SONIC ENGAGEMENT

Sonic Engagement examines the relationship between community engaged participatory arts and the cultural turn towards audio, sound, and listening that has been referred to as the 'sonic turn'.

This edited collection investigates the use of sound and audio production in community engaged participatory arts practice and research. The popularity of podcast and audio drama, combined with the accessibility and portability of affordable field recording and home studio equipment, makes audio a compelling mode of participatory creative practice. This book maps existing projects occurring globally through a series of case study chapters that exemplify community engaged creative audio practice. The studies focus on audio and sound-based arts practices that are undertaken by artists and arts-led researchers in collaboration with (and from within) communities and groups. These practices include—applied audio drama, community engaged podcasting, sound and verbatim theatre, participatory sound art, community-led acoustic ecology, sound and media walks, digital storytelling, oral history and reminiscence, and radio drama in health and community development. The contributors interrogate the practical, political, and aesthetic potentialities of using sound and audio in community engaged arts practice, as well as its tensions and possibilities as an arts-led participatory research methodology.

This book provides the first extensive analysis of what sound and audio brings to participatory, interdisciplinary, arts-led approaches, representing a vital resource for community arts, performance practice, and research in the digital age.

Sarah Woodland is a researcher, practitioner, and educator in applied theatre and participatory arts, with a particular focus on engaging with communities and groups from diverse social and cultural backgrounds, and those with experience of the criminal justice system. She is Dean's Research Fellow at the Faculty of Fine Arts and Music, University of Melbourne, Australia.

Wolfgang Vachon has been creating with and supporting children and youth through theatre and other arts practices for three decades. His work has primarily been with people who are street involved, homeless, 2SLGBTQ+, survivors of trauma, and those living in state care. Wolfgang teaches Child and Youth Care at Humber College in Toronto, Canada.

Routledge Advances in Theatre & Performance Studies

This series is our home for cutting-edge, upper-level scholarly studies and edited collections. Considering theatre and performance alongside topics such as religion, politics, gender, race, ecology, and the avant-garde, titles are characterized by dynamic interventions into established subjects and innovative studies on emerging topics.

A Sourcebook of Performance Labor
Activators, Activists, Archives, All
Joey Orr

Sonic Engagement
The Ethics and Aesthetics of Community Engaged Audio Practice
Sarah Woodland, Wolfgang Vachon

The Immersive Theatre of GAle Gates
Daniella Vinitski Mooney

Strategies for Survival at SIBIKWA 1988–2021
Landmarks of South African Performance History
Phyllis Klotz, Smal Ndaba

Instruments of Embodiment
Costuming in Contemporary Dance
Eric C. Mullis

Hamlet's Hereditary Queen
Performing Shakespeare's Silent Female Power
Kerrie Roberts

For more information about this series, please visit: https://www.routledge.com/Routledge-Advances-in-Theatre--Performance-Studies/book-series/RATPS

SONIC ENGAGEMENT

The Ethics and Aesthetics of Community Engaged Audio Practice

Edited by Sarah Woodland and Wolfgang Vachon

Routledge
Taylor & Francis Group
LONDON AND NEW YORK

Cover image: Biosphere Soundscapes, Mexico, Sian Ka'an Biosphere Reserve

First published 2023
by Routledge
4 Park Square, Milton Park, Abingdon, Oxon OX14 4RN

and by Routledge
605 Third Avenue, New York, NY 10158

Routledge is an imprint of the Taylor & Francis Group, an informa business

British Library Cataloguing-in-Publication Data
A catalogue record for this book is available from the British Library

Library of Congress Cataloging-in-Publication Data
Names: Woodland, Sarah, editor. | Vachon, Wolfgang, editor.
Title: Sonic engagement : the ethics and aesthetics of community engaged audio practice / edited by Sarah Woodland and Wolfgang Vachon.
Description: First Edition. | New York : Routledge, 2023. | Series: Routledge advances in theatre and performance studies series | Includes bibliographical references and index. |
Identifiers: LCCN 2022035177 (print) | LCCN 2022035178 (ebook) | ISBN 9780367758387 (Hardback) | ISBN 9780367758370 (Paperback) | ISBN 9781003164227 (eBook)
Subjects: LCSH: Performing arts--Social aspects. | Music--Social aspects. | Sound--Social aspects. | Community arts projects.
Classification: LCC NX180.S6 S6235 2023 (print) | LCC NX180.S6 (ebook) | DDC 701/.03--dc23/eng/20220815
LC record available at https://lccn.loc.gov/2022035177
LC ebook record available at https://lccn.loc.gov/2022035178

ISBN: 978-0-367-75838-7 (hbk)
ISBN: 978-0-367-75837-0 (pbk)
ISBN: 978-1-003-16422-7 (ebk)

DOI: 10.4324/9781003164227

Typeset in Bembo
by MPS Limited, Dehradun

CONTENTS

ILLUSTRATIONS

Figures

Table

CONTRIBUTORS

Mitchell Akiyama is a Toronto-based scholar, composer, and artist. His eclectic body of work includes writings about sound, metaphors, animals, and media technologies; scores for film and dance; and objects and installations that trouble received ideas about history, perception, and sensory experience. He is an Assistant Professor of Visual Studies at the University of Toronto.

Gary Anderson, Associate Professor of Drama at Liverpool Hope University, publishes on performance and its relationship to Spinozism. His most recent publication is 'Laruelle prefers heresy to revolution: Non-philosophy and live art' in *Art Disarming Philosophy: Non-Philosophy and Aesthetics* (Rowman and Littlefield), edited together with Niamh Malone.

Sylvan Baker is a Senior Lecturer in Applied Theatre at the Royal Central School of Speech and Drama. He is co-founder and Co-investigator of TVF, and also of the Medical Research Council funded Re-Star, a multi-disciplinary practice research project working to support the mental wellbeing of neurodiverse young people.

Leah Barclay is a sound artist, designer, and acoustic ecologist whose work investigates the value of acoustic ecology as a socially engaged interdisciplinary practice. She is the Discipline Lead of Design at the University of the Sunshine Coast, Australia and Vice President of the World Forum for Acoustic Ecology.

Bianca Beetson (Gubbi Gubbi/Kabi Kabi) is a visual artist, curator, and arts leader with a background in community engaged arts practice. She is Director of the Indigenous Research Unit at Griffith University, Australia, and is a member of the Board and Indigenous Advisory Panel for the Queensland Art Gallery and Gallery of Modern Art.

Casey Burkholder (she/her) is an Associate Professor at the University of New Brunswick's Faculty of Education. Her research program centres on work with 2SLGBTQ+ youth and pre-service teachers to agitate for social change through participatory visual research approaches. Casey's recent research and projects can be seen at https://caseyburkholder.com

Christos Carras studied philosophy at Cambridge and the Sorbonne. He is the executive director of the Onassis Stegi (Athens) and an independent researcher focusing on contemporary sonic art, digital technologies in cultural production, the role of art within broader networks and fields, and the politics of cultural production and consumption.

Budhaditya Chattopadhyay is an artist, media practitioner, researcher, and writer, holding a PhD in sound studies from Leiden University, and an MA in New Media from Aarhus University. He is the author of three books, *The Nomadic Listener* (2020), *The Auditory Setting* (2021), and *Between the Headphones* (2021). https://budhaditya.org/

Rosemary (Rosa) Cisneros is a researcher, dancer, choreographer, and curator who works closely with the RomArchive and many NGOs. She is leading various EU-funded projects which aim to make education accessible to vulnerable groups and ethnic minorities, and coordinating cultural heritage projects that bring dance and digital technologies together.

Maureen Flint is Assistant Professor in Qualitative Research at the University of Georgia. Maureen's research centers the theory, practice, and pedagogy of qualitative methodologies, artful inquiries, and questions of social (in)justice in higher education. She graduated with a PhD in Educational Research from the University of Alabama in 2019.

Jill Halstead is a feminist researcher and composer. Her publications center on the body politics of musical creativity and participation. She is Professor at the Grieg Academy, University of Bergen and leader of the Grieg Research School in Interdisciplinary Music Studies, a regional consortium of five universities located on the west coast of Norway.

Maggie Inchley is a Senior Lecturer in Drama at Queen Mary University of London. She researches cultural audibility and the intersectional and aesthetic aspects of vocal performance. Inchley is Principal Investigator of the AHRC funded practice-based research project, The Verbatim Formula (TVF), which explores systems of education and care with care-experienced young people: http://www.theverbatimformula.org.uk.

Vadim Keylin is a sound art scholar and poet. He holds a PhD from Aarhus University and is currently a postdoc at the University of Hamburg, working on research into sound poetry in digital culture. He has previously published articles on experimental musical instruments, sound sculptures, and participatory sound art.

Tanyss Knowles holds a Master of Social Work from the University of British Columbia and a Bachelor of Arts in Anthropology from Simon Fraser University. She is a board member of The Vancouver Association for Restorative Justice (VARJ). Tanyss is passionate about community-building projects that meet local needs and promote community strengths.

Brandon LaBelle is an artist, writer, and theorist working with sound culture, voice, and questions of agency. He is editor of Errant Bodies Press, Berlin, and Professor at the Department of Contemporary Art, University of Bergen. He develops and presents artistic projects within a range of international contexts, often working collaboratively and in public. He is the author of Acoustic Justice (2021), The Other Citizen (2020), Sonic Agency (2018), among others.

Eric Lewis is professor of Philosophy at McGill University. He is the founder and president of AIM (Arts in the Margins), and LUC (Laboratory of Urban Culture). He is an active sound artist and improvisor and organises the annual Koumaria new media art residency program in association with Medea Electronique.

Katie MacEntee's research focuses on the use of participatory visual methodologies to address HIV and AIDS, gender-based violence, homelessness amongst 2SLGBTQ+ youth, and for the study of transactional sex, sexual and reproductive health, and education access. Currently, she is a postdoctoral fellow at the University of Toronto.

Niamh Malone, Associate Professor of Drama at Liverpool Hope University, publishes on her applied theatre practice in prisons and on working with people living with dementia. Most recent publication 'Beyond Judgement: Arts intervention in dementia care' in *Art Disarming Philosophy: Non-Philosophy and Aesthetics* (Rowman and Littlefield), edited together with Gary Anderson.

Dont Rhine co-founded the international sound art collective Ultra-red in 1994. Based in Los Angeles, Dont has collaborated with Clean Needles Now (now named L.A. Community Health Project) since 1992. He is an organiser with the Los Angeles Tenants Union and a part-time instructor at Vermont College of Fine Arts.

Vicki Saunders (Gunggari) is a public health researcher who uses arts-led and poetic enquiry in the fields of child protection and family wellbeing. She is a Senior Research Fellow in the Centre for Indigenous Health Equity at Central Queensland University, Australia.

Morgan Shiver is a graduate student in English at Kansas State University. Her studies focus on children's literature and media. She graduated with a Bachelor of Arts in English from the University of Alabama in 2018.

Frank J. Tester is Professor Emeritus, School of Social Work, University of British Columbia, Adjunct Professor of Native Studies, University of Manitoba, and the Kutenai Art Therapy Institute, Nelson, British Columbia. He is the president of

VARJ. Frank is a recipient of the Gustavus Myers Award for the study of human rights in North America.

Wolfgang Vachon has been creating, playing, and supporting children & youth through theatre and other arts practices for three decades. His work has primarily been with people who are street involved, homeless, 2SLGBTQ+, survivors of trauma, and those living in state care. Wolfgang teaches Child and Youth Care at Humber College in Toronto.

Syrus Marcus Ware is a Vanier Scholar, a visual artist, community activist, researcher, youth-advocate, and educator. For 12 years, he was the Coordinator of the Art Gallery of Ontario Youth Program. Syrus is also a core-team member of Black Lives Matter Toronto and Assistant Professor at McMaster University's School of the Arts.

Ryanne Whyte graduated with a Bachelor of Arts in Telecommunications and Film and African American Studies from the University of Alabama in 2019. She now lives and works in Atlanta, Georgia as a film/TV production assistant.

Sarah Woodland is a researcher, practitioner, and educator in applied theatre and participatory arts, with a particular focus on engaging with communities and groups from diverse social and cultural backgrounds, and those with experience of the criminal justice system. She is Dean's Research Fellow at the Faculty of Fine Arts and Music, University of Melbourne, Australia.

Nimalan Yoganathan is a PhD student in Communication Studies at Concordia University (Tiohtià:ke/Montreal). His research interests include soundscape studies, as well as decolonial and intersectional modes of environmental listening. Nimalan is also a sound artist who has conducted soundscape research in the Brazilian Amazon rainforest and the Inuit community of Inukjuak, Nunavik.

Toby Young is a Leverhulme Early Career Research Fellow at the Guildhall School of Music and Drama and previously Gianturco Junior Research Fellow at the University of Oxford. As a composer and producer, Toby has collaborated with artists including the Rolling Stones, Duran Duran, and the London Symphony Orchestra.

INTRODUCTION: DISTILLING AN INTERDISCIPLINARY APPROACH

Sarah Woodland and Wolfgang Vachon

We find ourselves in the midst of a sonic boom. Rapidly developing technologies are enabling the democratisation and consequent proliferation of sound and audio-based modes of creative production. The accessibility and portability of affordable home studio and field recording equipment are making audio a compelling and burgeoning mode of participatory, and community engaged forms of creative practice. These approaches are now increasingly being harnessed as a way of amplifying marginal stories and experiences and highlighting social injustices, through forms such as podcast, audio drama, and participatory sound art. Prior to the COVID-19 pandemic, sonic practices in art, performance, and media were already enjoying a steady rise in popularity, accompanied by a concurrent turn towards sound and listening in cultural theory that had begun to challenge the dominance of visual epistemologies in academic discourse (Bull, 2019, p. xvii). As a result of the pandemic, participatory arts—particularly those involving presence and performance—have faced a crisis in many parts of the world, and sound arts have been deployed to solve community participatory problems that arise from physical distance. This will arguably hasten a seismic shift in how we collaborate artistically that was already well underway and raises increasingly important questions about the ethics and aesthetics of community engaged sound arts.

Leading sound scholar Michael Bull (2019, p. xvii) describes the field of sound studies as a 'broad church', increasingly attended by artists and scholars from diverse disciplines who have found their way towards the amorphous yet compelling qualities of sound (Bull, 2019; Bull & Cobussen, 2020; Sterne, 2012). This provides the context for our volume. In curating this collection, we threw the net out deliberately wide, hoping to draw together practice and research from different disciplines that nevertheless have one thing in common: an emphasis on sonic arts practices that are undertaken by artists and arts-led researchers in collaboration with (or from within) specific communities and groups. While sound studies are

DOI: 10.4324/9781003164227-1

inherently interdisciplinary, there remains a dominance of music and sound art in examinations of the social and aesthetic dimensions of the work, and a tendency towards privileging 'high art' or the Avant Garde over more popular forms, that has only recently been directly addressed in Michael Bull and Marcel Cobussen's collection on sonic methodologies (2020). Beyond participatory public art works, there has similarly been little interrogation of how sound and listening are increasingly being used in community engaged practices. We hope to address this by offering a fresh perspective and multiple points of entry for artists and researchers undertaking a range of community engaged sonic arts practices.

For our own part, we have arrived at the idea of sonic engagement via applied theatre, drama, and performance studies, most often working in prisons (Sarah) and Child and Youth Care (Wolfgang). Making theatre in a women's prison in Australia, Sarah was looking for ways to amplify the incarcerated performers' voices beyond the prison walls—since they were not allowed to tour to the public, and public audiences were not allowed in. At the same time (in the 20-teens), Pan Pan Theatre Company toured their production of Samuel Beckett's radio play *All That Fall* (Quinn, 2014), inviting audience members to sit on deckchairs in a darkened room, lit on one side with a bank of blazing spots, and listen to the disembodied voices emerging from a sonically sparse world effected by minimal sound design. Sarah was intrigued by this novel rendering of radio drama and wanted to explore the possibilities inside the prison. This led to *Daughters of the Floating Brothel* (2015), an historical audio drama exploring the colonial legacy of female incarceration in Australia that was created with the women and played in a series of live listening events both in prisons and outside (Woodland, 2020). Following this, in 2016, Sarah started working at the Queensland Conservatorium Research Centre, a vibrant mix of traditional approaches to music making and musicology, and contemporary approaches to sound art, sound design, and acoustic ecology. Here she met Leah Barclay—a composer, sound artist, and acoustic ecologist (see chapter in this volume)—who talked about her work with First Nations communities in Arnhem Land, where field recordings of environmental soundscapes connected Elders powerfully to their Country,[1] even if they were located miles away. They talked about how this approach might be used with First Nations women in prison and *Listening to Country* (2018) was conceived: A pilot project where an interdisciplinary cross-cultural team of artist-researchers composed a soundscape with incarcerated women drawing on environmental field recordings from their various places of belonging and Country (see chapter in this volume; see also Woodland et al., 2019). The approach also incorporated the First Nations principle and practice of *dadirri* (deep, active listening) as introduced to the world by Miriam-Rose Ungunmerr-Baumann (2017); as well as performance, poetry, and visual arts. Sarah was then invited to work with Rand Hazou and a group of men incarcerated in Auckland Prison on a verbatim theatre project *Ngā Pātū Kōrero: Walls That Talk* (2019), exploring the men's experiences of imprisonment. Since producing a draft documentary podcast episode about the production, Sarah and Rand have begun to explore how these works might orient communities, institutions, and policy makers towards listening (as opposed to wilful deafness) when it comes to the cultural rights of those entrapped in settler colonial regimes of carcerality (Hazou et al., 2021).

After many years of making theatre with young people, Wolfgang was looking for a more accessible, ethical, and aesthetically satisfying creative process to use with people who had been marginalised, minoritised, stigmatised, and too often dismissed from conversations by professional care providers and politicians. The recent re-emergence of audio drama piqued his interest as a possible route to meet these needs. After speaking with other youth-working artists and young people with whom he had existing relationships, he applied for and received funding to develop a project related to transphobia in the homeless shelter system. His first collaborative audio drama project was *Transitioning Home*[2] a queer hope-punk infused story set in the near future about a utopic shelter for trans, queer, and racialised people born from the negative experiences of past shelter stays. All the creators involved in this project were people who had direct experiences related to this world. Young trans, queer, and racialized folx who had lived in shelters, or avoided them because they feared for their safety. These perspectives were augmented by professionals who worked with young people, some (such as Wolfgang) with experience working in shelters, others working in different ways.[3] This project led Wolfgang to use audio drama as the primary method for his doctoral research (see chapter in this volume), developing two audio drama series as part of his final dissertation submission,[4] meeting Sarah, and the two of us mapping the process of applied audio drama/audio drama inquiry, and eventually co-editing this volume.

We share our pathways towards the sonic here as a way of both situating our thinking for this volume and illustrating just how serendipitous and *un*disciplined our disciplinary approach can sometimes feel. For us, the unique qualities of sound and listening in community engaged arts cut across disciplines and categories to expand our understanding of ethics and aesthetics in artistic practice and arts-led inquiry. The contributions in this volume therefore represent a breadth of different practices and voices, across diverse community, cultural, and global contexts. They may orient themselves towards artistic, educative, activist, or research outcomes; and include diverse practices such as prison radio, podcast, audio drama, performance art, sound art, sound walks, sonic memes, and more. The collection deliberately steers away from music, as there exists a strong corpus of work in the field of 'community music' that speaks eloquently for itself (see Bartleet & Higgins, 2018). Yet as attested to by the contributions from Toby Young, and Jill Halstead and Brandon LaBelle, analysis of community engaged soundscape and sonic sociality regularly reaches into music's domain; and certain music focused theories contribute to our understanding of sonic ethics and aesthetics as discussed later. In this, we also acknowledge and not-so-deftly sidestep longstanding debates over the hierarchies of music and sound, or where 'raw' sound ends and music begins (or vice versa) (see Demers, 2010; Sakakeeny, 2015).

As such, we do not presume to situate this volume firmly in any one field but rather hope to engage with and extend fields such as applied theatre, sound art, qualitative inquiry, social sciences, and sound studies to place the emphasis on sound in community engaged participatory arts practice. In writing this introduction, we therefore follow our own 'sonic imaginations' (Sterne, 2012) from our own disciplinary and artistic perspectives; hopefully contributing to the strength of sound as a 'transversal discourse', that can 'interfere' with traditional categorisations between different disciplines (LaBelle, 2020, p. 55). With such a diversity of practices, and the capaciousness of terms such as *ethics*, *aesthetics*, and *community*, the theoretical terrain that we cover in introducing a volume such as this could be almost limitless. We therefore attempt to distil here ideas around aesthetics, ethics, and community engagement that draw from our own practice-led research, as well as some of the common themes that have emerged across the range of chapters represented. We are also inspired by the global historical moment—the pandemic, the dominance of divisive politics, the urgency of ecological collapse. As such, we provide a preliminary interrogation of ethics and aesthetics in community engaged sound arts, giving particular attention to how they might promote activism, justice, decolonising pedagogies, and ethical participatory research.

What do we mean by community engaged?

Our provocation to authors in this volume was to consider works that use audio- or sound-based approaches in engaging specific communities or groups in artistic practice and arts-led research, that are directed concurrently towards social and artistic goals. In our own projects we have found that working in audio enables a flexible approach to participation, where non-professional, often-inexperienced collaborators in sites such as prisons, schools, and community settings, can be involved in performance and production in a range of ways. While engaging in creative practice and critical conversations, collaborators learn new skills and can produce satisfying creative outcomes. Privileging the sonic offers unique opportunities for affective and aesthetic engagement that extend beyond the usual modes of understanding and meaning making that exist within, in our case, theatre, and drama. We have found that audio works also allow projects to reach wider audience (and for longer) than an ephemeral performance event, engaging audiences in different ways, and extending the potential for marginalised voices to be heard beyond prison walls, across borders, and between diverse communities and cultures.

Community can be an amorphous concept, which despite falling somewhat out of favour in recent years, continues to hang on in cultural practice, research, and policy. If we draw from sociology, we might think of 'community' as a group with shared commonalities, which might be anything from geography, culture, language, age, religion, profession, who one was raised by, hobbies and interests, or an infinite number of other possibilities. While drawing on acoustic ecology, new materialism, and Indigenous knowledges, we might understand 'community' as inclusive of non-human actors and entities within the ecologies and cosmologies that make up our lifeworlds. This leaves much open to interpretation. It is beyond our scope to unpack this in detail, but in this volume, we see all the above constructions of community represented, along with several others. In our discipline of applied theatre, we are wary of how the word might assign a false homogeneity to groups with which we make work, but it nevertheless remains a compelling construct to describe how performance practice can bring people together—to build community—in a shared creative endeavour; and to bring audiences into a state of *communitas* (Turner, 2012), a heightened moment of potentially transformative shared affect. This of course takes us towards the ethics and aesthetics of community engaged practice, which we'll come to later.

Understandings of engagement shift depending on discipline, project, and personal framing among other factors. As editors, we each entered this conversation with different perceptions of what engagement meant, and the difference between engagement and participation. Turning to the Oxford English Dictionary (2021) one must pass through multiple definitions of the verb engage (to pledge, to bind, to hire, to promise, to betroth, to urge, to win over, to charm, to entangle) before finding a meaning related to how we use it here 'to attract and hold fast' and this too has some uncomfortable connotations. These multiple understandings are perhaps why many of the authors in this volume responded to the call from a variety of perspectives. In theatre, engagement can mean anything from people enjoying the performance, to audience members coming up on stage, to productions developed with or by community members, with some productions including all three. In other socially engaged and participatory arts, there appears to be a similar diversity of interpretations that range from simply engaging with social issues, to inviting fleeting interactions from publics or audiences, to facilitating longer and more collaborative encounters between audience and artwork, to working in embedded ways with groups or communities to conceive and execute works 'from the inside'.

The projects and works described in these chapters therefore appear at various places in what could be seen as a continuum—or perhaps more appropriately a web—of community engagement. These might range from projects led by specific communities from within, to those where 'outsiders' with longstanding relationships with communities and groups are invited in to collaborate, to those where artists or researchers invite community members to collaborate or have input in projects. Depending on how such collaborations occur, who is involved, and what is created, each of these will have its own implications for ethics and aesthetics. Rosa Cisneros' chapter provides an example of community-led practice and research, with the production of *Yellow Couch Convos*, a podcast made by Rosa (a Roma artist and researcher) interviewing other Roma artists, with the goal of countering negative stereotypes and narratives of Gypsy, Roma, and Traveller (GRT) communities. This might be seen as a compelling example of insider research that progresses a decolonising ethos (Smith, 2012), providing an alternative to cultural appropriation and 'extractive' approaches in artistic practice and research that still today are frequently executed from outsider (and often privileged) standpoints (see Robinson, 2020 for a comprehensive analysis of this). There are also elements of a community-led process in Maggie Inchley and Sylvan Baker's chapter, where they were asked by an organization of 'care leavers' to come and develop a project with them. The insider element was also present in that at least one of the facilitators themselves had residential care experience. A variation on community-led practice also exists in Christos Carras and Eric Lewis's chapter. After developing a relationship with the townspeople where an annual artists-retreat was located, the community (townspeople) asked the visiting artists to do sound-based research (media walks) with them. These new projects were initiated by the community members after they recognised the skills that the visiting artists brought. Elsewhere on the community engagement web, yet somewhat close to these approaches, we see the work of Tanyss Knowles and Frank Tester. Their chapter describes a project where the authors, as members of The Vancouver Association for Restorative Justice (VARJ), sought to increase public awareness about restorative justice and hired a local Indigenous playwright to work with them to create a radio play. By engaging an artist who represents communities most affected by the issues, VARJ deferred to Renae Morriseau in representing the issues authentically and in a way that would be culturally safe and appropriate. All these approaches privilege the insider voice in slightly different ways, positioning communities at the centre of projects that will ultimately represent their experiences.

Elsewhere on the web of community engagement, an artist or arts-led researcher might invite members of a specific community to be involved in aspects of the project such as creating elements of the work, performing, (co)facilitating or being part of post-performance discussions. The purpose of this is often to invite audiences into the world of that community, and again to centre their voices in the representation of their experience for a wider audience. Casey Burkholder and Katie MacEntee's chapter provides an example of this. Each author recruited and worked with a group to create 'cellphilms' based on the group's identity (people who trade sex in one case and LGBTQ+ youth in another). The group members conceived and created the cellphilms with the support of the researchers. Beetson and colleagues' chapter also reflects elements of this approach, where the pilot work that informs their methodology for 'listening to Country' took place inside a women's prison, with First Nations women enlisted as co-creators and co-investigators in the development of a soundscape and culturally situated knowledge base around sonic engagement and belonging. Leah Barclay's chapter about her work in acoustic ecology reflects a similar approach, which seeks to engage communities actively in the use of technology and production of sound-based knowledge and artworks generated in (and from) natural environments. In this sense, we see 'outsider' artists/researchers seeking to work ethically with specific communities or groups to make a work that will amplify their voices and experiences to a wider audience, at the same time building technical capacity and skills within those groups that may be useful to them later.

Slightly further away from these approaches on the web, we see projects that might involve drawing on the stories of community members, through interviews or conversations, to develop a work or research project. An example of this can be seen in the chapter by Mitchell Akiyama in conversation with Dont Rhine and Syrus Marcus Ware. The artist-activist-authors from Canada and the UK developed a project with self-identified drug users examining the politics of drug use in Toronto, Canada. The project drew on the voices of people the artists spoke with, which was then presented in a gallery, among other locations. Again, this approach seeks to centre the voices and experiences of communities or groups who may be considered 'marginalised', inviting public audiences to engage more deeply with issues that might be hidden from view. While Akiyama, Rhine, and Ware outline a deeply ethical approach and a longstanding commitment to the activist potential of this kind of work, it is worth noting that this model of engagement can run the risk of being tokenistic or shallow. Some public artworks that espouse to be socially engaged for example can come off as gimmicky, providing little more than Instagram-ability in the way that they invite participation. Or they can imitate, even become complicit within unsettling logics of surveillance culture. Similarly, some arts-based research that involves community-input (for example in planning or policymaking) pays lip service rather than inviting meaningful participation in the conception and execution of the work, let alone its fundamental epistemological positioning.

In all the above framings, it bares mentioning that people may well sit inside some aspects of the community and be outside in others. For example, an older artist with experience of drug addiction or trauma working with at-risk youth, may sit both inside and out. Or outside artists or researchers may have deep relationships or knowledge that mean they become trusted custodians of community stories and experiences. Wolfgang Vachon's chapter has elements of this as he is a professional in the field (child and youth care), which he studied, and he had personal relationships with several of the research participants prior to embarking on the research. We also see aspects of this in the chapter by Maureen Flint, Morgan Shiver, and Ryanne Whyte, as all had been students at the University of Alabama where the research project discussed was exploring race and geography. This again has implications for the ethics of representation in sound-based projects, and the sensitivity with which artists and researchers must approach such works. Elsewhere in the volume, cultural studies such as those undertaken by Nimalan Yoganathan, Vadim Keylin, and Jill Halstead and Brandon LaBelle, are concerned with observing the aesthetic, ethical, and activist operation of community in sonic phenomena as they are created from within communities, without outside intervention. In the case of Halstead and LaBelle, a comparative study is set up between an outsider observation of these phenomena (LaBelle) and an artist-researcher's perspective on working with multisensory listening in dementia care (Halstead) to progress ideas of 'more-than-social listening'. For cultural theorists, observing the operation of community in such works can provide important insights into how sound-based practices impact assembly, participation, and activism to effect social and cultural change.

Few of the examples in this volume demonstrate only one, and not another version of a community engaged process, in that we see possibilities for cross-over in all these examples. Each approach serves particular purposes and we do not frame one as any 'better' than another, although we certainly think one may be better than another in particular circumstances. Beyond thinking about community in the works described here, one can also think of how *audiences* engage. In applied theatre, for example, while the process of creating works is perhaps foregrounded, considerations of audience are of great importance. Drawing from its origins in Theatre of the Oppressed (Boal, 1985), many applied theatre works involve some kind of audience interaction—at the very least, a post-performance conversation or Q&A is a hallmark of such productions. This stems from the fundamental purpose of the work as being to facilitate community dialogue, a pedagogical or political motivation to invite audiences into the stories and experiences being depicted on stage. For our purposes here, sonic engagement might therefore be thought of in terms of the dialogic space that is established through the sonic encounter. Indicators of this might include how audiences respond, how many listens a project gets, what happens during the listening, how artists utilise different live or online strategies for interacting with listeners, what listeners do after listening or any number of other considerations. Thinking more deeply about community and audience engagement, and again drawing on our applied theatre backgrounds, within this dialogic and representational space are situated multiple permutations of power and the potential for it to exploit and oppress. Dagron (2001) helpfully identifies 11 aspects of participatory communication towards social change, which includes considerations of power and identity; and approaches that are 'Horizontal vs. Vertical', 'Process vs. Campaigns', 'With vs. For' among others (pp. 34–35). In these categories rest important understandings of who is doing what with whom and towards what purpose, which we will discuss further below. For now, though, we draw out these ideas and different examples of community and audience engagement, not only to highlight what we mean by the term 'community engaged audio practice', but also to invite reflection from readers about where their own work sits in the web, how they approach engagement with communities, and what therefore might be the ethical and aesthetic implications of this.

Sonic relations: Considering ethics and aesthetics

We have already begun to approach ideas of ethics and aesthetics in the ways that different sound- or audio-based practices and works engage with communities and audiences or invite their participation. Engagement might be exploitative or empowering, either replicating or working against prevailing power structures. But what of the specific qualities that sound and listening bring to community engaged arts? Sound has the capacity to slip past, around, under, and through borders and boundaries that might otherwise reinforce the status quo. As Salomé Voegelin (2014) so clearly puts it, 'Sound is the invisible layer of the world that shows its relationships, actions and dynamics' (p. 2). Sound also operates in unique ways to 'attract and hold fast' our attention, to elicit affective and emotional responses, and to produce heightened, embodied, transformative experiences for producers, participants, and listeners. If we consider artistic practice as an act of culture-making, or even world-making, then Voegelin invites us to engage imaginatively in a 'sonic possible world':

> Listening offers another point of view, an alternative perspective on how things are, producing new ideas on how they could be and how we could live in a sonic possible world, and how we could include sound's invisible formlessness in a current realization and valuation of what we understand to be the actual world. (pp. 2–3)

The chapters in this volume either discuss how sound arts and audio practices engage communities and groups directly in the creation of these sonic possible worlds, or analyse how sound operates to draw communities and publics into new formations of relational possibility—or both. For us, this is an aesthetic endeavour. Aesthetics in contemporary arts and cultural studies is, as Joanna Demers (2010) suggests, somewhat of a 'dirty word', rooted in Western thinking about Western judgements of what 'good' or 'beauty' resides in largely Western artistic practices and artworks. In framing an aesthetics for sound studies, Neil Verma (2019) similarly suggests that 'sound studies has a fraught relation with the legacy of aesthetic philosophy' (p. 50), a discipline that traditionally separates art from cultural and political life—a by-product of the dominance of Western visual epistemologies (p. 44).

Thinking about the aesthetic in community engaged arts can be even more problematic, as the term still conjures images of elitism, 'high art', and disinterested judgement. Scholarship in community engaged arts most commonly eschews engagement with aesthetics in favour of more palatable debates stemming from critical theory. Verma (2019) sees a similar pattern in sound studies, which reflects 'an anxiety that opening the question of aesthetics is liable to force a decoupling with the social' (p. 44). He cites several sound studies scholars who can be seen to contribute to an aesthetic understanding of the field, 'Yet these and others are not quite taking on aesthetic disciplinarity directly' (46). We too have noticed that consideration of the aesthetic most often manifests in discussions of Avant Garde sound art practices, as opposed to podcast, radio drama, and audio drama; and even then, often only in passing reference to the artfulness or artistry of works rather than through deep analysis of how artists and audiences engage (see for example the works contained in Cobussen, Meelberg and Truax's collection, 2016). With the exception of Vadim Keylin and Wolfgang Vachon, the authors of this volume similarly approach aesthetics more obliquely, sometimes focusing on the artistry of works (Chattopadhyay); sometimes the creative approaches used (Knowles and Tester); and sometimes their sensory, affective, or imaginative possibilities (Anderson and Malone). Given the focus of the volume on community engagement, analyses understandably foreground the social, political, and pedagogical implications of works over tackling aesthetics head-on.

Analysis of aesthetics and aesthetic engagement in music has a strong legacy, having its roots in Greek philosophy and emerging with the movement towards 'aesthetics' as a discipline heralded by Alexander Baumgarten in the mid-eighteenth century.[5] Again, this more traditional approach has evolved in recent years. Tia DeNora's work in the sociology of music presents a musical aesthetics that accords in many ways with our thinking about sound in the projects described in this volume. DeNora (2016) presents music as a material for making rather than judging culture, focusing on the good that music can do in the world, rather than what constitutes 'good music'. Focusing on the doing good in the world takes us into ethical terrain. DeNora further suggests, 'Music is always music-plus' (p. 2), which accounts for, not only music as an object or outcome in and of itself, but also as an embodied action that nevertheless has its own qualities that are distinct from other activities (pp. 2–3). Music's goodness, she suggests, is relational (p. 126)—it is '[a] locally situated, grounded, property that is performed and emerges relationally' (p. 125). In this, she suggests a 'hyper-contextual aesthetic' (p. 126) for music, that we see reflected in studies of other forms of sound and listening (see Verma, 2019).

In trying to distil a set of ideas that informs this introduction, we therefore find it useful to figure aesthetics and ethics together, bound by notions of relationality. In this we take our cue from DeNora as cited above, but also from our own intellectual traditions in applied theatre, which emphasises an 'aesthetics of care' for others (Thompson, 2015), and from the emphasis on relationality that is evident in contemporary sound studies as stemming in part from sound's resistance to disinterested judgement (Verma, 2019). In *The Auditory Culture Reader*, Bull and Back (2020) outline what they see as being 'at stake' when it comes to deep listening:

- Sound makes us re-think the meaning, nature and significance of our social experience.
- Sound makes us re-think our relation to the community.
- Sound makes us re-think our relational experiences, how we relate to others, ourselves and the places and spaces we inhabit.
- Sound makes us re-think our relationship to power.
- Sound makes us re-think our relationship to the other senses and to embodiment itself (pp. 20–21).

Brandon LaBelle's recent work in sonic agency (2018) and acoustic justice (2020) develops ideas of 'acoustic relationality', asking, 'What pathways of relationality are made possible by way of acoustics?' and, 'What relational capacities or imaginaries are afforded by underscoring listening as multimodal, as facilitating forms of care, compassion, and understanding that extend beyond the strictly aural?' (2020, p. 77). Crucially, LaBelle's analysis of deafness and his attunement to Deaf critical scholarship draws other ways of listening and sounding into his project for acoustic justice—for example, the vibrational and the haptic—that do not privilege the hearing ear. Biserna (2021) discusses sound walking as a methodology for producing spatial knowledge that 'implies a relational way of experiencing or, better, of engaging with space' (p. 300), describing it as an 'acoustic conversation between self and environment [that is] also explored through extra-aesthetic or anti-aesthetic acts and gestures' (p. 303). This relational conception of sound and listening figures sonic arts as acts of self-, world-, and culture-making, reflecting a holistic view that recognises the 'more-than-aesthetic' (Robinson, 2020, p. 52) qualities of sound in its capacity to encompass spirituality, ethics, and politics. Such approaches are aimed at 'dislocating the fixity and goal-oriented teleology of listening with more flexible listening practices that ... situate listening as a relational action that occurs not merely between listener and listened-to, but between the layers of our individual positionalities' (Robinson, 2020, p. 58; see also LaBelle, 2020).

This holistic orientation also reflects a move in contemporary sound studies towards 'engagement with multisensory experience' as opposed to the traditional Western siloing of the senses (Bull & Cobussen, 2020, p. 7). Michael Bull and Les Back (2020) cite David Howes (2017) in suggesting that the 'ideology of sensory exclusivity may be seen as related to the way in which modern Western society is centred on the autonomous individual rather than the relations between individuals' (p. 20). Listening can therefore be seen as part of a 'composite of sensory experiences' (Demers, 2010, p. 152) that are equally vital in understanding how we engage with works and each other. This is not simply related to how sound might evoke visual imagery (as in audio drama's 'theatre of the mind') (McMurty, 2019), but also how it might touch us physically through vibrations and resonances; and how it might work communicatively and provocatively through and between us—from skin-to-skin (Eidsheim, 2015; LaBelle, 2020). As Michael Bull highlights,

> Th[e] reorganization of the senses, of which sound plays a significant role has impacted upon our relationship to others, the world, and ourselves. It has also acted to transform our understanding of distance, intimacy and our sense of place; public, private, and global. More recently it has changed our understanding of what it is to be a consumer and producer of cultural goods (2019, p. 269).

Hopefully the discussion so far leads us towards a conception of the aesthetic that is focused on subjective aesthetic experience (via multisensory perception or *aesthesis*), and its role in cultivating relations (human or otherwise)—a *process*. Verma (2019) goes some way towards articulating this in his proposal for 'media aesthetics' which focuses less on an aesthetics of value, and more on the role that mediation plays in aesthetic experiences. Rarely discussed in the context of sound studies, John Dewey (1934) and subsequent pragmatist aestheticians turned their attention towards the process of aesthetic engagement rather than artistic products. Dewey and others since recognise the embodied and relational nature of aesthetic experience, and the potential for aesthetic experience to reside in the everyday (see also Shusterman, 2000). They make no distinction between 'high art' and popular culture in their potential to elicit an integrative experience between cognition, imagination, and emotion that engenders meaning-making. Crucially, pragmatist aesthetics (and its more recent development, somaesthetics) conceives aesthetic experience as that which occurs in the embodied interaction between subject-object, self-world—an integration between doing and undergoing—that leads towards transformative insight. Keylin adopts this frame in our volume, to explore how participatory sound art can facilitate playful experimentation and 'emergent socialities' which are integral to the aesthetic experience. These contributions from pragmatist aesthetics also echo Voegelin's (2014) description of soundscapes, sound artworks, and musical pieces 'as sonic timespace places, as sonic environments, which we inhabit as phenomenological subjects, listening intersubjectively and reciprocally: generating ourselves and the world we hear through our being in the world' (p. 3).

Pragmatist aesthetics recognises the legacy of Eastern and Indigenous philosophies as conceiving aesthetics as an 'art of living' (Shusterman, 2000), which has inherently ethical implications. It also reflects other important developments in aesthetics that influence us here, for example the integration between 'responses' and 'responsibilities' that is embodied in feminist care ethics (Tronto, 2013) and more recently 'aesthetics of care' (Thompson, 2015). In resistance to what Robinson (2020, p. 7) calls 'aesthetic assimilation', there is now a resurgence of 'decolonial aesthetics' (Deere, 2020) and 'decolonial listening practices' (Robinson, 2020, p. 37), which seek to challenge the base assumptions upon which Western listening, art appreciation and critique are built (see also Beetson and colleagues' chapter in this volume). These echo threads of the conversation in media studies centred on the politics of listening (Bassel, 2017): Who listens? What structural elements inform their listening? What does it mean to listen with a white or settler ear? (De Souza, 2018). Or to 'listen across difference?' (Dreher, 2009). All these ideas inform us as we think about aesthetics in community engaged sound arts. There is a re-tuning from 'aesthetic listening' where one is listening for enjoyment or pleasure to an 'aesthetics of listening' whereby one is hearing how aesthetics are calling, and perhaps demanding, a response. In our own work, a relational approach invites us to consider, not just how we engage and represent communities, but how we 'be with' communities ethically through this process.

Drawing on Indigenous ways of knowing (Robinson, 2020; Simpson, 2020) and new materialism (Barad, 2003; Rosiek, 2018), we can think of community engaged sound practices as subject(s) to subject(s). When we read the pieces in this volume and listen to the work they describe, we also recognise that each of us brings our own 'social aesthetics' (Born et al., 2017) our subjective, biased, and interested perspectives. Stadler (2015) in his essay 'On Whiteness and Sound Studies' reminds us that 'our understanding of sound is always conducted, and has always been conducted, from within history, as lived through categories like race' (para. 12; see also Robinson, 2020; Stoever, 2016). And we recognise that there is a 'distribution of the heard' at play in how and where the works sit within (or resist) existing hierarchies of attention and possibility (LaBelle, 2020 after Ranciére, 2006). In this framing we understand the community and all its members as subjects, we understand the people helping to realise the sound project as subjects (the artists, facilitators, researchers, and technicians be they from within or outside the community), we understand the listener as a subject, we can understand the acoustic spaces and environments that sound inhabits as subjects, and we can understand the sounds themselves as subjects. This then positions sonic engagements as intersubjective and interdependent processes that require care for all the elements: community, place, sound, listener, and facilitator/artist/researcher/technician.

In this volume, we hear young people who live in state care (Inchley and Baker), Idle No More protestors (Yoganathan), people who trade sex (Burkholder and MacEntee), rural Greek villagers (Carras and Lewis), incarcerated men (Malone and Anderson), injection drug users (Akiyama, with Rhine and Ware), refugee young people (Young), among others. We also hear the environments, places, spaces, and Countries that sustain us and cry out for our care (Barclay; Beetson et al.). What are our ethical obligations as listeners when we enter the relational spaces created by these sonic worlds? How might we be altered or transformed in the process? And what actions does the experience call us towards, as we move on from the exchange?

Such questions draw us towards ideas of agency and action in community engaged arts, which, when considered in relation to interdisciplinary sound and audio practices described in this volume, we might consider in terms of performance and performativity. We see these ideas increasingly explored in sound studies (Bull & Cobussen, 2020; LaBelle, 2020), and theories of sound increasingly explored in theatre and performance studies (Bennett, 2019; Herrera et al., 2020), which invite us to think through how sound is performed, and what effects this produces. Questions of role and representation emerging from Erving Goffman (1969) might invite us to ask how 'protesters' or 'refugee youth' or 'incarcerated men' are sonically performed in such works, and to what effect on the listener; or how listening is performed in response; or what roles and dynamics might be enacted through collaborative or participatory exchange. But further to this, Trevor Pinch (2020) argues that sound should be considered as a material element, 'Which mediates and stages social interaction' (p. 806). Introducing their special issue of the journal *Performance Matters* entitled 'Sound Acts', Patricia Herrera and colleagues (2020) summarise a performative approach to sound: Doing sound shifts the sonic away from the auditory to emphasize that 'all sounds issue from vibrating bodies—whether those bodies are human, animal, inanimate, and/or abject/object. … Vibrating bodies perform, and that the sonic is a constellation of acts' (p. 3). LaBelle (2020) draws on Judith Butler to extend these ideas, considering acoustics as a form of performativity. He frames acoustics as being not just about the formation of buildings as influencing sound and listening, but towards 'questions of aurality' and how 'hearing and listening, and the arena of sound, in general, are affected by various forces and knowledges' (p. 31). As such, 'Acoustics is underscored as a performativity, a practice of composition embedded within struggles over recognition and justice' (p. 31). Performativity therefore draws us towards the potential for sound as material to action, where the aesthetics of listening is bound together with the politics of listening, and the aesthetics of *sounding* is similarly bound together with the politics of sounding.

Conclusion

In drawing together the ideas presented here, we have adopted the metaphor of distillation, taking selected, varied (or vaporous) concepts of community, engagement, participation, aesthetics, relationality, care, and ethics; and distilling them into what we see as being the *spirit* of this interdisciplinary volume. The volume invites a diversity of artistic practices to adhere to the notion of sonic engagement, and for readers to consider what good this might do in the world today. In the pages that follow, we present sixteen wonderfully varied chapters, grouped together under a set of themes that we have identified across the different practices and disciplines represented. We begin with the chapter by Beetson and colleagues, deliberately positioning 'First Knowledges First'—outside our thematic organisation of the volume—to go some way towards privileging Indigenous voices and disrupting Western epistemic and aesthetic dominance of cultural and artistic fields. Part two focuses on 'Sonic Knowing' and the meaning and resonance that can be drawn from sound and audio as an artistic research method. The focus of part three is 'Sonic Assembly' and the role of sound in building communities and publics. Part four, 'Sonic Disruptions' draws together chapters concerned with how sound and audio might contribute to creating auditory counter-narratives of those people who are considered marginalised, minoritised, or stigmatised. And finally, part four takes us towards 'Sonic Resistance' or the sounds of activism. We acknowledge that all the contributions here traverse between and across these themes, containing elements of many or all in their work. It is not our intention to silo them from each other, but rather to continue building our contribution to knowledge across these interdisciplinary fields. We editors therefore introduce each part with a brief summary of how these works might progress what we see as some of the principal functions of sonic engagement in artistic practice and research. We conclude the volume with a reflection and acknowledgement of how sonic engagement might reverberate towards justice, and what it might offer us in the context of current intersecting global crises, as we strive towards living ethically in relation with each other, and within the communities, ecologies, and cosmologies that support us to thrive.

Notes

1 The term 'Country' is capitalised to describe First Nations Country or Countries affiliation and belongingness. Country is an essential ontological concept for many First Nations Peoples, and it aids in sharing their relationship to land, place and a sense of belonging (Carlson, 2016).
2 To listen to Transitioning Home please visit: https://transitioninghomepodcast.podbean.com/
3 To read more about this project see Vachon and Woodland (2021).
4 See www.TuningIntoCYC.org
5 For a history of the study of aesthetics in music from 1750-2000, see Goehr and colleagues' contribution to *Grove Music Online* (2001). For a helpful summary of aesthetics in music from 2000-2010, see the introduction to Joanna Demers' (2010) analysis of aesthetics in experimental electronic music. Neil Verma (2019) also engages with some of this history in his framing of aesthetics for sound studies.

References

Barad, K. (2003). Posthumanist performativity: Toward an understanding of how matter comes to matter. *Signs: Journal of Women in Culture and Society, 28*(3), 801–831. 10.1086/345321

Bartleet, B. L., & Higgins, L. (Eds.) (2018). *The Oxford handbook of community music*. Oxford University Press.

Bassel, L. (2017). *The politics of listening: Possibilities and challenges for democratic life*. Palgrave Pivot. 10.1057/978-1-137-53167-4

Bennett, S. (2019). *Theory for theatre studies: Sound*. Bloomsbury Publishing.

Biserna, E. (2021). Ambulatory sound-making: Rewriting, reappropriating,'presencing' auditory spaces. In M. Bull & M. Cobussen (Eds.), *The Bloomsbury handbook of sonic methodologies* (pp. 297–314). Bloomsbury.

Boal, A. (1985). *Theatre of the oppressed* (Charles A. & Maria-Odilia Leal McBride, Trans.). Theatre Communications Group.

Born, G., Lewis, E., & Straw, W. (2017). *Improvisation and social aesthetics*. Duke University Press.

Bull, M. (Ed.) (2019). *The Routledge companion to sound studies*. Routledge.

Bull, M., & Back, L. (2020). *The auditory culture reader* (2nd ed.). Routledge.

Bull, M., & Cobussen, M. (2020). *The Bloomsbury handbook of sonic methodologies*. Bloomsbury USA.

Carlson, B. (2016). Striking the right chord: Indigenous people and the love of country. *AlterNative: An International Journal of Indigenous Peoples, 12*(5), 498–512. 10.20507%2FAlterNative.2016.12.5.5

Cobussen, M., Meelberg, V., & Truax, B. (Eds.) (2016). *The Routledge companion to sounding art*. Routledge.

Dagron, A. G. (2001). *Making waves: Stories of participatory communication for social change*. The Rockefeller Foundation.

Deere, D. T. (2020). The spacing of decolonial aesthetics. *Journal of World Philosophies, 5*(1), 89–98. https://scholarworks.iu.edu/iupjournals/index.php/jwp/article/view/3603

Demers, J. (2010). *Listening through the noise: The aesthetics of experimental music*. Oxford University Press.

DeNora, T. (2016). *Music asylums: Wellbeing through music in everyday life*. Routledge.

De Souza, P. (2018). What does racial (in) justice sound like? On listening, acoustic violence and the booing of Adam Goodes. *Continuum: Journal of Media and Cultural Studies, 32*(4), 459–473. 10.1080/10304312.2018.1488524

Dewey, J. (1934). *Art as experience*. Capricorn.

Dreher, T. (2009). Listening across difference: Media and multiculturalism beyond the politics of voice. *Continuum: Journal of Media and Cultural Studies, 23*(4), 445–458. 10.1080/10304310903015712

Eidsheim, N. (2015) *Sensing sound: Singing & listening as vibrational practice*. Duke University Press.

Goehr, L., Sparshott, F., Bowie, A., & Davies, S. Philosophy of music. *Grove Music Online*. Retrieved 31 Jan. 2022, from 10.1093/gmo/9781561592630.001.0001/omo-9781561592630-e-0000052965.

Goffman, E. (1969). *The presentation of self in everyday life*. Allen Lane.

Hazou, R., Woodland, S., & Ilgenfritz, P. (2021). Performing te whare tapa whā: Cultural rights and decolonising corrections. *Research in Drama Education: The Journal of Applied Theatre and Performance, 26*(3), 494–510. 10.1080/13569783.2021.1940121

Herrera, P., Marshall, C., & McMahon, M. R. (2021). Sound Acts, Part 1: Calling Back Performance Studies. *Performance Matters, 6*(2), 1–7. https://performancematters-thejournal.com/index.php/pm/issue/view/17

LaBelle, B. (2018). *Sonic agency: Sound and emergent forms of resistance* (Vol. 1). MIT Press.

LaBelle, B. (2020). *Acoustic justice: Listening, performativity, and the work of reorientation.* Bloomsbury Publishing USA.

McMurtry, L. G. (2019). *Revolution in the echo chamber.* Intellect.

Oxford English Dictionary (2021). *Engagement.* Retrieved from www.oed.com

Pinch, T. (2020). The sound of economic exchange. In M. Bull & L. Back (Eds.), *The auditory culture reader* (pp. 797–820). Routledge.

Quinn, G. (Director) (11–16 February 2014). *All that fall by Samuel Beckett [Radio Play].* Brisbane: Brisbane Powerhouse.

Ranciére, J. (2006). *The politics of aesthetics: The distribution of the sensible.* Continuum.

Robinson, D. (2020). *Hungry listening: Resonant theory for Indigenous sound studies.* University of Minnesota Press.

Rosiek, J. (2018). Art, agency, and ethics in research: How the new materialism will require and transform arts-based research. In P. Leavy (Ed.), *Handbook of arts-based research* (pp. 632–648). New York: The Guilford Press.

Sakakeeny, M. (2015). "Music". In D. Novak & M. Sakakeeny (Eds.), *Keywords in sound.* Durham, NC: Duke University Press.

Shusterman, R. (2000). *Pragmatist aesthetics: Living beauty, rethinking art* (2nd ed.). Rowman & Littlefield Publishers.

Simpson, L. B. (2020). *Noopiming: The cure for white ladies.* House of Anansi Press.

Smith, L. T. (2012). *Decolonizing methodologies: Research and indigenous peoples.* Zed Books.

Stadler, G. (6 July 2015). On whiteness and sound studies. *Sounding Out.* https://soundstudiesblog.com/2015/07/06/on-whiteness-and-sound-studies/

Sterne, J. (2012). Sonic imaginations. In J. Sterne (Ed.), *The sound studies reader* (pp. 1–17). Routledge.

Stoever, J. L. (2016). *The sonic color line: Race and the cultural politics of listening* (Vol. 17). New York University Press.

Thompson, J. (2015). Towards an aesthetics of care. *Research in Drama Education: The Journal of Applied Theatre and Performance, 20*(4), 430–441. 10.1080/13569783.2015.1068109

Tronto, J. C. (2013). *Caring democracy: Markets, equality, and justice.* New York University Press.

Turner, E. (2012). *Communitas: The anthropology of collective joy.* Palgrave Macmillan.

Ungunmerr-Baumann, M-R. (2017). To be listened to in her teaching: Dadirri: Inner Deep Listening and Quiet Still Awareness. *EarthSong Journal: Perspectives in Ecology, Spirituality and Education 3*(4), 14–15. 10.3316/INFORMIT.732386012034745

Vachon, W., & Woodland, S. (2021). Finding resonance: Applied audio drama, inquiry and fictionalising the real. *Research in Drama Education: The Journal of Applied Theatre and Performance, 26*(4), 638–655. 10.1080/13569783.2021.1891038

Verma, N. (2019). The return to sound aesthetics. In M. Bull (Ed.), *The Routledge companion to sound studies* (pp. 44–53). Routledge.

Voegelin, S. (2014). *Sonic possible worlds: Hearing the continuum of sound.* Bloomsbury Academic.

Woodland, S. (2020). 'Daughters of the Floating Brothel': Engaging Indigenous Australian women in prison through participatory radio drama. In C. McAvinchey (Ed.), *Applied theatre: Women and criminal justice.* Bloomsbury Methuen Drama.

Woodland, S., Barclay, L., Saunders, V., & Beetson, B. (2019). Listening to Country: Exploring the value of acoustic ecology with Aboriginal and Torres Strait Islander women in prison. *Soundscape: The journal of acoustic ecology,* June 2019. World Forum for Acoustic Ecology. https://www.wfae.net/uploads/5/9/8/4/59849633/soundscape_vol17.pdf

PART I

First Knowledges First

INTRODUCTION

Sarah Woodland and Wolfgang Vachon

We begin our volume with a deliberate positioning of First Nations voices and Knowledges first. While subsequent chapters are grouped around specific themes, the chapter presented demonstrates a holistic approach that is resistant to this kind of siloing. We also wish in our small way to acknowledge and recognise the sovereignty of First Nations peoples globally and the legacy and longevity of Indigenous Knowledges that traverse and transcend disciplines that we (in the West) refer to as 'philosophy', 'ethics', and 'aesthetics'. Building on the important essay by Gus Stadler (2015) 'On whiteness and sound studies' and Jennifer Lynn Stoever's (2016) *Sonic Colour Line*, sound studies scholars have begun to recognise the dearth of non-Anglo European voices represented in the field (see Bull & Cobussen, 2020). We had hoped to include more than just one contribution from First Nations perspectives in this volume, but it is perhaps unsurprising that the impacts of the COVID-19 pandemic hit our First Nations contributors the hardest, with two initial chapter proposals having to be withdrawn as events unfolded. Dylan Robinson's (2020) recent work *Hungry Listening* represents the first major contribution to sound studies from a First Nations perspective, and this has provided an important touchstone for us. Robinson describes, among other things, how Indigenous forms of listening are 'a form of attention in which we are attentive not just to sound but to the fullest range of sensory experience that connects us to place' (p. 72).

This contribution from Australia's Bianca Beetson (Gubbi Gubbi/Kabi Kabi), Vicki Saunders (Gunggari) and their non-Indigenous colleagues Sarah Woodland and Leah Barclay explores this terrain through their chapter: Listening to Country; a methodology encompassing acoustic ecology, arts-led research, poetic inquiry, and bu'ra'nga'man | dadirri | yimbilli—three of many words in Aboriginal Australian languages that signify listening, knowing, understanding, and research. Through their cross-artform, cross-cultural approach, the authors present a model

DOI: 10.4324/9781003164227-3

for holistic, community engaged sound practice that aims to facilitate wellbeing and cultural connection for First Nations peoples experiencing disconnection from Country (ancestral or tribal homelands) and places of belonging. With potential applications in prisons, hospitals, aged care facilities, schools, and other institutional spaces, the ethics and aesthetics of the work are intertwined with deeper cultural and spiritual implications, some of which defy description in conventional written form. Such an approach recognises the symbiosis between people and Country, where the wellbeing of one is dependent on the wellbeing of the other, and sound and deep listening act to mediate this essential relationship.

References

Bull, M., & Cobussen, M. (2021). *The Bloomsbury handbook of sonic methodologies*. Bloomsbury.

Robinson, D. (2020). *Hungry listening: Resonant theory for Indigenous sound studies*. University of Minnesota Press.

Stadler, G. (6 July 2015). On whiteness and sound studies. *Sounding Out*. https://soundstudiesblog.com/2015/07/06/on-whiteness-and-sound-studies/

Stoever, J. L. (2016). *The sonic color line: Race and the cultural politics of listening* (Vol. 17). NYU Press.

1

BU'RA'NGA'MAN | DADIRRI | YIMBALI: ECHOES OF LISTENING TO COUNTRY

Bianca Beetson, Vicki Saunders, Sarah Woodland, and Leah Barclay

Bu'ra'nga'man | Dadirri | Yimbali[1]

The sounds of the words chosen to begin this chapter convey our collective goals and the key outcomes we seek. Each of these words were carefully chosen after consultations with others, to echo something about our individual and collective relationships with the places from where these words originated. They echo something slightly different in each context about listening, knowing, thinking, and understanding; and also our research questions.

The focus of our collective project of inquiry, along with our associated research projects, can be expressed in many ways. In this context, in this chapter, our research focus can be expressed as the application of sound and sonic engagement, as research method, whereby attention is focussed on the process of listening as both an object and outcome of inquiry. Sonic engagement for our emerging, somewhat nascent research collaboration is that which listens for calls from the heart (like those expressed in the Uluru Statement of the Heart[2]) and to requests to listen (like those echoed in the voices of many First Nations Elders and most notably the voice of Australian Senior of the Year 2021, Aunty Miriam Rose Ungunmerr-Baumann[3]).

The focus of our collective project of inquiry can also be said in different contexts, to be about the role of the sounds of those unique places, held in our memories, our bodies, and our hearts, that echo our collective and individual experiences of social emotional wellbeing. It can also be said to be about the different relationships we have with the Countries,[4] the lands, and the places that centre many models of First Nations wellbeing and identity (Dudgeon et al., 2014).

Listening to country: Mapping the country that connects us

This image (Figure 1.1) represents the Country[1] where we gathered for the Listening to Country project in 2018 as women, artists, and storytellers—it is a

DOI: 10.4324/9781003164227-4

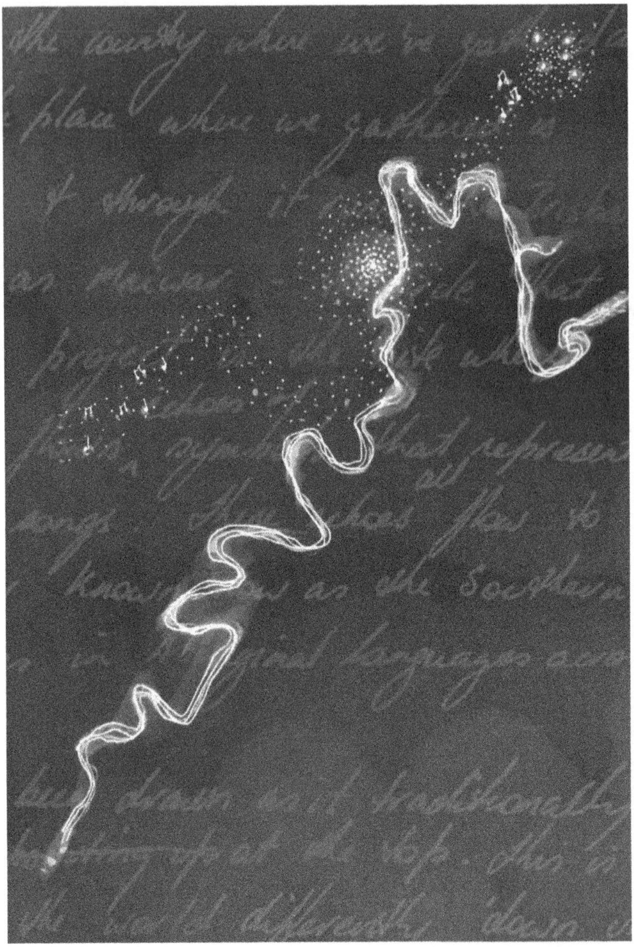

FIGURE 1.1 Listening to Country map: Vicki Saunders and Bianca Beetson (2018)

map of the place where we gathered to listen. This place is on Jagera and Turrbul Country and through it runs the Brisbane River, also known as Maiwar in some circles.

The circle of dots that mark a location along the river is the site where we gathered with the women at Brisbane Women's Correctional Centre. Into it and from it flow the sounds of many voices, mapped using symbols that represent sounds, music, and song. These sounds and their echoes all flow towards a point in the sky known as the Southern Cross (although it has many other names in Aboriginal languages across Australia).

Maiwar, the Brisbane River, has not been mapped as it traditionally is on guide maps for Brisbane, with North at the top. This is a reminder that we orient, see, and hear Country differently 'down under', under Southern stars.

Listening to country: The project

Listening to Country is both a pilot project that we conducted and a collaboration of like-minded researchers (both academic and community-based), all interested in exploring the value of Dadirri or deep active listening; and more broadly applying First Nations ways of, or approaches and methods to, listening in our work. Initially, Listening to Country was designed to explore the potential of acoustic ecology for promoting wellbeing with women in prison who were disconnected from the sounds of places of belonging. Along the way it began to respond to the need to create 'safer' spaces and to take up a more active space for Dadirri … for the 'unmarked forms of listening' (Robinson, 2020) in our research.

The Listening to Country project was delivered inside the Brisbane Women's Correctional Centre (BWCC) in 2018. It was created and funded to respond directly to a request from First Nations women at BWCC to create a 'culturally appropriate relaxation CD' and to raise awareness of the impact of the soundscapes of particular places—like the carceral contexts or institutions where bodies are confined.

Our intent for writing about this project is Dadirri; that is, our intent is to reflect and embody a deep active form of listening within the words we use to inscribe the project. This chapter is our first attempt to reimagine Dadirri—the bridge we have used in our work to amplify, scribe, transcribe, and echo the words of women (daughters, mothers, and grandmothers) living, and working far from their places of belonging. In many ways, however, our intent also responds to *binan goonj*—Mandandanji words from South West Queensland used to refer to people who 'can hear but don't listen'. This is the title of one of the first Aboriginal and Torres Strait Islander mental health care textbooks published in Queensland (Eckermann et al., 2010). These words encapsulate a critical element voiced by many First Nations leaders in mental health care reform in Queensland and one that is echoed in the word—Bu'ra'nga'man.

The words that entitle and begin this chapter are used here to first acknowledge our relationships with the traditional custodians of the Countries where these words belong/come from. They are also used to acknowledge the tensions and opportunities that attend the use and sometimes misuse of First Nations languages in copyrighted texts, narratives, and stories aimed for other audiences. This year (2022) marks the beginning of UNESCO's 'decade of Indigenous languages'. In its strategic recommendations for the decade, the Los Pinos Declaration organised by UNESCO and adopted internationally, emphasises,

> Indigenous peoples' rights to freedom of expression, to an education in their mother tongue and to participation in public life using their languages, as prerequisites for the survival of indigenous languages many of which are currently on the verge of extinction. The Declaration highlights the importance of enabling the use of indigenous languages in

> justice systems, the media, labour and health programmes. It also points to the potential of digital technologies in supporting the use and preservation of those languages.
>
> *(UNESCO, 2020, para. 4)*

In prison environments around Queensland, Aboriginal and Torres Strait Islander words are beginning to emerge in the titles of programs and units and in the vision statements for supporting First Nations staff and those in custody and in the community. An example is the Murridhagun Cultural Centre, which supported our pilot in BWCC, which was named from the collective term for Queensland First Nations peoples (Murri) and the Yuggera language word for earth, place, or Country (dhagun). For First Nations peoples, language is vital to resurgence, providing a linguistic (and sonic) anchor to First Nations sovereignty through the re-utterance of words that, in many parts of Queensland, the colonisers and settlers sought to erase. This becomes part of our project for listening (and sounding) in Listening to Country, as do the other sounds of Country (dhagun) that locate us geographically, culturally, and spiritually.

Echoing the sounds of belonging places

(Words from the women collated by the team[5])

My belonging place
Is safe
It's where you know
you're home

Where I feel peaceful
Where my ancestors' bones are
It's where I went swimming as a child
Where my children are, my mother

It's the beach, the weir, the river
The desert, the dirt

It's where all my ancestors are
Home for me is my culture,
that's the biggest thing
It's the land I've walked
I belong to the mountains, to the sea
To the place where wild lemons grow

Where everyone knows everyone

I've been living rough, doing it hard
I tried not to come back

but you know, it's here
where you guys …
my second family are

Listening to country: The collaboration

Our research collaboration builds on several years of engagement with Brisbane Women's Correctional Centre, delivering participatory drama projects, including audio drama (Woodland, 2021). The initial Listening to Country pilot brought together an interdisciplinary team of artist-scholars all interested in supporting local Elders in their efforts to promote cultural connection and wellbeing in prisons and all interested in developing a shared understanding of acoustic ecology, Dadirri and what sonic engagement means for First Nations communities.

As our first collaboration, Listening to Country was a relationship building exercise. It drew together our respective cultural standpoints, interests, and practices, which encompass acoustic ecology, visual art, performance, poetry and, of course, sound. For us, however, Listening to Country is not just about sound or research about sound: it is about ways of listening that are unique to Countries in Australia. We formed partnerships with Elders, who provide cultural guidance and wrap-around support to First Nations mothers, daughters, sisters, aunties, and grandmothers seeking connections to family, culture, and Country while in prisons and other institutions of care. We did this with the aim of supporting and contributing to their work.

For the authors of this chapter, working creatively at the nexus between such amorphous concepts as place, sound, acoustic ecology, sonic engagement, culture, belonging, and wellbeing within BWCC seeded a cross-cultural, cross-disciplinary research team and a novel approach and research methodology for promoting wellbeing in collaboration with local Elders.

What we heard

(Words from the women collated by the team)

It can send you womba[6]
It can keep you strong
Listening to how birds speak
of where we belong
A river runs through what we found
In the sounds beneath the sound

amplifying the silence,
silencing the noise …

While this project was about creating safe spaces for First Nations women in prison to share and to listen; it was also about the reciprocal benefits of listening and how the authors, as conduits, could amplify the voices of these women. The introduction of Dadirri and the use of deep active forms of listening in this project allowed us not only to hear the women's stories through yarning,[7] but also to hear the place or the context in which these stories were being told. When we use the term 'listening' in this chapter, we do not privilege the vocal sounds made by humans, or the hearing ear, but rather acknowledge the non-verbal, the sounds we can only hear with our eyes and feel with our bodies—these are the human and non-human terrestrial sounds in which we are constantly immersed. In today's world, there is no such thing as silence: noise pollution impacts our health and wellbeing and Country just like air pollution does (Doctors for the Environment Australia, 2020).

It can send you womba

From the women, we heard that prison soundscapes 'can send you womba, crazy'. The immutable, ever-present sounds of rattling keys, doors buzzing and slamming, coupled with the sounds of footsteps entering and exiting the room, walkie-talkies, muffled voices, and activated distress alarms were stressful and caused agitation and heightened anxiety. They triggered a constant state of hypervigilance, which is known to cause chronic stress and to lead to stress-related illnesses (see Moran, 2019). This statement was echoed by a mother and grandmother, who had lived in BWCC, on a particular site on Jagera and Turrbul land, for three years; her words emerged after drawing a picture, a memory-soundscape of a place that located a sense of peace, home, and belonging for her: 'What I miss most is … the sounds of night time … three years without the sounds of night time … it does things to your head, you know?'

At the start of the pilot project, during the initial listening activities; external noises and movements triggered agitation and a need for the women to remove themselves from the yarning circle. One of the key challenges we found working in the prison was that we were asking people to listen for relaxation in a place that is never silent. These sounds are the sounds of institutionalisation, and we also experienced them as a part of the carceral violence experienced by the women (Rice, 2016).

It can keep you strong

As the pilot project progressed, the women began to assert their agency through sound as they collaborated on the soundscape design. The soundscape gave the women an opportunity to disrupt the sonic violence of their day-to-day prison existence as they identified the potential for the soundscape's end-use—as something to which they could listen to in the morning to replace the sounds of women crying for their loved ones.[8] The goal was to transform the sounds of the institution from a place of trauma and violence into a place of peace, serenity and connection as they enjoyed the morning chorus of the natural world, albeit via digital audio file and headphones.

During the pilot, the Listening to Country team used various creative tools to collectively compose the soundscape; many of these activities evoked personal stories, reflections, and memories of their belonging place. The women's yarns helped to form and shape the story held within the soundscape as they talked about missing the sounds of night-time, the crackle of the fire, the sound of ceremony (songs, the clapping of hands, the stomping of feet, the deep bass sounds of the yidaki[9]), the sound of their men, the laughter of children, and the intimate sounds of human connection (breath, heartbeat, laughter).

Listening to how birds speak of where we belong

We listened collectively, deeply, to stories of birds—good and bad, the ones that herald new life, the ones that foretell the loss of life, and the ones that are just our ancestors checking in or passing by. They spoke of the healing and wellbeing that comes from connection to their belonging place, and the sounds they heard made them feel as if they were there. Country lives in your memory—you just need to close your eyes and it is still there (Figure 1.2).

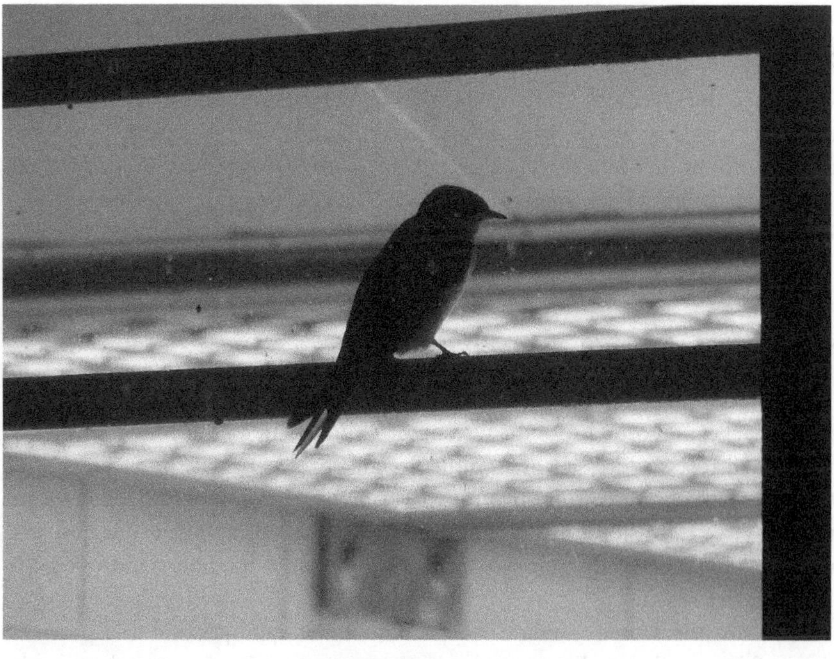

FIGURE 1.2 Listening to how birds speak: a swallow perches outside the workshop room at Brisbane Women's Correctional Centre. Image credit: Woodland (2018)

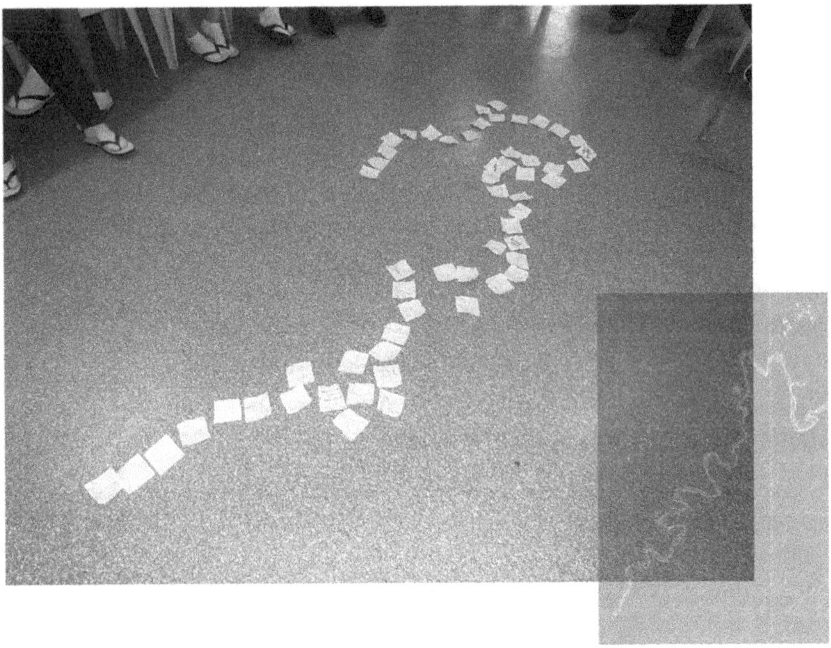

FIGURE 1.3 'The River of Words' workshop activity Brisbane Women's Correctional Centre. Image credit: Woodland (2018)

A river runs through what we found

We opened the space with yarns about belonging and connection, which saw the creation of a river of words, with the words taking the form of Maiwar,[10] the giant serpent forming the spine of Meanjin.[11] And, just like Maiwar, it connected the songlines and formed the spine of our soundscape. Out of this process, the following group poem was created, which was recorded and embedded into the layered soundscape created (Figure 1.3).

In the sounds beneath the sound

Towards the end of the project, amidst the toxic soundscapes of the prison and the playback of recorded sounds, we could no longer ignore a deafening silence that needed to be addressed in a culturally safe space. One of the challenges of working with a cross-cultural team is that sometimes the non-Indigenous team members need to step outside the camp so that Mob[12] can do business. Supported by the Elders, the two First Nations team members addressed the silence in the room.

We heard the women's concerns around cultural appropriation and potential misuse of the soundscape. For example, the women experienced angst at the thought of non-Indigenous women using the soundscape to do yoga. This concern centred around the soundscape containing sounds of cultural and spiritual significance to the women who created it, coupled with fears that non-Indigenous women would not respect or understand the cultural and spiritual importance of the sounds contained within the soundscape and pay no regard to the social, cultural, and political context in which the soundscape was created. The women's concerns were resolved by revisiting the project's research ethics and the publication outcomes of the research agreement. This resulted in the full soundscape being shared only with the participants.

The final product was a one-hour immersive audio work based on field recordings of natural environments and cultural rituals of various Countries in South East Queensland. However, the product that was created was more than just a soundscape or audio file: it enabled the women to exert a sense of sonic agency. It allowed them to disrupt the traumatic sounds of prison by using the soundscape as a non-violent transformative intervention, leaving them feeling spiritually alive and connected. In addition, the women could replace the prison soundscape with the soundscape of Country/belonging place via headphones when they needed relaxation, de-escalation, or to mentally escape. This enabled a reduction of institutional sound-induced trauma and anxiety, replacing it with an opportunity for transformational wellbeing through the sound of Country and its connection to memory and place.

Amplifying the silence, silencing the noise ...

(Words from the women collated by the team and arranged through group poetry process)

Like a Dreamtime story
Like they were children guiding me
Everything makes a sound
Always listen to the birds and the old people
The sounds of footprints going through country
It takes you back to the time before there were cities
My ancestors were there, telling me where to walk
When you hear it, you can feel it
Sitting in the dark, can't hear, no bird, no fire
I can't say the meaning
The true meaning is deep within us, in the dance
The only thing I hear is my heart beating
I feel warm, I feel myself
Centring your spirit

Echoes from the field

In Australia, as in other settler colonial nations such as Aotearoa, New Zealand, and Canada, First Nations peoples are being imprisoned at an alarming rate. Aboriginal and Torres Strait Islander women are the fastest-growing prison population in Australia. Representing nearly 30 percent of the prison population, they are currently 21.2 times more likely to be imprisoned than non-Indigenous women (Australian Law Reform Commission, 2017). The majority are mothers, experiencing the trauma associated with separation from family, community, and country; their incarceration creates a ripple that affects entire communities (Walters & Longhurst, 2017). This is part of the 'torment of powerlessness' (Referendum Council, 2017) experienced by First Nations people in Australia: it includes the continued legacy of forced removal, marginalisation and incarceration that began with Australia's establishment as a penal colony and the systemic racism that pervades our contemporary institutions (Fforde et al., 2013; Henry et al., 2004; Paradies, 2006).

Current prison programs are failing to address the specific needs of First Nations women, and research has shown a demand for holistic, innovative, and flexible approaches to engage and support women and their children before, during and after entering prison (Kendall et al., 2019). Connection to Country is central to most narratives of First Nations wellbeing. To increase wellbeing, the focus of many efforts in native title, health, education, environment, and cultural heritage lies in strengthening connections to place, belonging and Country. Therefore, the Listening to Country pilot at BWCC represented an original, experimental, creative approach to promoting cultural maintenance and wellbeing among mothers, daughters, sisters, aunties, and grandmothers living off Country in prison.

Creating an immersive audio work with the women was underpinned by the principles of acoustic ecology and Indigenist research. Acoustic ecology (Wrightson, 2000) formed the bedrock for producing the audio work and understanding the connections between listening to environmental soundscapes and the wellbeing of individuals and communities. In Australia, acoustic ecology has emerged as a socially engaged, dynamic, and interdisciplinary field concerned with the ecological, social, and cultural contexts of our sonic environments. Acoustic ecology evolved from research investigating the value of listening to natural environments and the negative implications of exposure to noise on our health and wellbeing (Schafer, 1977). Acoustic ecology incorporates other understandings around the value of listening, including Steven Feld's (1996) 'acoustomology', with sound as a distinctive medium for knowing the world, and Pauline Oliveros's (2005) 'deep listening', where sound facilitates expanded consciousness and healing with transformational changes in the body and mind.

The researchers were conscious of using sound intentionally to promote 'acoustical agency' (Rice, 2016), where participants might take control of the sonic environment and resist the oppressive industrial soundscapes of the prison. On a larger scale, Brandon LaBelle (2018) frames the burgeoning movement towards political resistance through sound and sonic cultures as 'sonic agency'. Key to the process were yarning and Indigenous storywork, research methods which are grounded in the idea that meaning-making occurs through storytelling, and our bodies *are* the story (Archibald, 2008; Barlo et al., 2021; Hughes, 2013). Working with story in Aboriginal health research means moving past toxic stories and telling our stories in ways that make us stronger (Abadian, 2006; Wingard & Lester, 2001). Arts-led and poetic forms of inquiry were central to our methodology. Creative acts such as drawing, sound maps, poetry, storytelling, and yarning enhanced and expanded our immersive listening and acoustic ecology processes.

Over a two-week period, the Listening to Country team used these principles and approaches with a group of women at BWCC, a reception and remand centre located just west of Brisbane, where Aboriginal and Torres Strait Islander women make up nearly a quarter of the prison population. The participant group fluctuated in size between one and 12, depending on a range of outside factors that included health or legal visits, transfer to other centres and release. The process involved listening to environmental soundscapes that we brought in from outside, yarning together about what feelings and memories these evoked, then exploring what sounds or environmental soundscapes might connect the women to their Country or their 'belonging place'. It was essential to include the broader idea of belonging for those women who did not know their country or ancestry. We then visited the locations the women selected and made environmental field recordings at different times of the day. This included dusk soundscapes with a crackling campfire, dawn chorus in the rainforest, laughing kookaburras by a river, local waterfalls, ocean waves, and dolphins underwater. The field recordings were edited into short soundscapes from each location (over 15 sites) and brought back to the prison to begin listening and collaborative composition with the women.

We also made recordings inside the prison of words, poetry, breathing, heartbeats, clapsticks, and footsteps, which were layered and sculpted with the environmental soundscapes. This process was not without its tensions. For security reasons, we could not bring a laptop into the prison. So the composition process in prison was done with drawing, brainstorming, and constructing 'sound maps' using paper printouts. This, in turn, created tensions around cultural safety and ownership, as the women felt they were experiencing less control over the creative process, as discussed earlier. In addition, the process of deep listening and recording was challenged by the industrial soundscape of the prison environment as described elsewhere, which disrupted the potential for quiet focus and reflection. Yet there were also many moments of meaningful engagement and connection. The emerging outcomes demonstrate potential for our acoustic ecology approach to be used for building evidence of what works in healing and rehabilitation programs within and beyond the prison walls and in communities (Marchetti et al., 2022).

The final work was produced in surround sound, using a quadraphonic array for playback in the prison to enhance the sense of immersion. The work also exists as a binaural mix (for headphone playback) and a stereo version for use in knowledge translation settings. The women provided the research team with permission to share the work outside the prison, but as the soundscape was produced for a very specific purpose inside the prison, the version that is shared publicly has been adapted, with changes including filters on voices so participants cannot be identified.[13] During future iterations of the work, we hope to make the soundscape as accessible as possible inside the prison, and we will shift away from headphone listening experiences to transducers in public areas to diffuse the sounds. This could include constructing 'listening walls', where leaning against the wall's surface will trigger the soundscapes to play through the listener's body and allow them to connect with Country in an immersive and embodied way. This would also open possibilities to shift away from a fixed media experience towards live streams and responsive soundscapes that facilitate more personalised listening experiences.

After analysing the data from the pilot and identifying recommendations for ways forward to enhance our approach, the work is gathering support from Elders and other stakeholders in the community. We plan to expand the program to other carceral and institutional contexts and sites, to explore how Listening to Country may be transferable to and beneficial for at-risk youth, Elders/seniors off Country in care or hospital, women transitioning from prison to the community and others. The pilot has informed the further development of a masterclass developed to respond to interest from different groups in the communities, with capacity development and knowledge translation activities planned throughout Australia to explore possibilities and questions raised by First Nations groups.

In 2019, we shared the soundscape with women from a different prison, the Southern Queensland Correctional Centre, situated about 80 kilometres west of BWCC on Yuggera Ugarapul Country.

This was the reflection the women shared as collated in another echo poem.

Echoing words from the field

(Words of the women collated by the team) Figure 1.4

I could see where we were
Every little sound
The crickets the birds
the old people
my kids, the grass, sand, the water
I heard everything
The raindrops,
the snake hiss or was that just me
I heard the riverbank, tea trees,
night time fishing, the bridge
Looking out at a full moon
I heard everything, water mainly
Aboriginal people dancing, red people, fire
A lot going on, love, gathering, dancing,
ancestors, children laughing
The birds, they were morning birds,
Following in footsteps
I could hear
The waterfall, the fire, the emu

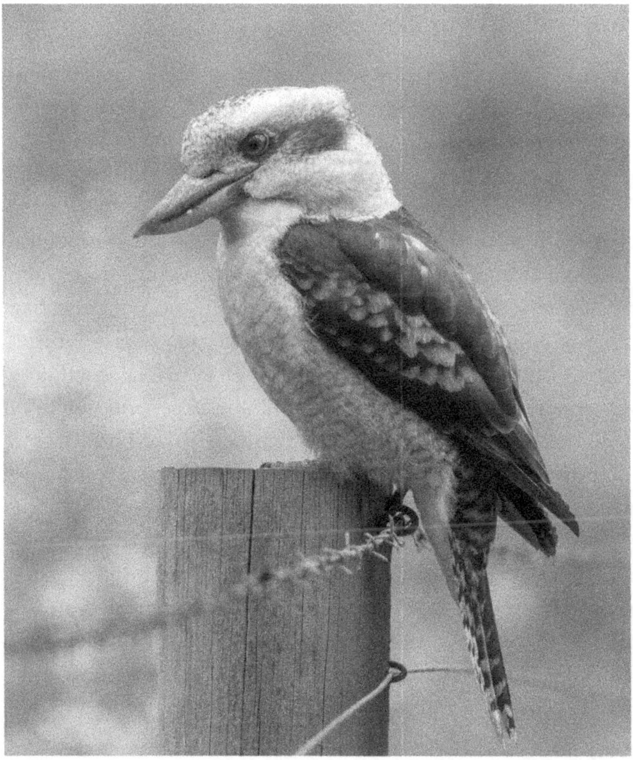

FIGURE 1.4 'Very Happy Kookaburra with Barbed Wire'

When I went to Highfields
The St George river
At night, the clapsticks
Kookaburra
My family and I fishing
Yarning around the fire
I read raindrops,
just a camp you know
my heartbeat,
hills
I heard cultural chanting
Walking through water
Footsteps in the dirt
Rainforest
Fire
My image is of a very happy kookaburra
And barbed wire

Concluding (but not ending)

Writing this chapter has sparked a preliminary interrogation of sonic engagement and the ethics and aesthetics of community engaged research in Listening to Country. Through our collaboration, we have invested time in growing the necessary relationships and partnerships that enhance and amplify cultural connection, maintenance, and wellbeing for ourselves and others. And we have tried to do so in ways that align with an aesthetic/ethic (made visible by the ontological principles embedded in Dadirri and First Nations ways of knowing, doing and being), and an emergent aesthetic/ethic (made audible in the creation of collective soundscapes, or landscapes of composed sounds, connected to particular Countries and places of belonging). Using the acoustic technologies available (including language and song), our aspirations were to amplify the soundscapes and voices of knowledge holders in contexts where First Nations women paradoxically find freedom in the carceral/institutional contexts that they inherit. In these places, we shared stories with mothers, daughters, sisters, aunties, and grandmothers living away from home, from family, and we listened. Doing this (applying the practices and methods of Dadirri, sonic engagement, acoustic ecology, and the immersive sonic experiences that emerged through our collaboration) has changed us all. While present only in written (silent) form here and anonymous in this chapter, we thank the women whose voices have shaped our understandings of what Bu'ra'nga'man | Dadirri | Yimbali mean and changed our understanding of research and carceral care.

My belonging place
Is safe
It's where you know
you're home

Acknowledgements

The Listening to Country pilot at BWCC, and the resulting knowledge translation, were supported with a grant from the Lowitja Institute for Aboriginal and Torres Strait Islander Health Research. The project team consists of two Aboriginal and two non-Aboriginal researchers, and two Aboriginal Elders. Aunty Melita Orcher and Aunty Estelle Sandow from the Brisbane Council of Elders have been working for several years as volunteers in Queensland prisons to reconnect incarcerated Aboriginal men, women, and young people with family, Country, and culture. The team was supported by an advisory group that included the Brisbane Council of Elders and Dr Claire Walker (Wiradjuri), Director of the Murridhagun Cultural Centre of Queensland Corrective Services.

The research discussed in this chapter was completed with support from Queensland Corrective Services. The views expressed herein are solely those of the author and in no way reflect the views or policies of Queensland Corrective Services.

Notes

1 **First words First:** For the authors of this chapter, putting first words first has become a practice that guides our work. It is a way of acknowledging first, the Countries that ground our work and the place, and the word, that drew our research interests together. It was our individual encounters with the word Dadirri, along Maiwar, the Brisbane River, on Jagera and Turrbul Country, in South East Queensland, Australia that first brought us together. Dadirri is a philosophy, practice and principle that is becoming more widely known and comes from the Ngan'gikurunggurr and Ngen'giwumirri languages of the Aboriginal peoples of the Daly River region in Australia's Northern Territory. Senior Australian of the Year (2021) Miriam-Rose Ungunmerr-Baumann first introduced this term to locate and frame her work and ways of being the world (Ungunmerr-Baumann, 2017). As a practice, Dadirri means listening with more than the ears and is a listening practice like listening to the sounds beneath the sounds of a flowing river. Yimbali is a Gunggari word from South West Queensland and was the first word we encountered that echoed the healing and spiritual elements of Dadirri foregrounded by Aunty Miriam Rose. Bu'ra'nga'man was the first phrase we encountered as we began writing this chapter; as we began thinking about what we had heard and where we are all now, and where we currently are not—at the Brisbane Womens Correctional Centre (BWCC), near Maiwar, along the Brisbane River that is flooding as we write. Bu'ra'nga'man encapsulates our listening approach as we begin to write this in the year that starts the UNESCO International Decade of Indigenous Languages (UNESCO, 2020). These words come from the Gubbi Gubbi/Butchulla languages of South East Queensland and mean understanding, knowing through two-way listening. We use a different font for these words to signify their situation outside and beyond Anglo-European ways of knowing and modes of expressing knowledge.

2 The Uluru Statement from the Heart arose from a National Constitutional Convention held in at Uluru in 2017, where First Nations Peoples gathered to develop a proposal for recognition, truth telling, and Makarrata (agreement-making) https://ulurustatement.org/the-statement/ See also Referendum Council, 2017.

3 https://www.miriamrosefoundation.org.au/about-miriam-rose-foundation/

4 The term 'Country' is capitalised to describe First Nations Country or Countries affiliation and belongingness. Country is an essential ontological concept for many First Nations Peoples, and it aids in sharing their relationship with land, place, and a sense of belonging (Carlson, 2016).

5 The poem inscribed here is something we call an echo poem—an echo poem aspires to embody Bu'ra'nga'man | Dadirri | Yimbali. This poem uses words captured verbatim—words that resonated with the women as we listened to each in the prison—yarning (talking together) about the sounds of the places that ground and centre our sense of home and belonging. The space for this yarn was held by two Elders from the BWCC Elders Group (Aunty Melita Orcher and Aunty Estelle Sandow) and revolved around their memories of the sounds of places that make them feel well. In talking about these sounds we began remembering the sounds of our own belonging places. These words have been collated as a memento, to condense and echo what we heard.

6 'Womba' is a word commonly used in Indigenous communities across Queensland and broadly means 'crazy'.

7 Yarning is a fundamental practice within Australian First Nations communities—a sharing of information, with a view to generating understanding (Bessarab & N'Gandu, 2010).

8 This was an actual sound identified by the women as contributing to the traumatic day-to-day prison soundscape.

9 This is a Yolŋu language word for the instrument that has become generically known as 'didgeridoo'.

10 The traditional Yuggera language name for the Brisbane River.

11 The traditional Yuggera language name for Brisbane.
12 'Mob' is a colloquial term used by First Nations Australians for those within specific kinship relationships, but also referring to fellow First Nations peoples in general.
13 To hear the public version of the soundscape, visit www.listeningtocountry.com

References

Abadian, S. (2006). Cultural healing: When cultural renewal is reparative and when it is toxic. *Pimatisiwin: A Journal of Aboriginal and Indigenous Community Health*, *4*(2), 5–28. http://www.pimatisiwin.com/online/?page_id=356

Archibald, J.-A. (2008). An Indigenous storywork methodology. In G. J. Knowles, & A. L. Cole (Eds.), *Handbook of the arts in qualitative research: Perspectives, methodologies, examples, and issues* (pp. 371–393). Thousand Oaks, CA: Sage.

Australian Law Reform Commission. (2017). *Pathways to justice: Inquiry into the incarceration rate of aboriginal and torres strait islander peoples*, final report no. 133. Sydney: Australian Law Reform Commission. https://www.alrc.gov.au/publications/indigenous-incarceration-report133

Barlo, S., Boyd, W. B. E., Hughes, M., Wilson, S., & Pelizzon, A. (2021). Yarning as protected space: Relational accountability in research. *AlterNative: An International Journal of Indigenous Peoples*, *16*(5), 90–98. https://doi.org/10.1177/1177180120917480

Bessarab, D., & Ng'andu, B. (2010). Yarning about yarning as a legitimate method in indigenous research. *International Journal of Critical Indigenous Studies*, *3*, 37–50. http://hdl.handle.net/20.500.11937/37083

Carlson, B. (2016). Striking the right chord: Indigenous people and the love of country. *AlterNative: An International Journal of Indigenous Peoples*, *12*(5), 498–512.

Doctors for the Environment Australia (2020). Submission into the 2019–2020 Independent Review of the Environmental Protection and Biodiversity Conservation (EPBC) Act. https://www.dea.org.au/wp-content/uploads/2020/12/2019-2020-Independent-Review-of-the-Environmental-Protection-and-Biodiversity-Conservation-EPBC-Act.pdf

Dudgeon, P., Milroy, H., & Walker, R. (2014). *Working together: Aboriginal and Torres Strait Islander mental health and wellbeing principles and practice*. Australia: Telethon Kids Institute, Kulunga Aboriginal Research Development Unit, Department of the Prime Minister and Cabinet.

Eckermann, A. K., Dowd, T., Chong, E., Gray, R., & Nixon, L. (2010). *Binan goonj: Bridging cultures in aboriginal health*. Australia: Elsevier.

Feld, S. (1996). Waterfalls of song: An acoustemology of place resounding in Bosavi, Papua New Guinea. In S. Felt & K. H. Basso (Eds.), *Senses of place*. Santa Fe, NM: School of American Research Press.

Fforde, C., Bamblett, L., Lovett, R., Gorringe, S., & Fogarty, B. (2013). Discourse, deficit and identity: Aboriginality, the race paradigm and the language of representation in contemporary Australia. *Media International Australia*, *149*, 162–173. 10.1177/1329878X1314900117

Henry, B. R., Houston, S., & Mooney, G. H. (2004). Institutional racism in Australian healthcare: A plea for decency. *Medical Journal of Australia*, *180*(10), 517–520. 10.5694/j.1326-5377.2004.tb06056.x

Hughes, K. E. (2013). Resilience, agency and resistance in the storytelling practice of Aunty Hilda Wilson (1911–2007), Ngarrindjeri Aboriginal Elder. *M/C Journal*, *16*(5). http://www.journal.media-culture.org.au/index.php/mcjournal/article/view/714

Kendall, S., Lighton, S., Sherwood, J., Baldry, E., & Sullivan, E. (2019). Holistic conceptualizations of health by incarcerated Aboriginal women in New South Wales, Australia. *Qualitative Health Research, 29*(11), 1549–1565. 10.1177/1049732319846162

LaBelle, B. (2018). *Sonic agency: Sound and emergent forms of resistance*. London: Goldsmiths Press.

Marchetti, E. Woodland, S., Saunders, V., Barclay, L., & Beetson, B. (2022). *Listening to Country*: A prison pilot project that connects Aboriginal and Torres Strait Islander women on remand to Country. *Current Issues in Criminal Justice, 34*(2), 115–170. https://doi.org/10.1080/10345329.2021.2018813

Moran, D. (2019). How the prison environment can support recovery. *Prison Service Journal*, (242), 44–49. https://www.crimeandjustice.org.uk/publications/psj/prison-service-journal-242

Oliveros, P. (2005). *Deep listening: A composer's sound practice*. New York: iUniverse.

Paradies, Y. (2006). A systematic review of empirical research on self-reported racism and health. *International Journal of Epidemiology, 35*(4), 888–901. 10.1093/ije/dyl056

Referendum Council. (2017). *Final report of the Referendum Council*. Canberra: Department of Prime Minister and Cabinet. https://www.referendumcouncil.org.au/sites/default/files/report_attachments/Referendum_Council_Final_Report.pdf

Rice, T. (2016). Sounds inside: Prison, prisoners and acoustical agency. *Sound Studies, 2*(1), 6–20. 10.1080/20551940.2016.1214455

Robinson, D. (2020). *Hungry listening: Resonant theory for Indigenous sound studies*. University of Minnesota Press.

Schafer, R. M. (1977). *The tuning of the world*. New York: Random House.

UNESCO. (28 February 2020). *Upcoming decade of Indigenous languages (2022–2032) to focus on Indigenous language users' human rights*. https://en.unesco.org/news/upcoming-decade-indigenous-languages-2022-2032-focus-indigenous-language-users-human-rights

Ungunmerr-Baumann, M-R. (2017). To be listened to in her teaching: Dadirri: inner deep listening and quiet still awareness. *EarthSong Journal: Perspectives in Ecology, Spirituality and Education 3*(4): 14–15.

Walters, A. & Longhurst, S. (2017). *Over-represented and overlooked: The crisis of Aboriginal and Torres Strait Islander women's growing over-imprisonment*. Melbourne: Human Rights Law Centre and Change the Record. https://www.hrlc.org.au/reports/2017/5/18/report-over-represented-and-overlooked-the-crisis-of-aboriginal-and-torres-strait-islander-womens-growing-over-imprisonment

Wingard, B., & Lester, J. (2001). *Telling our stories in ways that make us stronger*. Adelaide: Dulwich Centre.

Woodland, S. (2021). The art of listening in prison: Creating audio drama with incarcerated women. In H. Cahill & P. O'Connor (Eds.), *The Routledge companion to applied performance* (pp. 65–76). London: Routledge.

Wrightson, K. (2000). An introduction to acoustic ecology. *Soundscape: The Journal of Acoustic Ecology 1*(1), 10–13. https://ciufo.org/classes/ae_fl13/reading/Intro_AE.pdf

PART II

Sonic Knowing: Meaning and Resonance

INTRODUCTION

Sarah Woodland and Wolfgang Vachon

Much has been explored and written about how art and performance, both *in* and *as* research, can reflect the complexities of human experience and bring about embodied forms of knowing (see Arlander et al., 2017; Leavy, 2018). Returning to the idea of aesthetics that we framed in the Introduction, we hope that everyone involved in a work's creation and reception will achieve some level of transformative insight. We suggest that this is a goal (for want of a better word) that drives most community engaged art works. Whether political, pedagogical, celebratory, or ameliorative, these works are made to integrate affect and effect (Thompson, 2009), ideally privileging neither one over the other in generating meaning and bringing about change. In which case, we are led to wonder what insights can be achieved specifically through sound and listening in arts-based or arts-led research? What can we *know* from sonic engagement? And of particular interest for several of the contributors to this volume (including us), what unique affordances does sound or audio as a participatory arts-based research methodology bring to the process of inquiry? Of course, we recognise that no single theory or approach can contain the answers to these questions. As Jonathan Sterne (2012) observes,

> There is no a priori privileged group of methodologies for sound studies. Instead, sonic imaginations are guided by an orienting curiosity, a figural practice that reaches into fields of sonic knowledge and practice, and blends them with other questions, problems, fields, spaces and histories. (p. 6)

This diversity is further reflected in Michael Bull and Marcel Cobussen's (2020) recent *Bloomsbury Handbook of Sonic Methodologies* that considers 'research being done on and through sound' (p. 2). Many contributions to that volume attend to similar considerations that we find in ours, exploring sound arts as both medium

DOI: 10.4324/9781003164227-6

and method, and recognising the dynamism of sound in the creation of knowledge: 'Sound appeals not only to the cognitive and the rational but also gives space for associations, imaginations, and speculation as alternative forms of knowing' (p. 5).

The chapters we have grouped under this theme bring forward innovative approaches to qualitative research and pedagogy using sound, from audio drama and radio plays (Vachon, Knowles, & Tester, respectively) to 'participatory sound portraits' (Flint et al.) and 'cellphilms' (Burkholder & McEntee).

To begin, Wolfgang Vachon frames 'audio drama inquiry' as a novel form of arts-based research. He shares the approach he developed through his PhD project, *Tuning into Child and Youth Care,* an audio drama that explores the experiences of Child and Youth Care practitioners who have lived in residential placement. His central claim is that audio drama inquiry is a 'sonically dynamic' and 'ethically engaged, community responsive, and aesthetically enticing approach' to creating, analysing and disseminating research. Vachon acknowledges that ethics and aesthetics in audio drama inquiry are intertwined—what he explores through the framings 'aesthetics of ethics', 'aesthetic-ethics', and 'ethico-aesthetics'.

Casey Burkholder and Katie MacEntee discuss sound in cellphilm—the use of cell phone video as a research tool created by people to explore topics relevant to their own lives. Writing from Canada, the authors (both scholars) describe using this approach with 2SLGBTQ+ youth in New Brunswick to examine school-based experiences and oppression; and 'young cis and queer women who trade sex' in Toronto to explore harm reduction and access to health care. Through their reflexive study, the authors explore the micro politics of timbre (Eidsheim, 2019), DIY aesthetics and constructions of 'noise' and silence; concluding that 'Cellphilm audio aesthetics work to affirm the soundscapes of youth cultures'.

The contribution from Maureen Flint, Morgan Shiver, and Ryanne White asks, 'How might the process of composing sound portraits attune creators and listeners differently to the layered histories of place?' The authors draw on Tina Campt's (2017) notion of 'infrasounds', participatory listening, and Black feminist futurity, to explore two audio portraits of place related to a particular southern USA university. The chapter represents a compelling example of how research participants in ethical arts-led processes can, in some cases, become collaborators who then assume responsibility for the creation and transmission of knowledge.

Finally, Tanyss Knowles and Frank Tester considers a radio play—*A Mural of Social Change*—that was produced by the Vancouver Association for Restorative Justice (VARJ) in collaboration with Cree playwright Renae Morriseau, with the aim to 'inform the public about how restorative justice [RJ] works in real-world settings'. We suggest that the pedagogical potential of the work is strengthened by an 'aesthetics of listening' as we outlined in our Introduction. This operates on multiple levels to engage all involved in the generation of knowledge: the radio play was created through a collaborative process that mirrored RJ, it invites the public into an intimate space of listening and 'emotional potency', and the drama itself reflects an RJ process that is largely dependent upon respectful and generative listening.

References

Arlander, A., Barton, B., Dreyer-Lude, M., & Spatz, B. (Eds.) (2017). *Performance as research.* Taylor & Francis.

Bull, M., & Cobussen, M. (2021). *The Bloomsbury handbook of sonic methodologies.* Bloomsbury.

Campt T. M. (2017). *Listening to images.* Duke University Press Books.

Eidsheim, N. S. (2019). *The race of sound: Listening, timbre, and vocality in African American Music.* Duke University Press.

Leavy, P. (Ed.) (2018). *Handbook of arts-based research.* Guilford Publications.

Sterne, J. (2012). Sonic imaginations. In J. Sterne (Ed.), *The sound studies reader* (pp. 1–17). Routledge.

Thompson, J. (2009). *Performance affects: Applied theatre and the end of effect.* Springer.

2

AUDIO DRAMA INQUIRY: A TELLING METHOD OF RESEARCH

Wolfgang Vachon

Introduction

Child and youth care (CYC) is a practice-based field that supports children, youth, families, and communities through a 'developmental-ecological perspective [which] emphasizes the interaction between persons and the physical and social environments, including cultural and political settings' (National Association of Child Care Workers, 2014, para. 1). Canada, South Africa, and the United Kingdom have the most developed CYC infrastructures, including formal education at the diploma, Bachelor, Masters, and Doctoral level. However, with only one Doctoral and three Masters' programs in existence, CYC does not have a deep research literature. As such, many of the concepts and practices informing CYC are adapted from allied disciplines. With two of the three Masters programs opening in the past decade, there have been important discussions regarding which approaches to inquiry are suited to the field, the implications and impacts of different methodologies, and how these might be taken up by practitioners, students, and other scholars (Bellefeuille & Ricks, 2010; Vachon, 2021). One exciting, albeit contested, area of this discussion is the value of arts-based inquiry (ABI) in CYC.

Tuning into CYC[1] intentionally situates itself at the centre of these methodology discussions by using ABI in the generation and dissemination of the study (Leavy, 2018), drawing on approaches to inquiry informed by research-based theatre (Belliveau & Lea, 2016), ethics of care (Tronto, 2015), CYC scholarship (Gharabaghi & Charles, 2019), Mad Studies (LeFrançois et al., 2013), and aesthetic theory (Born et al., 2017). *Tuning into CYC* is a collaborative CYC ABI developed with two cohorts of child and youth care practitioners with placement histories resulting in two publicly available audio dramas. The first cohort collectively devised two scripts (only one of which has been produced so far due to COVID

DOI: 10.4324/9781003164227-7

restrictions) and the second participated in conversations with me which were subsequently turned into a verbatim audio drama.

Child and youth care practitioners from care (CYCPfC) have formal education in CYC and a history of living in a residential placement, such as group homes, foster homes, or semi-independent living as children or youth. Placement criteria did not include detention facilities, hospitals, shelters, or psychiatric institutions (although, several of the participants also had experiences living in one or more of these facilities). The study explores how being from care and working as a CYCP is understood, conceptualised, theorised, and performed by CYCPfC. The audio dramas present the findings of the study, which have been fictionalised and turned into narratives. Holstein and Gubrium (2012) write, 'Stories are discursive ways of doing or accomplishing something. The narrator or storyteller actively shapes or constructs narrative reality to achieve particular descriptive, rhetorical ends—be they personal myths or extraordinary events' (p. 7). Stories are a way that people make meaning of events in their lives. People (both individually and collectively) shape and construct their narrative reality, which leads to shaping their lives, our lives—our reality. In creating our stories, we create ourselves. Through audio drama inquiry, creators have the opportunity to manifest what might be.

Audio drama inquiry

Audio drama is an emerging approach to qualitative inquiry with few examples and little methodological, theoretical, or practical writing about the form. In the process of trying to understand this nascent method, Sarah Woodland and I introduced the term *applied audio drama* and sought to map the contours of this approach (Vachon & Woodland, 2021). As both of us work in applied theatre, we drew heavily on this literature to situate our audio drama practices. Applied theatre encompasses theatre creation with(in) community groups who may or may not have theatre experience. For example, much of Woodland's work is with incarcerated women (Woodland, 2020) and my own has frequently been with young people who have social service involvement (Vachon & McConnell, 2018). Applied theatre projects are generally created for audiences who are from, or interested in, the communities creating the productions. We also drew on O'Connor and Anderson's (2015) conceptions of applied theatre as research (ATAR), which they frame as 'an act of political and cultural resistance that creates through the fictional frame a set of propositions that are co-constructed, analysed and then re-presented to communities as a method of creating new knowledge and forging social change' (p. 49).

This construction by O'Connor and Anderson aligned with our own work in applied audio drama, which we saw as 'engaged with process as much as product … not limited to specific structures, lengths, or forms', 'aesthetics [being] entwined with ethics and politics', and the work frequently containing 'socio-cultural agendas and provocations' (Vachon & Woodland, 2021, p. 4). Both ATAR and applied audio drama have at their roots an engagement with community, situate the process itself as a possible site of inquiry (MacNeill, 2018), and function with a socio-political and critical theory analysis. There is much I like about ATAR, and I continue to use many of the ideas and techniques; however, since not everyone is familiar with ATAR or applied theatre, I prefer the phrase audio drama inquiry as a way to make explicit that the form is a research method.

In her introduction to the *Oxford Handbook of Qualitative Research,* Leavy (2020) recognises the 'essence of qualitative inquiry as a way of understanding, describing, explaining, unraveling, illuminating, chronicling, and documenting social life' (p. 1). Audio drama inquiry lends itself particularly well to all of these elements. This is certainly not unique to audio drama alone; many forms of performance can do this, which may be why performance as a mode of qualitative inquiry is garnering a lot of attention at the moment. For Denzin and Lincoln (2018):

> A performative project informed by research and inquiry, involves acting in the world so as to make it visible for social transformations. This is a postqualitative, post-research-inquiry-world. It is a world defined by risk taking, by textual experimentation, by ontologies of transformation, a world defined by acts of love, struggles, and resistance, a world shaped by dramatic radical acts of activism. (p. 11)

Audio drama inquiry is an experimental risk-taking form well suited towards understanding, chronicling, illuminating, and transforming, given its ability to not only document and reflect but in its capacity to imagine and refract.

Soltani (2018) provides a definition of 'audio-dramaturgy', which he sees as unique from other forms of dramaturgy, including that used for radio drama:

> Audio-dramaturgy then, is the praxis that turns a fiction into a dramatic presentation in sonic form, to be experienced as the 'here and now' of a world by a listener: the practices and processes that construct and configure the sound structure, the technologies involved in its creation, aesthetic paradigms, hypotheses about the audience's response, and so on (p. 192).

Functioning within the here and now for the listener is an approach that ABI frequently draws on. In the opening sentence of the introduction to *Handbook of Arts-Based Research*, Leavy (2018) writes, 'Art, at its best, has the potential to be both immediate and lasting' (p. 3). This is in direct contrast to how research is often written about, for example, the American Psychological Association's 7th edition of their publication manual instructed writers that the reporting of results must be written in the past tense (American Psychological Association, p. 118). This automatic archiving forces a distance between the reader and research.

Audio drama inquiry brings the research experience into listeners' present life and invites them to engage by hearing *now*, at times in first-person, creating a form of inquiry immediacy and intimacy. Listening to an audio drama inquiry is a present tense experience complicated by the recognition that what is being heard is not actually happening, it happened. And it is not actually me. Further, audio drama is fiction which both invites intimacy through the structures of storytelling and creates distance through the acceptance that 'this is not real'.

Listening is a verb, an active process that requires attention, negotiation, and integration into/with self in relationship to that which is being listened to. In discussing 'world music' Roshanak Kheshti identifies the listening subject entering the 'space/time of *others*' (Brooks & Kheshti, 2011, p. 331, italics original). Given the troubled past of research, particularly its abuses of people (Smith, 2012), the idea of entering the space/time of others holds importance in audio drama inquiry, as well as all forms of qualitative inquiry. When listening to research as though being present (experiencing it as happening now), there are ethical and aesthetic implications which have the potential to simultaneously engage and 'implicate' the listener. This is a relevant, and evocative, shift for those who encounter the research. Audio drama inquiry is not only about what was, but also about what is, how it is, and the potential of what might become. The present and future framing of audio drama inquiry implicates and thus asks for accountability from all parties—researchers, listeners, and others.

Research-based theatre

Tuning into CYC draws upon several forms of research-based theatre (RbT), including Applied Theatre as Research (ATAR), playbuilding, verbatim, and ethnotheatre. Saldaña (2011) writes that ethnotheatre:

> Employs the traditional craft and artistic techniques of theatre or media production to mount for an audience a live or mediated performance event of research participants' experiences and/or researcher's interpretation of data. The goal is to investigate a particular facet of the human condition for purpose of adapting those observations and insights into a performance medium. (pp. 12–13)

Saldaña (2011) distinguishes ethnotheatre from ethnodrama, the latter of which he defines as, 'A written play script consisting of dramatized, significant selections of narrative ... Simply put, this is dramatizing the data' (p. 13).

Norris (2009) uses a process called 'Playbuilding' and sees 'Playbuilding as research', which can happen at any stage throughout the inquiry process. He shows examples of playbuilding used in the developing (gathering), mediating (analysing), and communicating (disseminating) the materials (data). This he presents in contrast to some other forms of research-based theatre, such as ethnodrama, 'Where data is traditionally collected, analyzed, and then disseminated through an "alternative" form of representation. With Playbuilding, data is generated and interpreted in a different manner, and, at times, these three phases are simultaneous' (p. 22). As will be shown in the 'Method' section below, audio drama inquiry was used through all phases of *Tuning into CYC,* not only as a form of dissemination.

Verbatim theatre can be thought of along a continuum from 'tight' to 'expansive' (Vachon & Slavatore, 2022). Tight verbatim is a 'word for word and gesture for gesture' (Salvatore, 2020, p. 1046) recreation of what was said and done by the people being enacted. We can glean a more expansive framing from Hammond and Steward (2008) who write:

> Verbatim refers to the origins of the text spoken in the play. The words of real people are recorded or transcribed by a dramatist during an interview or research process, or are appropriated from existing records such as the transcripts of an official enquiry. They are then edited, arranged or recontextualized to form a dramatic presentation, in which actors take on the characters of the real individuals whose words are being used. (p. 9)

For Hammond and Steward verbatim theatre is less a rigid set of rules to follow, and more a variety of overlapping processes, tools, and approaches based on archival or interview text used in creating a performance. They see it as a technique, rather than a form. Whether tight or expansive, when working in verbatim, one must think of the reasons for its use, its limits, intentions, and textual implications, among other considerations (Vachon & Slavatore, 2022).

RbT might be thought of as an umbrella term encompassing all the above theatre-based inquiry processes. RbT 'employ[s] creative means of data generation, analysis, and dissemination with stakeholders, community members, and other researchers … [As a] means of knowledge translation that uses embodied approaches to address a broad range of critical social issues' (Belliveau et al., 2021, p. 141). The idea of embodied approaches becomes provocative when considering audio drama. Audio drama inquiry is a participatory or community engaged process, and thus works with embodied subjects throughout the creation; further, it understands the end listener as an embodied subject, yet it may be heard or thought of as dis-embodied. It might be worth noting that when I had conversations with CYCPfC who resided outside of Toronto, these were most frequently held through telephone or internet without a video component. Voice is without a doubt connected to the body, although in audio drama the listener may hear it as disconnected from a particular body and is thus able to project it onto bodies of their knowledge or imagination.

Drama, be it audio, video, or live performance, exists predominantly in the realm of fiction. When working in fiction there is an invitation, 'All *dramatised* stories—that is all stories clothed in *fiction*—are conversation starters' (Prendergast & Saxton, 2015, p. 280, italics in original). Rather than definitively answering questions, dramatised stories can create space for dialogue and reveal opportunities to discover truths, through fiction, collaboratively. This participatory fictive aesthetic is a conscious resistance to traditional empirical claims of truth, such as those strongly associated with positivist traditions and purity claims to objectivity, rationality, and beliefs in the scientific method as well as the implication that once the 'ignorant understand', change will happen (Rosiek, 2018). Trustworthiness in drama is less attuned to the factual accuracy of the story and more attuned to what the story does, why it resonates, how the source material is refracted through the creative processes and received by the listener, the multiple ways research collaborators, creators, and listeners experience the story. Drama creates spaces for audiences to find their truths in the fiction.

The contours of audio drama inquiry

Audio drama inquiry is an aurally experienced research-based storytelling method, which exists in the digital sonic realm informed by, yet distinct from, radio, theatre, video, or film. It is commonly situated within podcast culture, yet holds a unique and narrow place. As the label indicates, it is fictionalised, while drawing on elements of documentary, memoir, or interview-based productions. Indeed, in *Tuning into CYC* many of the tropes of these non-fiction styles are present. Audio drama inquiry also holds similarities with (and owes some of its value to) radio drama, audio books, and audio-based 'entertainment-education' campaigns (de Fossard, 2005; Singhal et al., 2004). Like many forms of arts-based research, one might think of it as bricolage (Denzin & Lincoln, 2018, p. 11; Leavy, 2020, p. 9)—a method constructed from multiple and diverse elements which together create something new and different from the sum of its parts. For the rest of this chapter, I will lay out several guiding principles informing audio drama inquiry and discuss specific processes used while developing *Tuning into CYC.*

Inquiry led

Questions are central to all inquiry projects, and most studies begin with a set of well-honed research questions. These provide focus and represent the contours of what the researcher(s) will study. While all audio dramas develop through and are informed by questions asked during the script development, rehearsal, and recording process, not all audio dramas should be considered research studies. Audio drama inquiry has an explicit intention through the questions it is exploring and follows processes informed by established tenets of research—such as ethical protocols, being grounded in articulated theory, advancing knowledge claims, situated within a body of literature, using a methodical and public approach that is amenable to scrutiny and critique. Like all research, the depth, shape, and specific process of study for each individual audio drama inquiry will shift from project to project. For *Tuning into CYC* we began by considering the question, "What does lived experience of residential care do in child and youth care?"

Community responsive

Audio drama inquiry is a dialogic process which creates opportunities for multiple perspectives and participants. The level of participation will shift from project to project and all involved must be cognizant of the expectations and limits of the participation. As discussed in the introduction to this book, distinctions can be made between different levels of community participation, and these will depend on the desires, resources, and capacities of community members and researchers. There were multiple forms of community engagement present in *Tuning into CYC*.

Audio

Audio drama lives in recorded audio and exists sonically. The value of sonic engagement will be apparent to most readers of this book. I came to audio through theatre and what I perceived as the benefits of this form over other forms of applied theatre as it relates to participatory research practices. These include its engaging and accessible nature, the potential for large numbers of audience members to experience the production, the longevity of productions to exist on the web, its relatively low costs to produce, conceptions of sonic agency and the ability to resist coercive soundscapes, opportunities for imaginative intimacy, and the theoretical possibilities to explore ideas researched (Vachon & Woodland, 2021).

Truth through fiction

Audio drama inquiry breathes in the realm of fiction, while living in reality. Fiction allows emotional and intellectual distance, while simultaneously bringing the listener closer to the subject(s) investigated. Anyone who has done research has been faced with questions of what to include and what to omit; how to analyse the data and the tools used to do so. This is true whether working in qualitative or quantitative research. Different models and assumptions deliver different results. Nairn and Panelli (2009) claim, 'Academic writing, like "fiction" is a construction of "reality"' (p. 97), each develops narratives in order to present their truths of the story. Audio drama inquiry acknowledges this fictive construction and embraces it for what it can do.

Theoretically supported

Tuning into CYC is informed by the practices and ideas of RbT, CYC, care ethics, and Mad studies, the latter two of which will be discussed more below. In order to situate audio drama inquiry within scholarly literature one must be aware of and be able to articulate what theoretical lenses it draws upon. Arguably, all audio drama works with theory, but not all dramatists, dramaturges, actors, directors, and sound artists can identify the theory utilised, nor coherently articulate it. As this project is part of a doctoral program, the requirements for understanding and working with theory were higher than they might be if I was not being overseen by an academic institution, for example if all of my funding came from arts grants.

Aesthetics of inquiry

It is possible to dismiss aesthetics as irrelevant to inquiry; however, opening up inquiry to the scrutiny of aesthetics provides multiple benefits. This begins with recognising that aesthetics are grounded in theory as much as subjective experience. In addition to notions of beauty, aesthetic considerations include form (Bourriaud, 2002), purpose (Kant, 2007), context (Born et al., 2017), quality (Thompson, 2015), engagement (Berleant, 2016), and the principles that underlie these concepts (Vachon, 2021). Aesthetics are a useful guiding frame for anyone embarking on an inquiry process, be it arts-based or otherwise. Many academics strive towards beauty in their research and writing, if we accept beauty as 'that quality of a person or thing which is highly pleasing or satisfying to the mind; moral or intellectual excellence' (OED, 2021). The other elements of aesthetics are equally relevant: we have intention (purpose) when we engage in research; we seek to locate and respond to the unique context in which we investigate; we consider the quality of our work both in the sense of excellence and in the sense of its characteristics, conditions, and features; and there is a reciprocal or relational engagement to the work.

Ethically grounded

Audio drama inquiry is rooted in ethical principles, and positions ethics as a generative *process* towards understanding ways of engaging and participating which honours all those who participate. I will discuss ethics, alongside aesthetics, more fully at the end of this chapter.

The method used in tuning into CYC

Methodological choice for an inquiry depends on what is being studied, the intentions of the study, and the people engaged in the process. Inquiry is not static, and all elements may be required to adapt as situations change; this was certainly the case during *Tuning into CYC*. Factors such as participant numbers, COVID-19, funding, and logistical practicalities meant I shifted plans multiple times, some of which then required returning to the institutional ethics review board to request amendments. Fortunately, I began by exploring something that interested me and chose a method that resonated with who I am. Philosophically, it was also important to work in a way that was congruent with CYC practice, pedagogy, writing, and inquiry. These included being relational in my inquiry (Bellefeuille & Ricks, 2010; Vachon, 2021); seeing development of community as an intention and result of research (Etmanski et al., 2014); drawing upon productive ethical principles whereby ethics are considered a means to create possibilities, not only, or primarily, about sanctions (Faulkner & Tallis, 2009); pleasure as a criterion in considering methodological choices and research participation (Foster, 2016); understanding inquiry as acts of care between people and other materials (Barad, 2007); operating from an awareness of decolonising, anti-oppressive, and critical theory approaches (Amponsah & Stephen, 2020; de Finney et al., 2020; Vachon, 2020) and drawing on the skills and knowledge of all people involved (Sweeney et al., 2009). By starting from these positions, I recognise that research is not a matter of description alone, it is the manifestation of matter, and CYC inquiry is 'ontologically generative' (Rosiek, 2018, p. 637).

Material generation

The first audio drama, titled *ReFiled*, was collectively devised by a cohort of seven CYCPfC (participant researchers) and the second, *Hiring Care*, was based on recorded conversations with ten CYCPfC from five provinces across Canada. A third source of material included three 'reflective panels' with 13 CYCPs who watched and then discussed a staged video reading of excerpts from *Hiring Care*.

Participant-researcher group processes

Introductory and script building workshops

There were two distinct stages for generating materials with participant-researchers; introductory workshops and script building sessions. The two introductory three-hour workshops were attended by CYCPfC who expressed interest in the project, five people attended the first workshop and two the second. These sessions were designed to be an opportunity for people to participate in an arts-based data generating process allowing them to 1) contribute to the project; and 2) to experience what the multi-session audio drama inquiry process would be like. The initial workshops used audio, improvisation, conversation, and poetry to generate materials. These sessions were audio and video recorded.

After the initial workshops, we moved into the script development process—generating, mediating, and shaping the empirical materials. Over the course of 12 sessions, we developed two audio drama scripts. *ReFiled* is constructed as a personal narrative podcast investigation told by a CYCPfC working in a group home, who along with a colleague physically restrains a resident. This event catalyses a reconsideration of their role as a CYCP. While reviewing the 'Serious Occurrence Report' (SOR) submitted following the restraint, they see their own past SORs from when they were a resident at the same agency. This motivates them to retrieve their other child welfare and mental health files. The story comes out of direct experiences with participant-researchers as well as distanced and abstracted scenes which capture emotional states that CYCPfC articulated—for example, being shot at for trying to retrieve information about themselves. There is a clear departure in these nightmarish episodes from 'lived experience' representing, though a fictive construction, the traumatic emotional reality identified by CYCPfC.

The second script comprises a series of linked scenes around the themes of dogs, relationships, and care. The narrator is a talking dog who tells its life story, which mirrors some of the experiences of those in the group. Abuse, abandonment, living 'on the streets', going into placement, and then dedicating one's life to caring for others. The story is set in a fantastical near future dystopian world. The theme of dogs was unexpected for me and not an area I was remotely considering when I started this project. The storyline came out of conversations and improvisations with the group during which many participant-researchers identified that they had important relationships with dogs prior to, during placement, and for some continuing to this day. Unfortunately, due to COVID safety concerns, this script has not yet been produced.

Generating materials

We used a wide array of processes to generate and consider materials during the workshops. These included, but were not limited to, sharing experiences through conversation, improvisation, creating soundscapes, developing interview questions, tableau work, poetry, drawing, vocal explorations, constructing conceptual frameworks of the inquiry, letter writing exercises, and others. As part of the creation process, I occasionally brought in outside materials to catalyse the generation of group material, such as transcripts from the conversations I had with other CYCPfC. I quickly learned the group preferred to work with their own stories rather than outside materials. This made sense as using transcripts to glean relevant information can be time consuming, detail oriented, individual work, and may occlude people sharing their own experiences in the process of voicing other perspectives.

Mediating materials

While developing and mediating (analysing) materials, there was an aesthetic engagement (Berleant, 2016) constituted through an iterative praxis. Participant-researchers would improvise around themes (emanating from discussions), which I documented through notes, audio recordings, and video recordings (the latter two of which were then transcribed). I used these materials to write scripts that were brought back to the group to read, analyse, discuss, perform, improvise, and rework—the mediation process. Several times participant-researchers would write text or scenes which would then follow the same mediation process. These scenes were often quite different in tone and style, and thus an important and meaningful contribution to the work (such as the scene described above about being shot at). Other scenes were explicitly autobiographical which allowed people to have specific important elements or aspects of themselves and their stories inserted into scripts, in their own words. After 12 workshops we had the basic shape of two scripts, and we moved into the rehearsal phase.

Rehearsal

The rehearsal phase focused on working with, refining, and exploring the content we had developed. I came to understand these considerations as embodied enactments or dialogues 'between data and ideas' (Coffey, 2018, p. 25). Scripts were acted out scene by scene with different people rotating through different characters. After each scene there would be a discussion to assess if it was reflecting what the group wanted to say. If not, we would do more improvisations and additional changes were made. I would then rewrite, revise, and return. This process went back and forth until we were all satisfied with the scripts. In the fall of 2019, we completed *ReFiled* and had a 3rd draft of what we expected to be 'season two'. Season two (the dog script) elicited more discussion and contrary positions than season one, so we decided to record *ReFiled* and continue working on season two which we planned to record in the spring of 2020.

Recording

Once we completed and cast *ReFiled*, we recorded the audio-drama. The cast was a mixture of participant-researchers who were part of the creative process as well as trained actors. Before recording each day, I led a warm-up followed by run-throughs of the scenes. We then recorded, usually doing several takes where the actors had the scripts in front of them and then two to three takes 'off book'. We recorded in a closed makeshift studio as well as several different locations. One of the many benefits of audio drama, is the ability to have scripts present, which is particularly useful for non-actors, time constraints, or those who have difficulty memorising.

Conversations with CYCPfC

Recorded conversations

Due to lower than hoped for turnout of participants to the initial workshops (seven) and concerns about attrition, I expanded recruitment beyond workshop participants and held conversations with 10 CYCPfC from five Canadian provinces. Drawing on the work of Doucet (2019), I use the term conversation rather than interview in recognition of how some people with child welfare involvement have negatively experienced 'interviews' through their time in the system. As the project was an audio drama inquiry, I made the decision to only record the audio and not video of people.

I wrote a verbatim script based on these conversations. The script is centred around an idea stated by one of the conversationalists that they had considered writing a booklet for organisations who want to hire people from child welfare to work in their organisations. This is the frame used in the script to bring a collection of CYCPfC together to have conversations about 'being from, and working in, care'. Over seven episodes they discuss ideas related to stigma, care, racism, empathy, systemic barriers, education, cultural violence, and the benefits of their lived experiences in CYC work, among other topics related to having placement experience.

I approached the transcripts as an artist-researcher, crafting the verbatim script through finding narrative lines, looking for relationships and conflicts (of people, ideas, and words), and listening for moments of engagement. It was a personal, subjective, and biased process, rife with my own prejudices and agendas. In this approach, I drew on Leavy's (2018) arts-based data analysis procedures which includes, 'Having an internal dialogue', 'Using theory and or literature well', and at the workshop stage 'Garnering feedback from peers' (pp. 578–579). When listening to the conversations, transcribing them, and then re-reading the transcripts, I paid attention to ideas that were similar to and distinct from the collectively created script developed previously. I looked for possible narrative lines considering the aesthetics elements listed above. Informed by theoretical and empirical literature, I tuned into topics that were present or absent in the literature (theories of care, self-identity, notions of benefits and limitations of lived experience, trauma and healing, stigma, among others). I placed conversationalists into 'dialogue' with each other, seeing what was revealed when I had the characters speak to one another. I sought out the sonic possibilities, verbal textures, and what Campt (2017) calls 'infrasounds', those sounds that 'like a hum, resonates in and as vibration' (p. 7). Working this way was generative, expansive, rigorous and creatively satisfying. Once I had a functioning draft, I cast, rehearsed, revised, workshopped, re-revised, re-cast, and then repeated the whole process again.

Working without an established traditional research coding process was more challenging than I anticipated. There is comfort as a researcher in having a step-by-step set of tools to follow from which one can expect particular outcomes. Outcomes that are accepted, recognised, and that fall within conventional academic norms. Ones that are less likely to be challenged because they are familiar to academic audiences, despite their limitations. Qualitative inquiry has been in a continuous process of unsettling research norms (Denzin & Lincoln, 2018), and more recently in unsettling settler frames (Huaman & Martin, 2020). Drawing on ethics of care, Mad studies, and CYC as orienting epistemologies, I sought to be relational, ethical, responsive, care-filled, and aesthetically engaging through(out) the inquiry. To this end, between the two cohorts of CYCPfC, multiple, and cohort specific ways were developed to realise these intentions. With the collectively devised scripts, ethical guidelines were created with the group, the materials were developed and considered alongside the people whose experiences were being studied, and there were multiple opportunities to add, alter, or withdraw materials. With the verbatim cohort, transcripts and the audio recordings were sent to them to verify, they were invited to remove anything that they did not want used in the project, they were offered drafts of the script, invited to workshops/readings of the script, and were asked for feedback (in group, one-on-one, and/or written). With both groups people were compensated for their time; professional artists, editors, and sound designers were hired; and I have developed or enhanced ongoing relationships with many of the people involved in the project, and several of them with each other.

Reflective panels

Reflective panels (adapted from Fenge & Jones, 2012) were developed to solicit input from CYCPs about the verbatim script. Three panels were held, each comprised of 4–6 people with a total of 13 people attending to listen to scenes from *Hiring Care* (one hour) and then partake in a facilitated conversation (one hour). I sought a range of CYCPs regarding geography, type of workplace, ethnicity, and gender. I also wanted input from people with and without placement experiences, so along with selected CYCPs, I invited participant-researchers and those I had conversations with, whose words had been used to develop the script. Conversationalists were invited to watch the video in advance so that they had time to reflect and take care of themselves in case any difficult emotions came up. I also offered to meet with them individually to discuss anything they wanted; two of the three who came to the reflective panel watched in advance, none decided to speak with me one-on-one. All participants were sent the conversation prompt questions in advance. Those who were represented in the script had additional questions sent, which they could respond to privately in writing or in person. Some who attended the panels disclosed participation in the study, others did not.

Based on feedback from the reflective panels I reworked the script, hired a dramaturge, and then did a workshop through New York University's Verbatim Performance Lab with actors, academics, and theatre artists who have experience in verbatim theatre. I was particularly interested in hearing how the play flowed, gaining feedback from an aesthetic perspective (as opposed to primarily a content perspective), and insight into how the piece landed outside of a CYC context. The consensus was that the piece was mostly ready for production, it was relevant and engaging for those outside of CYC, and that it held up as a viable verbatim audio drama.

Casting

I made the decision to work exclusively with trained actors for *Hiring Care*. This was based on a combination of aesthetic and ethical choices. Verbatim performance requires particular skills and approaches unique to the process (Hammond & Stewart, 2008; Saldaña, 2011; Salvatore, 2020, Vachon et al., 2019). Not all actors possess these skills, let alone people without theatre training. I made several choices in hiring, such as ensuring that actors mostly aligned in race, religion (where stated), age, and gender with their characters. All but one of the original conversationalists had been female, and this presented a sonic dilemma for recording. Several of the characters were composites of two conversationalists—I made one of these composite characters male although one of the conversationalists had been female. I also cast one of the female parts with a male actor. While the actor is male, the character remained female. These two casting choices brought more range to the vocal timbre of the performance.

Rehearsing and recording

Some of the people I had conversations with did not want their identity to be known and requested that I change identifying content. This presented some challenges (such as not being able to adhere to a 'tight verbatim' recreation and the omission of content) and some opportunities. I decided to embrace what might be generated by utilising an 'expansive verbatim' approach (Vachon & Slavatore, 2022). Thus, I invited actors to bring themselves, their perspectives, histories, and experiences to the character. This was done as a way to create distance from the original conversationalists, for aesthetic reasons, and as an analytical tool.

From an ethical perspective I continue to struggle with this choice. Wondering how it might distort conversationalists' experiences. If actors do not have placement experience, have not worked in CYC, or live in a different country than where the study took place; how does that change the material presented? These are relevant questions for audio drama inquiry researchers to consider when casting. I think for *Hiring Care*, the casting choices allowed for different readings and understandings of the text.

It is common for researchers, particularly at the PhD level to analyse text alone. By inviting actors to contribute their experience, social-locations, training, theoretical lenses, and history, I invited in different readings of the text. These readings are heard in the performances and added insights for me into the transcript text. This approach also aligned with the relational, community oriented, and intersectional principles informing the inquiry. It is not unusual when an inquiry process is collaborative that multiple people will analyse the materials looking for commonalities and unique understandings. These can take on many different iterations, such as 'codes' in the case of many qualitative inquiry processes (Saldaña, 2016), or themes and images in a Forum Theatre context (Boal, 2002), to provide just two examples. Regardless of the intention, there is a coming together in recognition that each person brings a different understanding to the materials. The readings and perspectives provided by Indigenous, racialised, and female actors went beyond my understanding of the characters. They brought insights that I had not considered, such as what might be behind a seven-second silence after a question from a white man to a woman of colour about race; such as how lines that I glanced over might be coded language about racism; about what might not be being said; and, about how my own biases might be showing up in the script. I also think that this approach to working with verbatim text engages the actor in a way that goes beyond the dismissive moniker of 'talent' and actually honours their wisdom.

Aesthetic of ethics

Aesthetics have long been, and continue to be, central to discussions of arts-based research (Cole & Knowles, 2008; Leavy, 2018; O'Connor & Anderson, 2015; Saldaña, 2011). For *Tuning into CYC*, I considered aesthetics through conceptions of beauty, form, context, purpose, and quality (Vachon, 2021). These elements all overlap and each respond to the others, thus they cannot be understood in isolation.

Aesthetic discussions in ABI risk a reductionist collapse into whether or not the *art* or *research* is subjectively 'good'. The separation into two categories (art and research) is a distinction made by critics and creators alike, revealing, perhaps an unwillingness to comprehend them as a whole. The notion of 'good', while useful if tied to specific criteria, may become shorthand for a personal affective response, and consequently fail to grapple with the multiple contextualising components that constitute such aesthetic pronouncements. As Leavy (2018) points out, through providing multiple criteria for which to evaluate ABI, arts-based inquiry attempts to achieve more than an affective response and simple reductions to notions of good, neglect to understand the complexity of the endeavour. Leggo (2008), in writing about poetic inquiry, suggests asking not, 'Is this a good poem?' but rather, 'What is a poem good for?' (p. 7). Considering purpose, while a direct challenge to Kant's (2007) notion of aesthetic 'purposiveness', is an essential question for community engaged ABI. Understanding aesthetic purpose when working with/in community recognises what Berleant (2016) calls aesthetic engagement, the shift away from art as object to art as experience. By neglecting aesthetic engagement, neglecting to ask what is the experience of all parties involved and what is the ABI 'good for', we risk removing the ethical from the aesthetic.

Born et al. (2017) write that social aesthetics hold an awareness of the central tenets of aesthetics as well as a critique of those very notions. Social aesthetics recognises that aesthetics are socially constructed and imbued with all the contextualising factors that make-up culture. Aesthetics, according to Born et al. (2017), are not objective, neutral, or disinterested—they have a purpose. The purpose is tied to the value that the community places on the object under consideration: 'A social aesthetic argues for, and investigates the details of, the many ways in which our interactions with art participate in or serve an array of political orientations and social and cultural processes' (p. 3). This reaches into the heart of ethico-aesthetic sonic engagements with(in) communities. ABI, and by extension audio drama inquiry, must grapple with the aesthetics of ethics. Where does the 'good' exist, and who determines this?

The obvious answer to the above question is: the community impacted, and (problematically at times) those impacting the community, in the case of audio drama inquiry researchers and artists. Here we might think of Thompson's (2015) call to an aesthetics of care, recognising that 'art making takes place in a series of relational acts, some more explicit and intentional than others' (p. 436). Considering how care is enacted through aesthetic decisions is a sharp turn away from purposelessness, re-framing the notion of 'good art'. *Tuning into CYC* sought multiple avenues for participant and co-researcher involvement, including in the realms of aesthetics, ethics, and structure. Coming from a field that places 'care' within its very name (child and youth care), holds relational practice as a defining element, and doing research though an arts-based approach on an industry that also borrows the care nomenclature,[2] considering aesthetics of care, aesthetic engagement, and relational aesthetics (Bourriaud, 2002) makes sense. This means being explicit in relational inquiry acts with a 'focus upon how the sensory and affective are realised in human relations fostered in art projects' (Thompson, 2015, p. 436). Concretely, for *Tuning into CYC*, this involved developing ethical protocols, sharing experiences, generating materials, analysing materials, writing content, discussing what was and was not included, performing the audio dramas, presenting at conferences, and co-authoring publications. This engaged approach requires recognising the multiple skills, experiences, and identities each person brings to the inquiry.

For *Tuning into CYC*, the people recruited for the project were emerging or existing professionals and simultaneously people with residential placement experience. In trying to understand this dual-identity dynamic I read mental health literature and in particular work coming from people who have experiences of residential mental health placement. This led me to Mad studies literature (LeFrançois et al., 2013; Voronka, 2016) and studying residential experience through a 'system user/survivor' lens (Sweeney et al., 2009). While not all CYCPfC involved claimed such identities ('survivor', 'system user', 'from care') or were even public about their care experiences, this literature was instrumental in developing ethical approaches for *Tuning into CYC*. When considering community engaged practices, Mad studies and system user/survivor research discourse, has much to offer qualitative researchers and artists who work in communities. This body of work begins with a stance recognising user/survivors' 'first-hand experience as a source of knowledge … [with] value and integrity … [and] as a unique source of understanding' (Beresford & Rose, 2009, p. 13).

There is a distinction made in the literature between user-led and user-involvement in research. Sweeney (2009) cites evidence indicating user-led research is perceived more positively by 'users' themselves, than user-involvement with 'the latter potentially creating tokenism and marginalisation' (p. 30). Sweeney also recognises the many institutional, funding, individual skills, and other barriers that exist which may limit or prevent user-led projects. Thus, she suggests considering user-led and user-involvement along a continuum with 'three distinct but continuous categories: *control*, *collaboration* and *consultation*' (Sweeney, 2009, p. 30, italics in original). This trifecta may also be a useful frame for community engaged sonic arts-practice, with collaborators contributing along a continuum from high (control) to low (consultation) in regards to both material development and art creation. Participation levels can be determined by community members availability, interest, and skills, recognising that it is potentially damaging to try and 'force' community members to do things that they have no desire or capacity to do.

Research done by Alison Faulkner looked at 'what principles and practices those of us involved in doing survivor research have developed' (Faulkner & Tallis, 2009, p. 54). From her study she identified ethical principles to consider when doing user/survivor research; I present these here augmented by interpretations from Sweeney (2009).

- *Clarity and transparency* with all parties involved, including protocols for informed consent and confidentiality.
- *Respect* for participants' views, ensuring views are 'heard', and flexibility regarding people's needs to take breaks or leave.

- *Equal opportunity*, including for diverse and minoritised populations, and adequate funding to support these opportunities.
- *Commitment to change,* doing research that leads to change, including implementation of the findings.
- *Empowerment*, shown by challenging stigma, fostering full participation, scaffolding participation to peoples' skills while teaching more if required or desired, support and supervision for all involved in the project, and ensuring funding to pay for these empowerment protocols.
- *Identity*, with discussions regarding shared identities.
- *Theoretical approach,* including transparency revealing, and discussion regarding, the theories informing research.
- *Accountability* with research participants, the wider society, and other people as required.

Notable in these principles is their productive, rather than restrictive, nature. Meaning, they are structured to create ethical opportunities and possibilities, as opposed to prohibitions. Thinking of ethics as generative territory works to resist bordering ideas which inform many discussions of ethics. While in each, considerations to do no harm exist, in the former this is done by considering how ethics can be a place of growth and possibility. There is a tendency to situate ethics as actions to be avoided rather than opportunities to be fostered. Part of this framing, no doubt, has to do with the differences between 'ethical practices' and 'ethical reviews' (Faulkner & Tallis, 2009). While institutional review protocol must be adhered to, generative practices can also be included in research design. For *Tuning into CYC*, I met the institutional ethics protocol, during which I described the rationale for community members to contribute to ethical considerations. This resulted in multiple generative ethical discussions with those who partook in the project and informed the practices for how we worked together. In a literal form of aesthetic ethics, we created a visual representation of how we were going to engage ethically which was shared with each new person in the group, with the ability for them to add or alter the image(s).

Conclusion

Audio drama inquiry has the potential to be an ethical, aesthetic, responsive, and community engaged process to generate, analyse, and disseminate research. As an emerging form it will be developed over the coming years by those who partake, theorise, and listen to the results. Undertaking it as a method of inquiry requires understanding the multiple shifting ways that information can be studied and disseminated while adhering to a process rooted in theory, ethics, and oversight, whose aims include advancing knowledge through aesthetic engagement.

Notes

1 To hear the audio dramas, visit www.ReFiled.ca, to read additional publications related to the project, visit www.tuningIntoCYC.org

2 When children are removed from their families it is frequently called 'going into care' or some variation on that depending on the country, and the discourse within the industry is rife with the use of the word care (care entrants, care homes, foster care, group care, children in care, care leavers, etc.).

References

American Psychological Association (2020). *Publication manual of the American Psychological Association: The official guide to APA style 7th ed.* American Psychological Association.

Amponsah, P., & Stephen, J. (2020). Developing a practice of African-centred solidarity in child and youth care. *International Journal of Child, Youth and Family Studies, 11*(2), 6–24. 10.18357/ijcyfs112202019516

Barad, K. (2007). *Meeting the universe halfway: Quantum physics and the entanglement of matter and meaning.* Duke University Press.

Bellefeuille, G., & Ricks, F. (2010). Relational inquiry: A child and youth care approach to research. *Children and Youth Services Review, 32*(10), 1235–1241. 10.1016/j.childyouth.2010.04.013

Belliveau, G., Cox, S., Nichols, J., Lea, G. W., & Cook, C. (2021). Examining the ethics of research-based theatre through Contact!Unload. In T. Prentki & A. Breed (Eds.), *The Routledge companion to applied performance* (Vol. 1) (pp. 141–148). Routledge.

Belliveau, G., & Lea, G. W. (2016). Introduction. In G. Belliveau, & G. W. Lea (Eds.). *Research-based theatre: An artistic methodology.* Intellect.

Berleant, A. (2016). *Aesthetics beyond the arts: New and recent essays.* Routledge.

Boal, A. (2002). *Games for actors and non-actors.* Routledge.

Born, G., Lewis, E., & Straw, W. (2017). *Improvisation and social aesthetics.* Duke University Press.

Bourriaud, N. (2002). *Relational aesthetics* (S. Pleasance, F. Woods, & M. Copeland, Trans.). Les presses du Réel.

Beresford, P., & Rose, D. (2009). Background. In A. Sweeney, P. Beresford, A. Faulkner, M. Nettle, & D. Rose (Eds.), *This is survivor research* (pp. 11–21). PCCS Books. https://www.pccs-books.co.uk/products/this-is-survivor-research

Brooks, D., & Kheshti, R. (2011). The social space of sound. *Theatre Survey, 52*(2), 329–334. 10.1017/S004055741100041X

Campt T. M. (2017). *Listening to images.* Duke University Press Books.

Coffey, A. (2018). *Doing ethnography.* London: SAGE Publications.

Cole, A. R., & Knowles, J. G. (Eds.) (2008). *Handbook of the arts in qualitative research: Perspectives, methodologies, examples, and issues.* Sage Publications.

de Fossard, E. (2005). *Writing and producing radio dramas: Communication for behavior change.* SAGE Publications.

Denzin, N. K., & Lincoln, Y. S. (Eds.) (2018). *The SAGE handbook of qualitative research* (5th ed.). Sage Publications.

Doucet, M. (15–18 May 2019). *Relationships matter: Examining the pathways to supportive long-term relationships for youth exiting care in Canada* [Paper presentation]. US: 15th International Congress of Qualitative Inquiry.

Etmanski, C., Hall, B. L. and Dawson, T. (Eds.) (2014). *Learning and teaching community based research: Linking pedagogy to practice.* University of Toronto Press.

Faulkner, A., & Tallis, D. (2009). Survivor research: Ethics approval and ethical practice. In A. Sweeney, P. Beresford, A. Faulkner, M. Nettle, & D. Rose (Eds.), *This is survivor research* (pp. 53–62). PCCS Books.

Fenge, L-A., & Jones, K. (2012). Gay and pleasant land? Exploring sexuality, ageing and rurality in a multi-method, performative project. *British Journal of Social Work, 42*(2), 300–317. 10.1093/bjsw/bcr058

Foster, V. (2016). *Collaborative arts-based research for social justice.* Routledge.

Gharabaghi, K., & Charles, G. (2019). *Child and youth care across sectors: Canadian perspectives* (Vol. 1). Canadian Scholars.

Hammond, W., & Steward, D. (2008). *Verbatim verbatim.* Oberon Books

Holstein, J. A. and Gubrium, J. F. (2012). Introduction: Establishing a balance. In J. A. Holstein & J. F. Gubrium (Eds.), *Varieties of narrative analysis* (pp. 1–12). SAGE Publications. 10.4135/9781506335117.

Huaman, E. S., & Martin, N. (2020). *Indigenous knowledge systems and research methodologies: Local solutions and global opportunities.* Canadian Scholars' Press.

Kant, I. (2007). *Critique of judgement.* Oxford University Press.

Leavy, P. (2018). *Handbook of arts-based research.* The Guilford Press

Leavy, P (Ed.) (2020). *Oxford handbook of qualitative research* (2nd ed.). Oxford University Press. 10.1093/oxfordhb/9780190847388.013.9

LeFrançois, B. A.; Menzies, R., & Reaume, G. (Eds.) (2013). *Mad matters: A critical reader in Canadian mad studies.* Canadian Scholars' Press.

Leggo, C. (2008). Astonishing silence: Knowing in poetry. In A. L. Cole & J. G. Knowles (Eds.), *Handbook of the arts in qualitative research: Perspectives, methodologies, examples, and issues* (pp. 165–174). SAGE Publications. 10.4135/9781452226545

MacNeill, C. (2018). *Applied theatre with looked-after children: Dramatising social care.* Peter Lang.

Nairn, K., & Panelli, R. (2009). *Using fiction to make meaning in research with young people in rural New Zealand. Qualitative Inquiry, 15*(1), 96–112. 10.1177/1077800408318314

National Association of Child Care Workers (2014). *Child & Youth Care.* http://naccw.org.za/international-child-and-youth-care-definition

Norris, J. (2009). *Playbuilding as qualitative research: A participatory arts-based approach.* Left Coast Press Inc.

O'Connor, P., & Anderson, M. (2015). Research in a post-normal world. In P. O'Connor & M. Anderson (Eds.), *Applied theatre: Research: Radical departures* (pp. 3–94). Bloomsbury.

Oxford English Dictionary. (OED) (2021). *Beauty.* https://www.oed.com/

Prendergast, M., & Saxton, J. (2015). Seduction of the real: The significance of fiction in applied theatre. *Research in Drama Education: The Journal of Applied Theatre and Performance, 20*(3), 280–284. 10.1080/13569783.2015.1059747

Rosiek, J. (2018). Art, agency, and ethics in research: How the new materialism will require and transform arts-based research. In P. Leavy (Ed.), *Handbook of arts-based research* (pp. 632–648). The Guilford Press.

Saldaña, J. (2011). *Ethnotheatre: Research from page to stage.* California: Left Coast Press.

Saldaña, J. (2016). *The coding manual for qualitative researchers* (3rd ed.). London: Sage.

Salvatore, J. (2020). Scripting the ethnodrama. In P. Leavy (Ed.), *The Oxford handbook of qualitative research* (2nd ed.). Oxford. 10.1093/oxfordhb/9780190847388.013.5

Singhal, A., Cody, M. J., Rogers, E. M., & Sabido, M. (2004). *Entertainment-education and social change: History, research, and practice.* Lawrence Erlbaum Associates, Publishers.

Smith, L. T. (2012). *Decolonizing methodologies: Research and Indigenous peoples.* London: Zed Books.

Soltani, F. (2018). Inner ears and distant worlds: Podcast dramaturgy and the theatre of the mind. In D. Llinares, N. Fox, & R. Berry (Eds.), *Podcasting new aural cultures and digital media* (pp. 189–208). Palgrave Macmillian.

Sweeney, A., Beresford, P., Faulkner, A., Nettle, M., & Rose, D. (Eds.) (2009). *This is survivor research*. PCCS Books.

Sweeney, A. (2009). So what is survivor research? In A. Sweeney, P. Beresford, A. Faulkner, M. Nettle, & D. Rose (Eds.), *This is survivor research* (pp. 22–37). PCCS Books.

Thompson, J. (2015). Towards an aesthetics of care. *Research in Drama Education: The Journal of Applied Theatre and Performance*, *20*(4), 430–441. 10.1080/13569783.2015.1068109

Tronto, J. C. (2015). *Who cares? How to reshape a democratic politics*. Cornell University Press.

Vachon, W. (2020). Queering child and youth care. *International Journal of Child, Youth and Family Studies*, *11*(2), 61–81. 10.18357/ijcyfs112202019519

Vachon, W. (2021). Child and youth care aesthetics and the beauty of relational inquiry. In H. Modlin, J. Freeman, C. Gaitens, & T. Garfat (Eds.), *Aspects of relational practice* (pp. 117–207). PreText Publishers.

Vachon, W., Hossain, R., Ramsay, N., Moore, M., & Milo, M. (2019). Frictions and actions: The actor's role in ethnodrama. *Qualitative Inquiry*, *25*(2), 118–129. 10.1177/1077800417746424.

Vachon, W., & McConnell, T. (2018). Allies, not accomplices: What youth work can learn from trans and disability movements. In P. Alldred, F. Cullen, K. Edwards, & D. Fusco (Eds.), *The SAGE handbook of youth work practice* (pp. 426–439). London: SAGE Publications.

Vachon, W. & Salvatore, J. (2022). Wading the quagmire: Aesthetics and ethics in verbatim theatre: Act 1. *Qualitative Inquiry* (published online, pp. 1–10). https://doi.org/10.1177/10778004221098988

Vachon, W., & Woodland, S. (2021). Finding resonance: Applied audio drama, inquiry and fictionalising the real. *Research in Drama Education: The Journal of Applied Theatre and Performance*, *26*(4), 1–18. 10.1080/13569783.2021.1891038

Voronka, J. (2016). The Politics of 'people with lived experience': Experiential authority and the risks of strategic essentialism. *Philosophy, Psychiatry, & Psychology*, *23*(3), 189–201. 10.1353/ppp.2016.0017.

Woodland, S. (2020). Daughters of the floating brothel: Engaging indigenous Australian women in prison through participatory radio drama. In C. McAvinchey (Ed.), *Applied theatre: Women and criminal justice* (pp. 79–104). Bloomsbury Methuen Drama.

3

WHAT DOES A CELLPHILM (CELLPHONE + FILM PRODUCTION + INTENTION) SOUND LIKE? THE ETHICS AND AESTHETICS OF CELLPHILM METHOD

Casey Burkholder and Katie MacEntee

Introduction

Cellphilms are an emerging participatory visual research method where participants use cellphone technologies to create short films that respond to a prompt (MacEntee et al., 2016, 2019). Cellphilms reinscribe participants' everyday media making practices—taking videos on mobile technologies—and turn these everyday practices towards a specific issue or prompt to bring insight to a research question. The accessibility of the mobile-based visual method is paramount. Participants bring a range of experiences in consuming and interpreting visual culture and they use this to inform the cellphilm production and the creation of new knowledge in the research context. Cellphilms are regularly analysed—and often co-analysed with participants—in relation to the video content, as well as visuals and aesthetics (Mitchell et al., 2017). However, the ethics and aesthetics of sound within cellphilms have been largely ignored within the methodological literature. To address this lacuna, we are drawn to reflexively study sound by revisiting cellphilms produced in two participatory visual research projects for social action.

The first study we explore was led by Casey and engaged 2SLGBTQ+ youth in New Brunswick, Canada through cellphilm method to consider school-based experiences and oppressions. Casey is a white femme bisexual Associate Professor who explores gender, sexuality, and schooling through participatory visual approaches with queer, trans, and non-binary youth. In the second study, we considered Katie's work with a racially diverse group of young cis and queer women who trade sex (exchanging intimacy—and sometimes sex—for money, gifts, food, drugs, etc.) in Tkoronto (Toronto), Canada who created cellphilms that explored pathways towards harm reduction and access to healthcare. Katie is a White cisgender woman who does not define her sexuality and who, at the time

DOI: 10.4324/9781003164227-8

of the research, was a post-doctoral fellow in Tkoronto—a territory that is cared for by the Anishinabek Nation, the Haudenosaunee Confederacy, and the Huron-Wendat. Katie's scholarship explores the use of participatory visual methodologies to study issues of gender-based violence, sexual and reproductive health, and HIV and AIDS healthcare with youth internationally. In both projects, we hear mostly voice over, some music, and some live audio. Some audio is grainy, some is clear. Some cellphilms are totally silent. When recording sound alongside images, the sound quality is often degraded—you can hear that it is a cellphilm, rather than a participatory documentary using a DSLR, for example. We note how the participant's cellphilm soundscapes and silence impact the media's emotional content (see also: Reason et al., 2016). In his book, *Sound Design Theory and Practice*, Leo Murray (2019) suggests:

> Sound choices—whether they are about the clarity of a line of dialogue, the choice and timing of music against a sequence of images, the suggestive possibilities of the background ambiences, or the realistic portrayal of an event—are all concerned with how they help to tell the story. (p. 6)

We wondered about Murray's assertion in our own work. To what extent are our sound choices conscious and unconscious? In what follows, we ask: what do cellphilms sound like? From our analyses of cellphilm sounds, we consider how sound—in its various formats, variations, and sometimes absence—contribute to cellphilms as a participatory visual approach for research intervention and social change. We discuss how cellphilm soundscapes recorded with mobile technologies echo and disrupt the mobile soundscapes of the digital media disseminated online everyday, and explore to what extent participants are heard especially when cellphilms are shared with public audiences—both online and in person (Britton, 2016; Burkholder & Rogers, 2020; Burkholder & MacEntee, 2016). Listening to cellphilms draws attention to the ways that personal technologies act as a 'metaphorical interface to practices of social and cultural formation more broadly' (D'Errico, 2016, p. 41). To the best of our knowledge, this is the first analysis of sound in cellphilms. By reanalysing a corpus of 18 cellphilms, four from a project with 2SLGBTQ+ youth who sought to articulate a response to societal and school-based exclusions through cellphilms and 14 from a project with participants who traded sex, we suggest that sound, music and silence must be listened to when analysing cellphilms. The sounds of cellphilms presents a rich entryway into a consideration of the possibilities of participant produced media as knowledge representation and knowledge dissemination.

Cellphilm method

Cellphilms are short films that reframe people's everyday media-making practices (taking cellphone video) and intentionally turn the practice towards a prompt for the purpose of research (MacEntee et al., 2016, 2019). In their discussion of the pedagogical promise of cellphilms to address issues of settler colonialism, gender-based violence, and social determinants of health, Sarah Flicker and her colleagues (2018) suggest, 'The process of making cellphilms encouraged students to move beyond representing the "damaged" relationships between land and bodies, and also point to the possibilities for new actions and reactions (the cellphilm already being one)' (Tuck, 2009, p. 41). Cellphilm method emerged from the work and scholarship of communication scholars, Jonathan Dockney and Keyan Tomaselli in South Africa in 2010 (Tomaselli & Dockney, 2018). Since that time, cellphilm method has been engaged to explore diverse issues including school-based research with teachers (Mitchell & De Lange, 2013); Indigenous language revitalisation in Oaxaca, Mexico (Schwab-Cartas, 2018); activism with ethnic minorities in Hong Kong (Burkholder, 2018); gender-based violence in South Africa (MacEntee, 2016; Mitchell et al., 2017); gender and water in the UK (Thompson, 2019); settler colonialism, gender-based violence and social determinants of health in Canada (Flicker et al., 2018); feminist tertiary education (Starr & Mitchell, 2018); and queer environmental futures (Lebel, 2022). In her consideration of the aesthetics of cellphilm production, April Mandrona (2016) describes the cellphilm method in relation to visual cultures, arguing that cellphilms encompass aesthetics employed in visual research, in art production, citizen journalism, and documentary style visuality. Although cellphilm method has been theorised in terms of aesthetics and visual cultures (Mandrona, 2016), the sound of cellphilms has yet to be examined.

Context

Our two studies are located in a research for social change framework—where the research knowledge we co-produce with participants seeks to make sustained local change in response to existing oppressions. Casey's project *Where Are Our Histories* was situated in the context of unceded Wolastokiyik territory—Fredericton, New Brunswick, Canada—and Katie's within the context of Tkaronto—Toronto, Ontario, Canada. Casey worked with six 2SLGBTQ+ young people to create short cellphilms that took up their experiences of belonging, spaces, and spoke back to the erasure of their communities' histories and experiences within schools, and social studies classrooms in particular (Burkholder et al., 2021). Participants worked collaboratively to produce four cellphilms over a series of three workshops. We held a number of community screenings to showcase our cellphilms and project findings: for the participants' parents, at a local queer film festival, and at a dedicated cellphilm screening at the public library organised by the Fredericton Feminist Film Collective. The *Where Are Our Histories* group has also shared these cellphilms on our YouTube Channel[1], which acts as the project's digital participatory archive (Burkholder & Thorpe, 2019). Casey has also shared the cellphilms amongst academic audiences, through her teaching, and disseminating research on cellphilm method and 2SLGBTQ+ youth experiences at academic conferences.

Katie's study worked with 14 young women in Tkoronto who have experience trading sex. The goal of the study was to learn about the harm reduction practices and access to healthcare for this highly marginalised and stigmatised population. The participants met one-on-one with Katie to engage in an interview about their experiences trading, their harm reduction practices, and their access to healthcare (MacEntee et al., 2022). The cellphilm prompt—What is important for people to know about trading sex in [Tkaronto]?—was developed by members of the research team—Katie MacEntee and Sarah Flicker—to guide the participants' cellphilm production and link the media to a common thread of communicating a message to others about trading sex. The goal was to support participants in actively producing and directing project findings and creating media that could be shared in the dissemination of the results. The resulting cellphilms highlighted the participants' key concerns, such as: staying safer while trading, mental health and access to mental health care, stigmatisation, and self-determination. With participants' consent and involvement, the cellphilms were compiled into a composite video[2] called *Celling Sex*, which reports on study findings and advocates for ending stigma towards young women who trade sex. In co-facilitated events involving participants and the research team, the video was screened to numerous different stakeholder audiences around Tkaronto to raise awareness, stimulate discussion and spark community change (Kendrick, MacEntee, Wilson, & Flicker, 2022). Like Casey, Katie has also screened all or portions of the *Celling Sex* composite video and a few individual participant cellphilms (with participant's consent) for university undergraduate and graduate classes and at academic conferences.

Theorising sound in participatory media production

To better articulate the importance of sound and noise in cellphilms, we have learnt from preeminent sound theorists who conceptualise the affect of noise in a situated fashion. Influential early film theorist and critic Béla Balázs (1985), for example, describes the potency of sound in film:

> Every sound which is really produced somewhere must of necessity have some such space-quality and this is a very important quality indeed if use is to be made of the sensual reproducing power of sound! … One of the most valuable artistic faculties of the microphone is that sounds shot at the point of origin are perpetuated by it and retain their original tonal coloring. A sound recorded in a cellar remains a cellar sound even if it is played back in a picture theatre, just as a film shot preserves the viewpoint of the camera, whatever the spectator's viewpoint in the cinema auditorium may be. (p. 8)

When producing cellphilms, participants are usually conscious of the sounds (or lack thereof) that are being recorded in order to support their production's overall message or intention. They decide whether or not to include dialogue, music or other audio pick-ups. What we hear in cellphilms is (mostly) the intentional, time capsules of place and time, that transmit the participants' message and lends further meaning to their visual recording.

However, not all recorded material in a cellphilm is intentional. Not all projects have the use of professional recording devices—for example, external microphones—and it is difficult to control environmental sounds that the cellphone microphone may record. And while participants may be familiar and even adept at making videos using their cellphones, it is rare that they have specific training on sound technology. Therefore, it is not unusual (or necessarily unwanted) to hear the sound of a hand brushing past a microphone, (which is placed in close proximity to the record button), or wind (when recorded outside), the sudden scrape of a chair on floor (when another participant off camera gets up to join a scene). These sounds can obscure the narrative and other audio the producer desired to capture.

In fact, much of the environmental sounds that are captured in cellphilm audio, might be better categorised as noise. Marie Thompson suggests that the use of noise in music has typically been articulated in terms of 'crossing a line between the musical and the extra-musical, the wanted and the unwanted' (2017, pp. 5-6). However, she also argues that noise must be theorised in a complex way:

> That there is much more to noise than unwanted sound, and to fail to recognise this is to fail to recognise the crucial role noise plays in auditory culture and in material culture more generally. There is no music, no mediation, no *son* without noise. (p. 3)

Thompson writes, 'Recognising that noise has the potential to be "good" as well as "bad", to have positive as well as negative effects, the ethico-affective definition thus allows more fully for noise's generative capacity in artistic contexts' (p. 4). It is this space between wanted and unwanted sound that we use as an entry point in our analysis of participants' cellphilms. In our future cellphilm facilitation practice, we will explicitly centre the ways in which recorded sounds are wanted and unwanted in the cellphilms produced by our participants.

What about silences?

Although a great deal of literature on cellphilms and participatory video describes the ways that these modes allow participants to disrupt silences around their experiences (Kindon, 2016; Milne et al., 2012), what is actually heard and what is unheard in these participant-produced films has been largely overlooked in the academic literature. Some sound theorists consider the idea of the unheard through silence. Balázs (1985) notes in his reflection on sound and film:

> Silence is when the buzzing of a fly on the windowpane fills the whole room with sound and the ticking of a clock smashes time into fragments with sledgehammer blows. The silence is greatest when we can hear very

> distant sounds in a very large space. The widest space is our own if we can hear right across it and the noise of the alien world reaches us from beyond its boundaries. (p. 2)

He goes on to note that complete silence can be disorienting if not unintelligible:

> A completely soundless space … never appears quite concrete, and quite real to our perception, we feel it to be weightless and unsubstantial, for what we merely see is only a vision. We accept seen space as real only when it contains sounds as well, for these give it the dimension of depth. (p. 2)

Balázs describes silence as disorienting. Contrastingly, in her book seeking to theorise noise, Thompson (2017) writes:

> This auditory politics [of silence] is underpinned by a dualist aesthetic moralism, which positions noise as bad to silence's good … Where silence is associated with a positive affectivity—it has the capacity to calm, revive and rejuvenate—noise is defined by its negative affectivity—it damages, destroys, deafens and harms. (p. 5)

Building on these understandings of sound, in the following sections we consider the interplay between silence and noise in participant produced cellphilms.

Data collection

We approached the analysis of cellphilm sounds by thinking about the extent to which we centred sound and silence in our cellphilm facilitation practices and workshops. In a study with asylum seekers and labour migrants in Ireland, Alexandra (2017) describes the ways in which research practitioners and facilitators impact media production:

> I have implicitly argued for the benefit of cultivating a reflexive curiosity about one of the most pivotal aspects of our work—its collaborative, dialogical essence. Instead of conceptualising the role of the educator, media mentor, or "story specialist" as a "hands off" position in which one "leaves no fingerprints," we would benefit from recognising the ways in which we influence, shape, and inform the process, as well. (p. 352)

When we have given cellphilm workshops, we have spent time guiding participants through a process of deciding on the content of their cellphilm and its visual format, as well as highlighting options for participants to negotiate their own and others' consent and anonymity in relation to image production. Our emphasis on sound has usually been focused on providing tips on the challenges of recording sound on a mobile device (e.g. we have often reminded participants that it is important to be physically close to the phone's microphone in order to be heard and to avoid filming in windy outdoor locations). While we have sometimes provided an introduction to the concepts of 'framing the shot' and asked participants to consider what is left out of the visual, we have overlooked the potential significance of what they might choose to audio record, and the use of silence and other soundscapes to support the communication of the cellphilm message. Trained as participatory visual researchers, we have a nurtured awareness of what visual production offers to research as a means of both knowledge production and representation. Up until now, we have quite unconsciously overlooked the potential of taking an equally appreciative critical analysis to the qualities and productive nature of sound.

Listening to the *where are our histories and celling* sex cellphilms

In total, we considered the soundscapes of 18 cellphilms made by youth between the ages of 14 and 21. In Table 3.1, we provide an overview of each of the productions, a brief description of its content and a synopsis of the sound characteristics that they employ. In this section, we provide a more in-depth description of these characteristics and how we feel they detract from or contribute to the overall cellphilm aesthetic and our understanding of how sound contributes to the knowledge production and knowledge representation of cellphilm method. We summarise these characteristics into three themes: 1) youthful cellphilm sound aesthetics, 2) DIY sound qualities as accidental reveals of the cellphilm producers, and 3) sound ethics in cellphilm method.

Youthful cellphilm sound aesthetics

We noticed several aesthetic qualities that showed up in both projects' cellphilms. Producers make conscious sound decisions. For example, we often noted the use of pre-recorded music—or music that has been saved on one phone being recorded on another phone—narration, scripted, or improvised dialogue between characters, and strategies that encourage visual but not auditory anonymity. We identified the tinniness of re-recorded music in some of the cellphilm productions (e.g. 'Nackawic Needs a GSA Now!!', see Table 3.1) and Katie likened this to the ways that young people can use their phones to publicly listen to music together or to create living soundtracks to their lives. Building on the distinct youth-like quality of cellphilming (MacEntee & Flicker, 2019; Mandrona, 2016), when a producer requested that Katie choose a soundtrack for one of their cellphilms ('Know Your Options'), she chose Cardi B's 'Drip'. It had a good beat and it was a hit at the time. She was also motivated to use one of Cardi B's songs because the artist is open about having engaged in transactional sexual activities, such as exotic dance. Cellphilm audio aesthetics in this way work to affirm the soundscapes of youth cultures.

TABLE 3.1 Cellphilm overview

Cellphilm title	*Cellphilm synopsis*	*Cellphilm sound characteristics*
Celling Sex Cellphilms		
Do You	Sex work is work, there are good days and bad days, everyone's experience is different.	Narration with no other identifiable characteristics. Changes in voice over tone due to distance of cellphone from narrator. Sound of producer manipulating props.
It's Not All Sugar	Maintaining Sugar Daddy relationships. Earning money. Impact on personal relationships.	Narration with no other identifiable characteristics. Post-production voice changes for confidentiality.
El Paseo	Experience of trading sex, family finds out, stigma based on her job as dominatrix.	Narration with no other identifiable characteristics. Narrator rambles. Sound off camera of someone Laughing at something the narrator says.
Stay Safe	Girl arranging and going on a date, strategies to stay safe.	Accidental capture of off camera voice of director saying 'Action'. Live action voice of individual playing a character with limited view of other identifiable characteristics. Voice changes marking gendered and hetero-natured communication between woman and man versus woman and woman.
A Message to My Younger Self	How to go on a transactional date, strategies to stay safe while trading sex, goals of trading sex.	Narrators with no other identifiable characteristics. Narrator's accent and words suggests her race. Background noises as potential interruptions (ladder banging, traffic, birdsong).
Transactional Sex Montage	Juggling trading sex and everyday life.	Sound of computer mouse being used to scroll through sugar daddy profiles on a transactional sex dating website. Everyday noises (e.g. laundry machine). Culturally specific noises (e.g. preparing ink for use in traditional Chinese calligraphy and brush strokes).
Catfish Revealed	Experience of being catfished, strategies for avoiding being catfished.	Soundtrack of synthesizer music.

(*Continued*)

TABLE 3.1 (Continued)

Cellphilm title	*Cellphilm synopsis*	*Cellphilm sound characteristics*
I am More	A person trading sex can have two stories. When providing support, listen to both stories, see the whole person.	Narration with no other identifiable characteristics. Silent interludes in narration to allow viewers to take in images on screen.
The Surmount	Sugar Daddy stalking, police intervention is ineffective.	Narration and view of face. Audio is quiet.
R-Rated Material	Mental health worker shames person for trading sex. Strategies so support people who trade sex.	Silence for first 42 seconds. Soundtrack of piano music.
Becoming an Independent Woman	Strategies to stay focused on goals and get ahead of life and not have to rely on trading sex for money.	Narration with no other identifiable characteristics. Researcher heard clearly from off camera at end.
I Am More Than What My Body Has To Offer	Experience of trading sex described, using sex to fill an emotional voice can be harmful to oneself, healthy relationships can help one heal.	Narration with no other identifiable characteristics. Background potentially disruptive (e.g. wind, traffic). Researcher heard quietly off camera prompting participant with a question for consideration in her narration.
Know Your Options	Trading sex is providing a luxury service. Have goals, manage your money, stay focused.	Participant approved researcher's choice of Nikki Minaj 'Drip' soundtrack.
Convenience?	Don't develop emotions for the people you trade with, slut shaming is hurtful, young women who trade sex are not alone.	Narration with minimal identifiable characteristics (hands). Slight accent suggests participant's nationality. Sound of participant manipulating props.
Where Are Our Histories?		
Nackawic Needs a GSA Now!!	The cellphilm shows what Gender-Sexuality Alliances (GSAs) offer for queer, trans, and non-binary youth in school spaces, and was produced in response to being told that they could not form a GSA in their middle school.	Grainy recording of songs. Aside from the music, there is no narration or dialogue. Music was recorded live from a phone to a phone so we capture several environmental noises throughout. (e.g. 0:57: we hear live recording of the sound of a person writing with a permanent marker).

		Sad song cuts abruptly to happier music, which was also recorded from a phone. More instances of background noise—the phone being dropped on a table, something making a staticky sound—perhaps it is a finger or hand placed over top of the cellphone's microphone. The cellphilm ends with music recorded phone-to-phone while hand drawn credits roll. Then, silence.
Coming Out	The cellphilm discusses what it feels like to come out as a young queer person.	Sound of statics while the speaker narrates through a voiceover. Speaker has a feminine sounding voice and that they are a young person. We hear that the narration has been recorded in a quieter setting—fewer environmental noises are captured in the sound. Static and narration are present throughout the cellphilm until the final credits roll when the filmmaker uses an instrumental piece of music recorded phone to phone, and which has a staticky quality to it.
Gendered	Gendered was produced to highlight the ways that school spaces are highly gendered, and agender youth experience gender policing through teacher language, physical spaces, and interactions with peers.	Sound echoes in the cellphilm. Dialogue between speakers was recorded live on a participant's cellphone. Some of the dialogue is muffled. We hear environmental sounds, like shoes squeaking on the floor, the auditorium chair squeaking, some unscripted laughter. Sometimes these noises obscure the planned dialogue. We hear people sighing. We hear other environmental sounds like: air circulating, doors closing, steps shuffling on cement. The sound of a heartbeat ends the cellphilm.
Making & Breaking Friends	This cellphilm highlights the ways in which relationships and friendships can be thorny and fraught in the context of middle school. Producers describe what makes and breaks friendships.	The cellphilm begins with silence—broken by a voiceover. A conversation between four voices—one adult and three young people—discussing what friendship feels like in the context of middle school, how schools encourage gender-based violence through oppressive policies, and what it feels like when a friendship is broken. No music used in the cellphilm. The cellphilm ends with static.

DIY aesthetics reveal the cellphilm producers

We think that one of the most powerful qualities of a cellphilm is in its DIY aesthetics. Not all cellphilms are slick media productions, but some are, perhaps because of the everyday nature of production in spaces like Tiktok. The shaky, hand-held recorded visual quality is further enhanced by unclear and unintended recorded sounds. For example, dialogue that is poorly balanced with environmental sounds or noises that are off camera disrupt the narration. In the *Celling Sex* cellphilm, 'It's Not All Sugar', we hear the creaking chair as the camera person shifts her weight. In 'Nackawic Needs a GSA NOW!!', there is the staticky sound of the filmmaker's finger over the cellphone's microphone. In the *Celling Sex* cellphilm, 'Do You', a cut from one scene to the next changes the voice quality of the narration as the producer moves the camera closer to her than in the previous shot. Music is used in background sometimes obscures the live recorded sound. Without careful mixing, the originally recorded voices can be lost. These seemingly unintentional sounds break the film-studies proverbial 'third wall' to remind the viewer of the cellphilms producers themselves. Through these inadvertent and perhaps amateur production errors, the viewer is reminded that there are people off camera controlling the production. These people are not professional filmmakers, so much as 'everyday' people who claim expertise through the production-as-research process. When we have screened these cellphilms we have noticed that audiences are deeply moved by the DIY and emotional quality—the rawness—of the cellphilms produced by real people who are also often present at the screenings and can speak further about their knowledge and what needs to change in society to promote social wellbeing (Burkholder & Rogers, 2020; Kendrick et al., 2022). We argue, these accidents contribute to the DIY quality that, while at times distracting or disruptive to the content of the cellphilms, also give the cellphilms power as they remind us that people typically marginalised or overlooked in more traditional research have capacity and important insights to contribute to the knowledge production process.

Sound ethics and aesthetics in cellphilm method

Participants create impactful media through the mix of sound, silence, noise, the incorporation of music, narration, and the use of the visual. We suggest that there are specific sound ethics within these decisions and practices. In her discussion of audience attention in a participatory video project, Britton (2016) argues that aside from specific visual aesthetics, producers also curated the use of 'sound (that of rivers, birds, etc.) [which] were all critical in gaining audience attention' (p. 197). When we take the time to listen and take seriously the choices participants have made in terms of the sound, music, noise, and silence that they curate in their cellphilms, we hear the explicit choices they make. The narration and music are intentional. But, the environmental sounds, and other participants' voices captured in recordings are not. We hear these sounds in a new way when we listen carefully beyond dialogue, narration, and soundtrack.

We also note that we hear participant identities, for example in relation to race and gender. As listeners, we make assumptions about the producers based on their voices—as one of our editors remarked, these are the micro politics of timbre (Eidsheim, 2019). In the *Celling Sex* cellphilm soundtracks we hear the 'race of sound' (Eidsheim, 2019) in the narrators' voices. To Katie, the audio analysis made her wonder how these sonic clues to producers' racial and cultural backgrounds impacted the audiences. This question picks up on comments from, for example, audience members and chapter reviewers about the ethical concerns of the project's composite video. For example, the sonic clues that may identify some participants might reinforce stigma associated with racialised female bodies and sex work and put participants at future risk of discrimination. It reminds Katie also of the feedback from different audience members and participants who reflected on the benefits of sharing media that disrupts the mainstream representations of transactional sex that is dominated by White, blond, blue-eyed female college students[3]. These participants and audience members celebrated seeing (and hearing) their race and culture reflected back to them in the different cellphilms. Thus, the reception of these micro politics can be diverse.

Thinking through facilitating strategies in cellphilm and composite video screenings—that disrupt social stigmas—is paramount given the potential risk to participants, and the practice of ongoing and informed consent (Kendrick et al., 2022; MacEntee, 2016). Casey felt the need to be explicit about producers' gender identities which might be different from the assumptions that a listener might make in hearing the voices. In 'Making & Breaking Friends', for example, we hear an agender person, an adult cis femme, and a cis teenage boy and teenage girl. Without this supported reading from Casey, we acknowledge that people's assumptions about gender and voice might lead to participants being misgendered in the listening. In an attempt to create gender affirming screening practices, Casey intentionally uses producer's pronouns when setting up a Q&A. Casey might say, 'This cellphilm was produced by a young person who sought to describe their experiences in school. Their work really resonated with me. What were your (the audience's) responses to their cellphilm?' We want participants' identities to be honoured and represented and thus feel beholden to intervene, if not pre-empt, audience problematic and discriminatory assumptions about participants.

We also wonder about participants' thoughts about the mistakes and unintentional sounds that are recorded. When do these accidents support the DIY affect and therefore the emotional content of the cellphilm? And when might these accidents distract or detract from the participant's intentions? Equally interesting to think through are the moments when the cellphilm includes a sound unintentionally, but actually supports the video's message. An example from the cellphilm 'Gendered' is when we hear background talking during recording that makes the scene feel more like a school (which is often noisy and other conversations become background noise). From the *Celling Sex* project, in 'I Am More Than What My Body Has To Offer' we hear the sound of traffic and wind as the producer walks along a busy Toronto street telling her story of trading sex.

While the traffic and wind can sometimes disrupt the producer's narration, it brings the viewer into the moment like a friend accompanying the producer on a walk. Thinking these practices through, we argue that these unintentional sound practices are merits of the cellphilm artistic process. In future research, it would be interesting to learn to what extent the decision participants make in post-production editing notice or ignore these unintentional sounds. For example, a participant might decide to use a particular take because a sound accident was cool and they likely could not make it happen again (for example, the sound of a marker writing during the otherwise silent 'Nackawic Needs a GSA Now!!')

In the area of participatory visual methods, the issue of visual ethics is much discussed (Mitchell, 2011; Warr et al., 2016). And, as visual researchers, research ethics boards (especially those that are unfamiliar with the approach) can be apprehensive if they assume a participant cannot maintain their anonymity as a result of the use of a visual method (Flicker et al., 2007). Interestingly, when applying for institutional ethical review, neither Casey or Katie have ever had any explicit questions around the capture of participant voices and the potential risk this poses to their confidentiality. Given our discussions (above) about the potential to misgender participants, we acknowledge that there is an ethical risk in screening participants' cellphilms. In listening to the collection of cellphilms from *Celling Sex*, we hear that several participants used creative methods to obscure identifiable characteristics such as their faces but were comfortable with recording their voice for narration. One participant asked that their narration be augmented in order to obscure their identity. This was done post-production via iMovie, which has a function that automatically heightens or lowers the audio's pitch and effectively renders the participant's voice more unrecognisable. However, three participants made completely anonymous cellphilms as they chose to neither audio record their voice nor film themselves. As facilitators of cellphilm method, we assume that all participants have the capacity, with the support of transparent and frank conversations throughout the cellphilm process about the risks and benefits of identification, to make decisions over how much of themselves they want to put into their cellphilms (MacEntee et al., 2022). The project's method followed previous research that highlighted the capacities of young female participants to navigate an ongoing and active consent process that considered not just the making but the public showing of cellphilms (Young Indigenous Women's Utopia et al., 2021). But our sound analysis has led us to consider if we might in future make more of an effort to understand participants' comfort with including their voice over other characteristics in their media.

Less often, but no less importantly, we might also discuss the use of third-party sounds and music. When participants use a video editing app, for example, they may gain access to certain pre-recorded sounds and music meant for public use. We offer tips on how to engage with and upload music produced and shared within the Creative Commons. Some participants integrate music into their cellphilms, and to this we would discuss the impact of copyright infringement and the limitations of mixing third-party media on the dissemination of their cellphilms to wider audiences. To return to the example from the *Celling Sex* project, one participant approved Katie's suggestion of adding 'Drip' by Nikki Minaj as the soundtrack for her cellphilm. Thus, the researcher can also be in a position to invoke copyright infringement in the decisions they are allowed to make around a participant's work. To this, we add that the particular cellphilm under question was made to be completely anonymous and it has not been shared online or used in public dissemination activities with this soundtrack in place so as to protect the participant from negative judgement or legal recourse.

Commonalities and tensions in cellphilms' use of sound and silence

We have noticed that sound choices in cellphilm production shift the emotional content of the cellphilm, and shape the way that producers make choices about how they want the audience to interpret their films. We found that the cellphilms shared some characteristics in terms of sound, noise, and silence, and what we have called 'DIY aesthetics that reveal the producers'. As April Mandrona (2016) theorised in her discussion of cellphilms as visual material culture, there is an inherently DIY aesthetic to the sound quality of the cellphilms we surveyed. Graininess. Picking up environmental noise. The sound of a hand or shirtsleeve brushing past a microphone. The sound of wind and static. These sounds emerged again and again in the cellphilms. When participants included music in their cellphilms, they often recorded music from one phone to another (editing on the fly) instead of uploading a 'clean' version of a piece of music. Still, there's something to this DIY sound aesthetic that we think is worth noticing, or listening to. When the quality is lower, or less professional, the overall emotional content is not diminished. There is instead something more true-to-life in listening to these aesthetic choices. These are auditory cues that mark for the listener the production moment (Balázs, 1985) and the producer. We interpret this as particularly meaningful when working with individuals and communities whose voices have typically been excluded or overlooked from the topic at hand. The power of a cellphilm comes in part from the DIY reminders that this is not a highly stylised, commercial media production. This is a DIY video made by a person with lived experience and expertise that is worth listening to and learning from.

Darcy Alexandra (2017) describes the ways in which sound and creative practice are central considerations in participatory media production and engaged scholarship:

> Great care was given to mentoring participants as emergent media practitioners and to critically exploring the diverse elements of their documentary essays—visual montage, sound design, script development, and video editing. The audio-visual production process—the opportunity to be creative in new ways—contributed to unprecedented engagement

> among research practitioners … [who] interrogated their daily circumstances negotiating migration policy and revealed the structural violence of asylum and migrant labour regimes. (p. 336)

The cellphilm production process is central to the creation and sharing of important stories. In our work, this means speaking back to school and health systems (Casey) and health and justice systems (Katie).

Who is heard? Who is listening?

One of the main claims in the power of participatory visual research production—including cellphilms—is that they provide participants with space and avenues to speak back to existing societal oppressions and to be heard in spaces where they may not normally be heard, for example, by policymakers, by using the internet to disseminate, by holding public screenings where they are positioned as knowledgeable about their experiences and about the issues that their cellphilms take up (MacEntee et al., 2016, 2019; Milne et al., 2012; Mitchell and De Lange, 2013). However, in listening to the cellphilms explicitly, we suggest that facilitators must take an engaged role in facilitating the listening process. In our analysis, we also learned to identify our own assumptions about gender identity and voice as we heard in our discussion of 'Making & Breaking Friends'. This gives us—as cellphilm facilitators who screen our works for unfamiliar audiences—some clues about the kind of explicit support we need to provide audiences in listening to participant produced cellphilms. Going forward, we seek to engage this kind of listening-forward screening with audiences, and in doing so, we provide audiences with information before listening so that they can affirm participants' identities in relation to gender, sexuality, race, and ability—some things that may and may not be heard in listening to participant cellphilms.

Conclusion

What does a cellphilm sound like? In reviewing a corpus of 18 cellphilms from two distinct projects, we notice that there is an inherently DIY quality to cellphilm aesthetics—this is as true for their audio content as it is for their visuality. We have highlighted three commonalities in our analyses: youthful cellphilm sound aesthetics, DIY accidents that reveal the producers, and sound ethics. In listening deeply to cellphilms that were previously produced, we also wonder how listening more intently to cellphilms may shift our facilitation practices in future cellphilm projects. We suggest that deep listening is important in cellphilm production and also in cellphilm screenings. How can we guide audiences into listening to participant cellphilms while simultaneously making space for their identities to be affirmed in relation to aspects such as gender, sexuality, race, ability, age, class, religion? How will we centre sound, silence, noise, music, and the audio in more explicit ways when we are going forward? It might be valuable to include a cellphilm 'listening' before engaging participants in a cellphilm viewing in cellphilm production workshops. This would mean that participants could listen to cellphilms and discuss what they hear, before re-watching cellphilms with both audio and visual put together. By explicitly centering sound and silence in this way, participants would be supported in considering the environmental and contextual noises that are included within these cellphilms as important materialities. We suggest that there is a clear aesthetic to cellphilms—both in terms of their audio and visual components—and we think that thinking more explicitly about sound, noise, and silence would productively expand the practice and theorising of participatory visual research.

Notes

1 The *Where Are Our Histories* cellphilms can be viewed at: https://www.youtube.com/channel/UCXORsJs60OVKJ7TSDnEl6xg

2 Composite videos are video texts that draw together multiple videos produced in research studies and that provide larger context for the individual visuals produced, and which are often screened in communities outside of that where the research have taken place (Mitchell et al., 2017). Composite videos are a mode of knowledge dissemination that seeks to provide additional context that might be lost when single cellphilms are disseminated (MacEntee, 2016).

3 See for example Sandler, D. (2020). *Thousands of Students Attend Uni Thanks to Sugar Daddies* [Press release]. Retrieved from https://www.seeking.com/p/sugar-baby-university-2020/canada/#gref

References

Alexandra, D. (2017). Implicating practice: Engaged scholarship through co-creative media. In G. Jamissen, P. Hardy, Y. Nordkvelle, & H. Pleasants (Eds.), *Digital storytelling in higher education* (pp. 335–353). Palgrave Macmillan.

Balázs, B. (1985). Theory of the film: Sound. In E. Weis & J. Belton (Eds.), *Film sound: Theory and practice* (pp. 116–126). Columbia University Press.

Britton, K. (2016). *Rhetorical visions/affective images: Participatory video as aesthetic event* [Doctoral dissertation, The University of New South Wales].

Burkholder, C. (2018). DIY media-making for social change: Hong Kong's ethnic minority youth speak back to exclusion and call for social action through cellphilms. In S. Bastien & H. Holmarsdottir (Eds.), *Youth as architects of change: Global efforts to advance youth-driven innovation for social change* (pp. 155–174). Palgrave MacMillan.

Burkholder, C., Hamill, K., & Thorpe, A. (2021). Zine production with queer youth and pre-service teachers in New Brunswick, Canada: Exploring connection, divergences and visual practices. *Canadian Journal of Education* [online], *44*(1), 89–115. https://journals.sfu.ca/cje/index.php/cje-rce/article/view/4535. (Accessed 29 May 2021).

Burkholder, C., & MacEntee, K. (2016). Exploring the ethics of a participant-produced cellphilm archive: The complexities of dissemination. In D. Warr, M. Guillemin, S. Cox, & J. Waycott (Eds.), *Ethics and visual research methods: Theory, methodology and practice* (pp. 211–224). Palgrave Macmillan.

Burkholder, C., & Rogers, M. (2020). Moving beyond celebration toward action: Screening cellphilms and participatory videos for different audiences. In M. Carter, C. Mitchell, & H. Mreiwed (Eds.), *Art as an agent for social change* (pp. 10–20). Brill/Sense.

Burkholder, C., & Thorpe, A. (2019). Cellphilm production as posthumanist research method to explore injustice with queer youth in New Brunswick, Canada. *Reconceptualizing Educational Research Methodology, 10*(2–3), 292–309. 10.7577/rerm.3680

D'Errico, M. (2016). *Interface aesthetics: Sound, software, and the ecology of digital audio production* (Publication No. 10112120) [Doctoral dissertation, University of California at Los Angeles]. https://escholarship.org/uc/item/0mv9v64c

Dockney, J., & Tomaselli, K. G. (2010). Third screens, third cinema, third worlds and triadomania: Examining cellphilm aesthetics in visual culture. *Communitas, 15*, 97–111.

Eidsheim, N. S. (2019). *The race of sound: Listening, timbre, and vocality in African American music*. Duke University Press.

Flicker, S., Sayde, A., Hedlund, K., Malivel, G., Wong, K., Owino, M., & Booy, S. (2018). Teaching and learning about the relationships between land, violence and women's bodies: The possibilities of participatory visual methods as pedagogy. *Agenda, 32*(4), 32–44. 10.1080/10130950.2018.1544436

Flicker, S., Travers, R., Guta, A., McDonald, S., & Meagher, A. (2007). Ethical dilemmas in community-based participatory research: Recommendations for institutional review boards. *Journal of Urban Health, 84*(4), 478–493.

Kendrick, C. T., MacEntee, K., & Flicker, S. (2022). Screening stories: Methodological considerations for facilitating critical audience engagement. In C. Burkholder, F. Aladejebi, & J. Schwab-Cartas (Eds.), *Facilitating community research for social change: Case studies in qualitative, arts-based and visual research* (pp. 240–257). Routledge.

Kendrick, C. T., MacEntee, K., Wilson, C. L., & Flicker S. (2022). Staying safe: How young women who trade sex in Toronto navigate risk and harm reduction. *Culture, Health & Sexuality, 24*(7), 920–934.

Kindon, S. (2016). Participatory video as a feminist practice of looking: 'Take two!' *Area, 48*(4), 496–503. 10.1111/area.12246

Lebel, S. (2022). Facilitating queer art in the climate crisis. In C. Burkholder, F. Aladejebi, & J. Schwab-Cartas (Eds.), *Facilitating community research for social change: Case studies in qualitative, arts-based and visual research* (pp. 195–207). Routledge.

MacEntee, K. (2016). Facing responses to cellphilm screenings of African girlhood in academic presentations. In K. MacEntee, C. Burkholder, & J. Schwab-Cartas (Eds.), *What's a Cellphilm? Integrating mobile phone technology into participatory visual research and activism* (pp. 135–152). Brill/Sense.

MacEntee, K., Burkholder, C., & Schwab-Cartas, J. (2016). What's a cellphilm? Introducing the methodology. In K. MacEntee, C. Burkholder, & J. Schwab-Cartas (Eds.), *What's a Cellphilm? Integrating mobile phone technology into participatory visual research and activism* (pp. 1–15). Brill/Sense.

MacEntee, K., Burkholder, C., & Schwab-Cartas, J. (2019). Cellphilms. In P. Leavy (Ed.), *The Oxford handbook of methods for public scholarship* (pp. 419–442). Oxford University Press.

MacEntee, K., & Flicker, S. (2019). Doing it: Using participatory visual methodologies with youth to study sexual health research. In S. Lamb & J. Gilbert (Eds.). *The Cambridge handbook of sexual development: Childhood and adolescence* (pp. 352–372). Cambridge University Press.

MacEntee, K., Kendrick, C., & Flicker, S. (2022). Quilted cellphilm method: A participatory visual health research method for working with marginalised and stigmatised communities. *Global Public Health, 17*(7), 1420–1432.

Mandrona, A. R. (2016). Visual culture, aesthetics, and the ethics of cellphilming. In K. MacEntee, C. Burkholder, & J. Schwab-Cartas (Eds.), *What's a Cellphilm? Integrating mobile phone technology into participatory visual research and activism* (pp. 183–198). Brill/Sense.

Milne, E. J., Mitchell, C., & De Lange, N. (2012). *Handbook of participatory video*. Rowman & Littlefield.

Mitchell, C. (2011). *Doing visual research*. Sage.

Mitchell, C., & De Lange, N. (2013). What can a teacher do with a cellphone? Using participatory visual research to speak back in addressing HIV&AIDS. *South African Journal of Education, 33*(4), 1–13. http://www.scielo.org.za/scielo.php?script=sci_arttext&pid=S0256-01002013000400010

Mitchell, C., De Lange, N., & Moletsane, R. (2017). *Participatory visual methodologies: Social change, community and policy*. Sage.

Murray, L. (2019). *Sound design theory and practice: Working with sound*. Routledge.

Reason, M., Jola, C., Kay, R., Reynolds, D., Kauppi, J., Grobras, M., Tohka, J., & Pollick, F. (2016). Spectators' aesthetic experience of sound and movement in dance performance: A transdisciplinary investigation. *Psychology of Aesthetics, Creativity and the Arts, 10*(1), 1–47. https://ray.yorksj.ac.uk/id/eprint/1294/1/final_PACA.pdf

Schwab-Cartas, J. (2018). Keeping up with the sun: Revitalizing Isthmus Zapotec and ancestral practices through cellphilms. *Canadian Modern Language Review, 74*(3), 363–387. https://utpjournals.press/doi/abs/10.3138/cmlr.4056?journalCode=cmlr

Starr, L., & Mitchell, C. (2018). How can Canada's feminist international assistance policy support a feminist agenda in Africa? Challenges in addressing sexual violence in four agricultural colleges in Ethiopia. *Agenda, 32*(1), 107–118. https://www.tandfonline.com/doi/full/10.1080/10130950.2018.1427692

Thompson, J. (2019, July 23). "To drink or not to drink?" Exploring drinking water practices using cellphilming. Faculty of Education Colloquium Series, University of New Brunswick, Fredericton, NB.

Thompson, M. (2017). *Beyond unwanted sound: Noise, affect and aesthetic moralism*. Bloomsbury Publishing USA.

Tomaselli, K. G., & Dockney, J. (2018). Cellphilms, visual culture and new narratives, *English Studies, 132*(1), 142–161.

Tuck, E. (2009). Suspending damage: A letter to communities. *Harvard Educational Review, 79*(3), 409–428. https://meridian.allenpress.com/her/article-abstract/79/3/409/31956/Suspending-Damage-A-Letter-to-Communities

Warr, D., Guillemin, M., Cox, S., & Waycott, J. (2016). *Ethics and visual research methods*. Palgrave Macmillan.

Young Indigenous Women's Utopia, MacEntee, K., Altenberg, J., Flicker, S., & Wuttunee, K.-D. (2021). Cellphilming and consent: Young Indigenous women researching gender-based violence, In R. Moletsane, L. Wiebesiek, A. Treffre-Goatley, & A. Mandrona (Eds.), *Ethical practice in participatory visual research with girls and young women: A focus on rural, Indigeneity and transnationality* (pp. 110–133). Berghahn.

4

COMPOSING PLACE: CREATING PARTICIPATORY SOUND PORTRAITS AND COMPILATIONS

Maureen Flint, Morgan Shiver, and Ryanne Whyte

Introduction

This chapter explores the journey of composing audio portraits as an entry point to understand the layered histories of place through an attention to the affective and resonant possibilities of sound. In educational research, Gershon (2013a, 2013b) argues that sounds are aural, vibratory affects that are intrinsically bound with the production of space and place and make possible representations of multiple contexts and encounters. Sounded methods approach the use of sound in research from a variety of perspectives—from ethnography (Arnfred, 2015; Gershon, 2013a; Wozolek, 2015), to narrative (Baker et al., 2020; Droumeva, 2015; Flint, 2021b; Gershon & Van Deventer, 2013), and geography (Duffy et al., 2016; Waitt et al., 2020; Wood, 2010). Sound has also been approached as a methodological tool, for example, attending to soundscapes as enacting a slow ontology (Ulmer, 2017), considering the signification of rhythm and intonation in transcription and oral history (Madison, 1993), or as a practice of listening and attunement (Bessette, 2016; Flint, 2021a; Gallagher et al., 2017). In ethnographic research, sound is often put forward as a means to

> attend to the sensorium in ways that move away from ocular privileging and toward a representation of the senses that are entwined with the lived experiences … creating a space for what was said but not heard in the data.
> *(Wozolek, 2015, p. 4)*

This chapter explores the possibilities of sound as a participatory research method that engages participants in the process and production of research. Within this chapter, we invoke the term participatory to include both the deliberate engagement and inclusion of research participants in the production, analysis, and representation

DOI: 10.4324/9781003164227-9

of an inquiry (Ballinger, 2008) as well as the sense of co-implication, and engagement produced through listening to or encountering visual and sounded representations of such an inquiry. We argue that sounded methods enable engagement with research that moves beyond a passive encounter of findings or outcomes to one that invites participatory thinking and engagement.

Examples of participatory sound studies include Bayne et al.'s (2014) study of distance students' experience where participants were asked to create 'digital postcards' that combined sound and images, Snowball and McKenna's (2017) student-created podcasts, or soundscapes or portraits created by or in collaboration with participants (Baker et al., 2020; Gershon & Van Deventer, 2013; Stevenson & Holloway, 2017). Through participatory research methods, we are also interested in sound as an embodied and haptic experience (Campt, 2017). In other words, we are interested in sound for its possibility to create moments and encounters of connection. Through this study, sound and audio create the possibility for a relational and connected experience that spans temporalities. For example, there is the moment the audio was created, on walking tours that Maureen took with both Ryanne and Morgan. There is the temporality of re-listening, first by Maureen as she created transcripts, then by Ryanne and Morgan as they re-listened to the audio to create clips. These temporalities of listening during the creation of the sound portraits form a current under the later listening and engagement of attendees at an art installation where these portraits were displayed, and then again now, for you the reader of this chapter.

The layered temporality of sound and the sharing of sound through these portraits offers a felt, embodied, and emotional experience. Listening to them is a participatory experience that is not simply about hearing to comprehend the content of the words spoken, but is a fundamentally haptic experience, an experience of touch, perception, connection, and sense (Campt, 2017). In this chapter, we consider the listener as the three of us as we created the audio compilations, the attendees at an art installation where the clips were heard (along with images and text), as well as you, the listeners of the audio clips and readers of the chapter. We theorise the capacity of sound and the practice of listening as participatory experiences with the work of Tina Campt (2017). Campt's work explores images of the Black diaspora and conceptualises sound as:

> An inherently embodied process that registers at multiple levels of the human sensorium … Sound can be listened to, and equally powerful ways sound can be felt; It both touches and moves people. In this way sound must therefore be theorized and understood as a profoundly haptic form of sensory contact. (p. 6)

To 'listen', for Campt, is participatory—beyond a one-way physical transmission of waves through an eardrum towards an embodied attunement to what else might register 'how they move, touch and connect us to the event' (p. 9). Importantly, for Campt, listening and hearing are practices that are intimately connected with futurity—a Black, feminist futurity that 'moves beyond a simple definition of the future tense as *what will be* in the future … to the tense of possibility that grammarians refer to as the future real conditional or that *which will have* ***had to*** *happen*' (p. 17, emphasis original).

The imperative of the future real conditional tense, Campt argues, necessitates a practice of listening that moves beyond what *is* to what *must be*, a practice of attunement to the haptic, quiet, quotidian practices that are embedded as affects and intensities. This is a practice of refusal that Campt operationalises through her exploration of a series of images. Refusal, Campt (2017) writes, is a practice that, 'highlights the tense relations between *acts* of flight and escape, and creative *practices of refusal*—nimble and strategic practices that undermine the categories of the dominant' (p. 32, emphasis original). We think with her theorisation of listening and practices of refusal in multiple ways. First, we explore how a participatory practice of creating sound portraits takes up multiple sensory levels, evoking an 'ensemble of seeing, feeling, being affected, contacted, and moved' that connects speakers and listeners across space and time (p. 42). Second, we consider the 'infrasounds' of events in our analysis, attending to the quiet, the interstices, the unsaid. This attunement moves us to consider the ways in which white supremacy becomes embedded in everyday practices of the university and the experiences of students in higher education. In addition, we argue that an attentiveness to infrasounds moves us as co-creators, and hopefully moves you as the listener and reader, to listen and orient to the future real conditional tense. As Campt (2017) describes, this is:

> A responsibility to create one's own future as a practice of survival. The future real conditional is an essential component of a black feminist practice of futurity as an existential grammatical practice of grappling with precarity, while maintaining an active commitment to the labor of creating an alternative future. (p. 116)

Our process of creating audio portraits came out of a shared desire to linger in moments and stories that exceeded categorisation, thematisation, or abstraction. Simultaneously, we offer these audio portraits as a passageway or threshold for listeners to linger with us—to be moved by the unsaid, the interstices, the affective capacities of the stories we share through audio and images. We offer this as a kind of participatory thinking. Through creating and listening, we can orient together towards an alternative future. In what follows, we first present the context of the study, the timeline, and methods of our research before moving to focus on the audio and visual portraits created by Ryanne and Morgan in collaboration with Maureen, and our presentation of these portraits as part of a community art installation.

Context, timeline, and methods

An interest in sound with this project began out of the involvement in a participatory arts organisation that the three of us were involved with as fellows,

Maureen as a graduate student and Morgan and Ryanne as undergraduates. The organisation, known as Creative Campus, had previously been an undergraduate focused group which put together creative events and activities for the campus. In years past, the organisation had brought in community artists and hosted maker events around crafts such as bookbinding, soapmaking, and screen-printing. The year that the three of us joined, however, the organisation was in a transition from an events-oriented program to one that was focused on research-creation. Part of this shift to research-creation also involved including graduate students who were involved with research projects or dissertations involving the arts as fellows.

Our project began as an offshoot of Maureen's dissertation work, which was interested in the sociohistorical context of race and racism on campus, and students' navigations of campus history and legacies of white supremacy. Maureen made a presentation to the larger Creative Campus group about her project and some ideas she had for exploring her research questions artfully. Other members of Creative Campus joined her research team based on this presentation. The team was led by Maureen, and five Creative Campus undergraduate students (including Ryanne and Morgan) took part as members of the team. As members of the team, Ryanne and Morgan participated in the study and also helped conceptualise the artful re-presentations of the project. During the writing of this paper, when asked why they initially participated, Ryanne reflected that she became involved because when

> we first started I liked [the] vision for the project ... [and] think of place in a lot of different ways with the way I grew up (moving around every couple years) so I think place is a worthwhile thing to delve into.

Morgan reflected that she 'thoroughly enjoyed that as a member of the research team, I also got to work with the data we collected in a creative way—I had never done research like this before, so it was a valuable experience for me!'

Our interest in audio specifically was inspired by a moment from Maureen's pilot study from the previous fall. She had been embarking on walking interviews on campus with students as part of her interest in place and sociohistorical contexts (Burke et al., 2017; Evans & Jones, 2011; Kuntz & Presnall, 2012). During one of these walking interviews, as she and another student approached a confederate monument on campus, a campus tour walked by. As the campus tour guide shared a narrative of test scores and student success, the student she was with shared their experiences with the history and placement of the confederate monument they stood in front of (Flint, 2018). Maureen was interested in how these stories and narratives of place contradicted, intertwined, and congealed in past-present-future *tellings* and *rememberings*. The audio of that encounter produced a kind of 'indexicality of voice and place [that began to] grasp something at a sensuous level that is considerably more abstract and difficult to convey in a written ethnography' (Feld & Brennis, 2004, p. 465). In other words, the layering of these histories and tellings became visceral in the audio, as the campus tour guide discussed Honors College admission rates and scholarship programs and the student reflected on the contested history of the name of the Honors College and his own advocacy work to bring attention to it in relation to the confederate monument. In the audio, the resonance of these disparate voices, the way they paused, caught, overlapped, and repeated, evoked the histories of place in a vibrant and three-dimensional way that seemed to become flattened and linear in the transcripts of the same encounter.

Maureen had played the audio from this encounter with the confederate monument for the research team with Ryanne and Morgan, and audio was on our mind. And then, around the same time that we were conceptualising the form of the project, we visited a local art installation as a group with Creative Campus. *The Selma Portrait Project* by Kathryn Mayo (2017) combined oral history interviews with large format wet plate collodion ambrotype, an antique photographic process. Walking through the gallery, the audio from interviews played as viewers walked between 11×14-inch glass plate portraits of Selma residents. As Mayo (2017) wrote, 'the reflective, almost three-dimensional quality of the glass plates allow the viewer to see themselves in each portrait, bringing to mind questions of personal responsibility, obligation and accountability' (para. 4). The pitch, tone, and vibration of the storytelling all combined to trouble linear narratives of history and place, resonating in the body, inviting the listener to feel and experience along with the speaker. In this way, the exhibit offered a kind of 'participatory thinking' where gallery visitors, simply through their presence, became implicated and part of the stories and narratives of Selma.

Following this exhibit, we wondered how we might approach the audio collected for our project in a way that evoked a similar type of reflective participation in the listener. The University of Alabama, where our study took place, is a site that is contested in many ways, complicated through and with multiple histories of white supremacy, settler colonialism, and civil rights activism. These histories are not located in a time-since-past but are embedded in place as infrasounds that continue to resonate in the present. As Campt (2017) described, infrasounds are the quotidian *buzzings* that continue to linger and be felt, long after physical sounds have faded away. Infrasounds are the buzzings of racism and colonialism, and they are also the potential for resistance and subversion. The period of time that Maureen, Ryanne, and Morgan were all connected to the University of Alabama marked a range of contradictory events that entangled with the campus's history, part of the infrasounds of place. In 2013, when Maureen was a second-year master's student and Morgan was applying to the college, a university newspaper article exposed the systemic practises of segregation within white Panhellenic sororities (Crain & Ford, 2013). This led to a series of protests and activism on campus, including a march from the main library to the administration building, touted as 'The Final Stand in the Schoolhouse Door'. This event mapped back to a 'stand at the schoolhouse door' that had occurred 50 years earlier, when the then-governor of the state had stood to bar access to two Black students, James Hood and Vivian Malone, seeking to enrol on campus. The spring following the protests around sorority segregation, Elliot Spillers was elected to the Student Government (SGA) presidency, the first Black student elected to the office in over 40 years. His successful election broke the decades-long grip of the Machine,[1] a voting bloc of 'Old Row' (historically white) Panhellenic and Intra Fraternity Council organisations with a history of intimidation and violent practises against Black students and candidates.

These moments that happened before we came together as a research team layered with events that occurred throughout the project. For example, around the time we recorded our walking interviews, the university dedicated a plaque on campus to honour Autherine Lucy Foster, the first Black student to attend the University. Later that same semester, there were several incidents involving sorority women posting racist and anti-Black Instagram posts. To think sound and listening with Campt (2017) attends to these happenings as infrasounds—buzzings that are said and unsaid in the audio portraits created by Morgan and Ryanne. In what follows, we move through our process of data generation and creation of the audio portraits.

Data generation and creation

In the fall of 2017, a year after Maureen's initial encounter at the confederate monument that inspired this research, Morgan and Ryanne both participated in a focus group with two other students in Creative Campus, and then both individually went on walking interviews with Maureen as part of her dissertation research, and as members of the Creative Campus research team. The question guiding Maureen's dissertation research asked how students navigate the socio-historical context of race on campus, and how the past and present context of white supremacy mattered in the production of place. Maureen also conducted two additional focus groups, went on seven other walking interviews, and had three individual interviews with students who could not make a focus group. Maureen transcribed the focus group and walking interviews. Then, she pulled out transcribed sections from walking interviews, interviews, and focus groups that connected to the overall research question, and brought in the excerpts to a Creative Campus team meeting that Morgan and Ryanne attended. Guided by a participatory ethic where we sought to produce the data as co-researchers, we read through the excerpts together, marking, highlighting, and underlining resonant sections. Then, we began cutting and layering excerpts next to one another, placing voices, stories, and experiences in conversation (see Figure 4.1).

After this collaborative activity, Maureen composed a 6 minute and 12 second audio compilation[2] using a free open-source audio editing software, Reaper, to clip, trim, and layer segments of recording using the collaboratively created highlights and groups as a guide. The compilation brought together student narratives in dialogue, troubling, and contesting linear or easy histories of place, playing with pause and quiet and silence.[3] The narratives shared by students explored how histories of racism, settler colonialism, and white supremacy became visible or invisible, how students confronted or encountered those legacies, and how they mattered in the present. Maureen shared the compilation with the

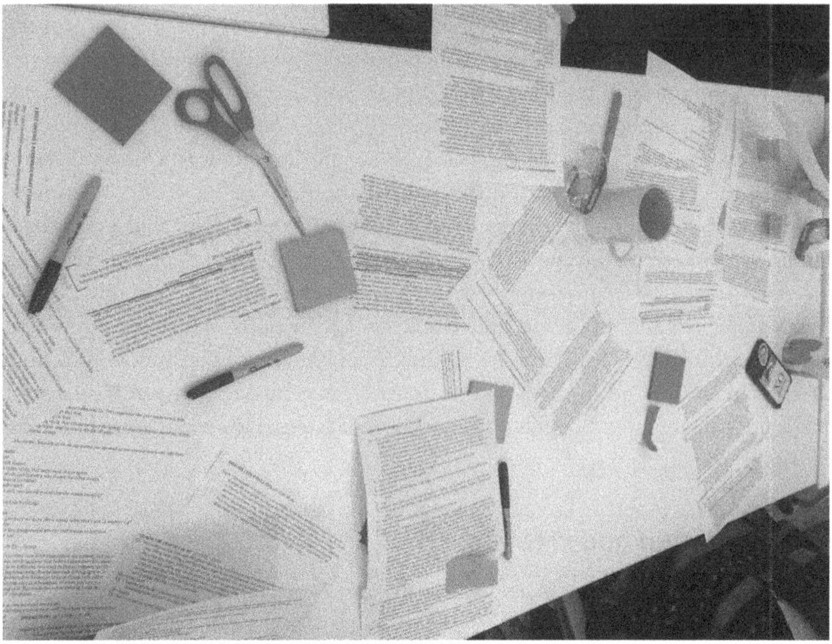

FIGURE 4.1 Participatory analysis

Creative Campus group (including Morgan and Ryanne) a few weeks later, near the end of the fall semester, for feedback and input.

As we returned to campus in the spring, Creative Campus announced that we would end the semester with a community art installation. Work from each of the research teams would be displayed, along with artefacts created during campus-wide events. Some students from our research team opted towards logistics and marketing, while Morgan and Ryanne expressed interest in taking a more active role in the representation of the project. This led to some reflection on the audio compilation Maureen had created the last semester. Our reflections included ethical questions such as: what does it do to cut, juxtapose, and layer the voices of someone else? How might Morgan and Ryanne's representation of their stories differ from that of Maureen's? Guided by a participatory ethic, the initial audio creation had included the undergraduate co-researchers in the ordering and grouping of segments of transcripts. Yet, those excerpts were ones that Maureen had selected as important or relevant to the research—they were her interpretation. Likewise, aesthetic decisions in the audio such as making some voices quieter or louder, placing voices in conversation with each other through layering multiple tracks—those were all representational decisions made by Maureen. We wondered about other ways of engaging in participatory inquiry, how Morgan and Ryanne as co-researchers might become more involved in the representation and analysis of their stories. Thus, Maureen invited Morgan and Ryanne to create their own 'audio portrait'. This practice was inspired by an article by Gershon and Van Deventer (2013) where Gershon interviewed Van Deventer, who was also a poet, creating a 'portrait' or narrative of Van Deventer through compiling clips of audio from a series of interviews. This practice, Gershon and Van Deventer wrote, was 'simultaneously a piece of sound/art, a narrative about a poet and his voice … and a piece of scholarship that utilizes a hybrid form of narrative research/inquiry' (p. 99).

To create the portraits, Maureen shared the audio and transcript from their respective walking interviews with Morgan and Ryanne, asking them to listen to the audio and pull clips or moments that connected with them to begin to create a 'portrait'. They both took a few weeks to listen. Using Audacity, a free sound editing software, they created, labelled, and exported clips. Maureen then set up individual interviews with both Morgan and Ryanne, where they talked through the process of creating clips. In the following section, we move through the clips selected by Ryanne and Morgan, the conversations we had after the clips were created, and then explore the process of their installation at a local gallery exhibit as part of the culminating event for Creative Campus.

Creating portraits

Both Ryanne and Morgan approached the process of creating clips in similar ways. In our conversations after, they noted that they 'tried to keep the project in mind … trying to map the campus … an audio view of the campus' (Ryanne) and 'I kind of knew how you might fit them together and like just little short clips of things that could fit with what other people might be talking about' (Morgan). Both pulled a number of clips that they then titled with a summative statement or main idea. Morgan described this later as creating 'chapters, labelling the topics that we talked about' as she initially listened and then she 'would go back and say oh, yeah, I talked about that, that was interesting and listen to it and find what little chunk to pull out'. These 'chapters' created by Ryanne, and then Morgan, are discussed in the following sections.

Ryanne[4]

Awful freshman roommates
Black tarps at frats
Golden hour of squirrels
Reese Phifer's character
Maureen's wardrobe
Construction on campus
First time in a sorority house
Thinking about college experiences
Campus was my world
Making UA's 'party status' invisible
Freshman UAPD situation
Shelby is so beautiful

At the time of the study, Ryanne was in her third year of the Telecommunication and Film program. The clips Ryanne chose (which can be listened to by following the link in the endnotes) include mundane moments of campus life. For example, the 'golden hour of squirrels' contemplates the ubiquity of squirrels on the campus at particular times of day. In 'Shelby Hall is so beautiful' and 'Reese Phifer's character' she reflects on the architecture and lore of buildings on campus. In addition, she recounts how her navigations on campus changed after she moved to an off-campus apartment her sophomore year in 'campus was my world'. In other clips she reflects on the process of the study itself such as in 'thinking about my college experiences'.

A theme linking together the remaining clips is a question of visibility and invisibility in campus space. In 'black tarps at frats' Ryanne shared how fraternities are mandated to 'hide' their parties from view (even as the existence of the tarps makes the parties hyper-visible), and the permissive behaviour of administrators and campus police towards the fraternities and sororities. In 'making UA's party status invisible' she recounted a story where a campus police officer at a campus education program told her that there were always undercover cops at the fraternity and sorority events (though as these events rarely got shut down, their job was presumably not to prevent underage drinking or other illegal behaviour). Lingering in the infrasounds of these clips about fraternity and sorority life is how whiteness functions as property right—manifested in the right to take up space on campus (a tenet of critical race theory [Ladson-Billings & Tate, 1995]).

Whiteness as the right to take up space continues to resonate in the clip titled 'Maureen's wardrobe'. This conversation happened while Maureen and Ryanne were walking on campus, and Ryanne had asked Maureen 'How do you feel walking around campus all the time?' In her asking this of Maureen (who is white) she draws attention to the fact that 'whiteness' is produced and reinforced through certain behaviours and actions—in this case, through wearing particular clothes, a 'uniform' which both Maureen and Ryanne eschewed. When we discussed the clips later, Maureen had asked Ryanne why she had included that clip, and she had said

> I just I think it's a part of like the campus's story that … like a big part of the uniform for campus especially for girls is shirts that cover your shorts that you're wearing and that's it and like tennis shoes … so the fact that you're like sort of the antithesis of that I think is so interesting.

In turn, Maureen had asked Ryanne how she felt she fit into that uniform and she shared:

> I remember one of the first days … I was going to [the Dining Hall] with one of my roommates. I wore the Bama Bound t-shirt that they gave us and shorts because it was a hot day. And I looked just like everybody else … [and] I looked down and I looked up and I was like, I can't do this again. I'm not going to do this again.

The 'uniform' that Ryanne describes was, at the time, a marker of 'belonging'—a standard of white femininity that was produced and disciplined by the Panhellenic sororities on campus. As Ryanne noted she 'refuse[s] to be like one of these—I can't do it'. At the time, Maureen had nodded along, agreeing that she too 'can't do it'. Listening to the infrasounds, the quiet buzzing, the unsaid between our assertions of 'I can't' there is a layering of multiple tenses of refusal. There are pasts, presents, and futures that operate from different subject positions—Maureen as a white graduate woman and Ryanne as a Black undergraduate woman. For both of us, 'can't' resonates with could and could not and can and could have, but in different ways that are crossed with racism and white supremacy. To utter 'I could have' is more possible for Maureen than Ryanne, for Ryanne to say 'I could not have' is an utterance that just four years before was demonstrably true. For Ryanne to remark that 'I just looked like everybody else' and 'I can't do this again' is a refusal of the logics of white supremacy that emphasise sameness and assimilation. And, at the same time as our 'can't' functions differently, it also points to the future, a refusal that is located not simply in rejection, but in an assertion of 'what will be' (Campt, 2017, p. 17).

The descriptions of the politics of whiteness of campus stand in sharp contrast to another clip included by Ryanne—in 'Freshman UAPD situation' she elaborates on a story that she had shared in a focus group where she and a group of friends, while on a walk around campus after dark, found a burning trashcan in front of the administration building on campus. Later, when Maureen and Ryanne discussed her process of making clips she shared that when they called the police:

> He didn't say a word, he drove up, got out of his car. He looked at us real, like sort of menacing, and then just turned around and didn't … and we were just kind of like, all right, I guess we're gonna keep walking and it was just, it was just weird like it was a little scary, but it was mostly just strange.

This recounting led to another story, a moment that had happened a week earlier (than the conversation between Maureen and Ryanne) where a white woman in Ryanne's Women's Studies class had shared how a campus police officer would check in on her and her roommates and offer them rides home when they were drunk. Ryanne continued:

> I was like, that's really cool. I mean, I'm sure that's … it's also like, my mind can't help but go, okay—well, that's also like white students, white female students, versus every *every* interaction I've seen or experienced on campus with UAPD and Black students has never been that kind of positive association.

This interchange resonates across time and space, buzzing with connection over the three years since it was recorded (and into the future). It is a refrain of Ryanne's experience, a refrain of the white student's inability to imagine a different experience could be the case.

The thread of whiteness as an infrasound, embedded, and unsaid in campus space was continued in the conversation between Maureen and Ryanne after Ryanne created the clips, where Ryanne shared another moment that had happened the previous week, when she was walking across campus after leaving the library:

> This woman and her daughter walked up to me, and they are touring the campus … [and after] I gave them directions [the mom] was like, "Okay, how do you like it here? Like as a Black student?" She and her daughter were Black … And I was just like, "Oh gosh, that's a complicated question." It's funny because [when they stopped me] I was talking to my friend on the phone, FaceTiming with him, and he heard the entire conversation too. So, basically, I told them, "UA has its problems, but you know, you find your community here. It's it all works out … " I did mention like the videos and all that stuff[5]. And then she asked me, "What's the deal with the sororities?" She [tells me] she's thinking about joining a sorority. So, I was like, "Yeah, that's a whole 'nother animal. I don't really know about, but I know they're not super accepting of Black women. It's been like a whole thing of them not really accepting Black women." Then [my friend who I was FaceTiming with] showed up right after they left. He found me and he was like, "Why are you lying to them Ryanne?"

At the time, Ryanne laughed and assured both Maureen (in the re-telling) and her friend (in the moment) that she 'hadn't lied, just to clarify'. In other words, her friend was saying she should have told them that the campus is more racist than she let on. Then she continued, 'Me personally I know I like being here. I like, you know, the people I've met, the experiences … I had a great like college experience overall'. Again, in this moment, Ryanne enacts in her tellings a politics of refusal to a neat and linear narrative of either subjection or utopia, but rather orients to relationality, to possibility. This is not a naïve or innocent positioning. As Campt (2017) writes:

> Futurity is, for me, not a question of "hope" though it is certainly inescapably intertwined with the idea of aspiration … [Futurity] is a tense of anteriority, a tense relationship to an idea of possibility that is neither innocent nor naïve … It is frequently quiet and opportunistic, dogged and disrupted. (p. 17)

Through the process of composition, through creating a portrait of place, Ryanne offers a complicated and non-linear telling of what it means to belong, how to belong. Attuning to the composition of Ryanne's clips, there is a future real conditional tense of what must be, a tense that does not disregard the past, but imagines and lives *what must be* if we are to imagine a radically different future. This tense, importantly, comes through in the practice of listening, both initially, for Ryanne, as she created the portrait, for Ryanne and Maureen as they relistened together, and for the audience or listener who engages with the clips she selected. In what follows, we continue to think this concept of future-real-conditional with Morgan.

Morgan[6]

Choose your own adventure
Knowing that it exists
Everyone's entitled to their own opinions
Looking around the classroom
I really hope not everyone's like this
My idea of the South
The sororities got in major trouble

At the time of the study, Morgan was a senior majoring in English. She noted after creating the clips with Maureen that she had kept the focus of the project in mind while creating them (to listen, follow the link in endnotes). Specifically, she pointed to the initial compilation they had co-created and that Maureen had shared the previous semester as inspiring the clips that she chose:

> I do think my process of it would be different than someone who wasn't involved in the project. I had already listened to some of the stuff you put together, so I knew what you were kind of going for.

Morgan's clips also explore the infrasounds of whiteness and how it functions on the campus. In 'everyone is entitled to their own opinions' and 'knowing that it exists' she maps between her own choices as a white student to acknowledge the history of campus ('yeah, I'm not a fan of that' and 'not sugar-coating it') while also noting 'everyone is entitled to their own opinions'. In 'choose your own adventure' she describes that how students come to encounter history 'really is different for every student … ' As a white student, Morgan's clips map the way that confronting and acknowledging the history of campus and the continuing climate of white supremacy is positioned as a choice. It is not included in curriculum, or campus tours. Students who want to learn about the history must do so by seeking it out, or come across it incidentally, through a class they take or in casual conversation with friends. Further, as a white student, she has to continually choose to position herself differently from the culture of campus.

As Maureen and Morgan met to talk through the clips, Morgan elaborated on several of the stories that she had gestured to in the initial walking interview. In particular, she shared more about the story she had titled 'I really hope not everyone is like this'. During the initial walking interview, as shared in the clip, she had not elaborated on the story of a 'horrible' experience during freshman orientation, and Maureen had not followed up (as the conversation had moved on easily to another topic). As they talked about the process of making clips, Maureen asked if Morgan wanted to say anything more about any of the stories she had shared in the clips, and Morgan pointed to that clip. She then shared that moving from Washington State, she 'didn't know anyone'. She had moved on campus early because of her involvement in the university marching band (band members come to campus several weeks before the start of the semester to practice routines for football games, which sometimes start before the first week of classes), and found herself at a lunch table with a group of women who were going through sorority rush (which also happens before the start of classes):

> And they asked me about my roommates, and I was telling them my three roommates' names, and I got to my one roommate's name and one of the girls asked me, "Oh is she Black?" and I was like, "Yeah. Yeah, she's Black." And the girl said, "Oh, I'm not racist, but I could never room with a Black person," and I was so horrified in that moment. And I think now being older and being the bolder person that I am I definitely would have said something back to her like saying, "Well you probably are racist if you have to say that 'you're not racist.'"

Through the practice of re-listening, re-telling, highlighting, and composing, Morgan expanded beyond 'horrible' to become more specific. Horrible was not simply that she had felt lonely or not connected to campus, rather horrible was a moment of 'horror' at the views of other students. This complicatedly maps back to other clips pulled out by Morgan, where she notes that everyone has the right to their opinion. There is a contradiction in the quiet between her clips denouncing racism and also allowing space for it. We might listen closely to the question mark that hangs quietly between these statements, a humming infrasound. What does it mean to have a right to an opinion, who claims that right, who is afforded that right in policy and law, for whom is that right dangerous? When re-listening during the writing of this paper, Morgan noted that 'Listening, re-listening, and creating something from my own audio clips forced me to reconsider my experiences and the way I presented them'.

Throughout Morgan's clips, she also refuses easy linearity between the geographic location of the South and the persistence of white supremacy. For example, in her clip 'the sororities got in major trouble' Morgan references the protests that occurred in the Fall of 2013 around anti-Black racist practises occurring in white Panhellenic sororities to maintain their segregation. Morgan noted that this happened right around the time she was applying to the university. She observed that 'a lot of people in the North will have that stereotype about the South … After I've tried so hard to convince people that no it's not like that'. Later in their conversation, Morgan tied back to that clip, and connected it to the clip she had titled 'My idea of the South'. In that clip, she had shared how she had been born in Georgia, a state also in the Southeastern United States and had been surprised coming on campus to discover how white it was. She mapped back to this as she and Maureen talked through her creation of the clips saying:

> I am so bothered by the stereotypes people hold for people in the South. I cannot stand it. The main stereotype being that everyone is racist. Everyone is super conservative. Everyone is stupid … you know just in movies and TV shows that's how people from the South are portrayed … it's important to let people know that "hey, there's racist people everywhere" and it's not everyone from Alabama that has those views or holds those views.

Later, as we wrote and revised this article, she noted another example of refusal of determinative geography, noting that the woman she encountered during orientation was from California.

At the same time, Morgan also noted in the conversation with Maureen after creating the clips how the linearity between geography and history did and could function, pointing to one of the recent incidents on campus, of a woman who had posted a racist Instagram video noting that the woman said, 'She came to Alabama because she thought that people's views would align with her own'. Morgan continued:

> We've done a lot of talk about racism and race in our interviews because there is such a deep history of that at the University. As there is with any University, and especially in a University in the South … and I think there's a lot of big generalisations made about the people.

As with Ryanne, listening in the quiet spaces, the unsaid, the affects of the clips created by Morgan produces another (not unrelated) but positionally different portrait of campus. Morgan works a different politics of refusal, one that troubles the easy lines between geography and race, one that negotiates and seeks a refusal of the legacy of white supremacy. Again, the multiple practises of listening compose different entry points to these refusals, different geographies and encounters in place. Listening, as Bessette (2016) writes, offers a 'dual meaning of understanding and hearing' that produces an 'openness to difference' (p. 73). Importantly, this refers to difference both as a resistance to dominant narratives, as well as a difference in experience. In the following section, we explore how listening to these portraits is a participatory and engaged act.

Installation: A coda

The process of creating, composing, selecting, and then discussing audio clips offered Morgan and Ryanne an opportunity to create their own audio portraits of campus—an entry point to explore the complicated and ongoing legacies of race and geography. As Gershon (2013b) writes, 'Representing sounds sonically allows participants give voice for/to themselves while retaining information lost when translated to text' (p. 259). Thinking with Campt, we offer that this practice also becomes a practice of futurity, a practice of orienting to what *must be*, thinking in complicated and non-linear tellings of place and belonging. As noted earlier, these portraits were created to be included as part of a community arts installation featuring the work of Creative Campus fellows over the 2017–2018 academic year. At the end of the conversations where they discussed the creation of the audio clips, Maureen noted that in the creation of the initial audio compilation, she had offered 'her analysis or representation of these encounters and the stories you're telling, but you might do that differently'. She invited both Morgan and Ryanne to think about how they would want a gallery attendee to engage with their portraits, and what other modalities they might want to include to attune or guide a listener through their stories. Ultimately, Ryanne sent Maureen a folder of pictures—some of which she had taken on the walking interview they had both gone on, and others she had downloaded off her phone (see Figure 4.2).

These images (lettered A-L from left to right) connect with the stories Ryanne tells in her clips—for example, Ryanne includes a picture of Shelby Hall (Image F, G), a hawk (Image J), and a squirrel (Image E)—all referenced in her clips. She also includes several photos taken from windows of Reese Phifer (Image A, B, K, L), the building where her program was located, and the subject of one of her clips. Other photos gesture to stories not told, moments and encounters that exceed the clips, exceed the talk of the interview. As Campt (2017) writes, 'listening requires an attunement to sonic frequencies of affect and impact. It is an ensemble of seeing, feeling, being affected, contacted, and moved beyond the distance of sight and observer' (p. 42). Listening to the images Ryanne created, along with her audio portrait, orients the listener to a future-real-conditional-tense. That must which have had to have happened. For example, one photo is of the Malone-Hood Plaza (Image H), a marker on campus honouring the legacy of James Hood and Vivian Malone at the site of the initial 'stand in the schoolhouse door'. Viewed while listening to the clips, this photo offers other stories, other visualisations of entry points to be and become on campus. Listening, there is possibility for joy and connection and legacy and a futurity which refuses a pathological or reductive reading of Ryanne's experience. One that 'enunciate[s] alternative accounts' (Campt, 2017, p. 5). The listener/viewer is co-implicated, connected, participating with Ryanne in her imagining of the future.

In contrast, Morgan chose to compose a nonlinear transcription of her clips (see Figure 4.3), creating columns and rows of excerpts from her audio. As with Ryanne's photographs, these clips of words zigzag in and between the audio. As the viewer listens, reading along with the text, there are moments of serendipitous alignment, when as Morgan shares a story, the viewer is reading the same words. There are also words that are not included in the stories shared in Morgan's clips, words and excerpts that Morgan took from the transcript of the interview, or from our conversation after making the clips that exceed the audio. 'Listening requires an attunement to sonic frequencies of affect and impact. It is an ensemble of seeing, feeling, being affected, contacted, and moved' (Campt, 2017, p. 42).

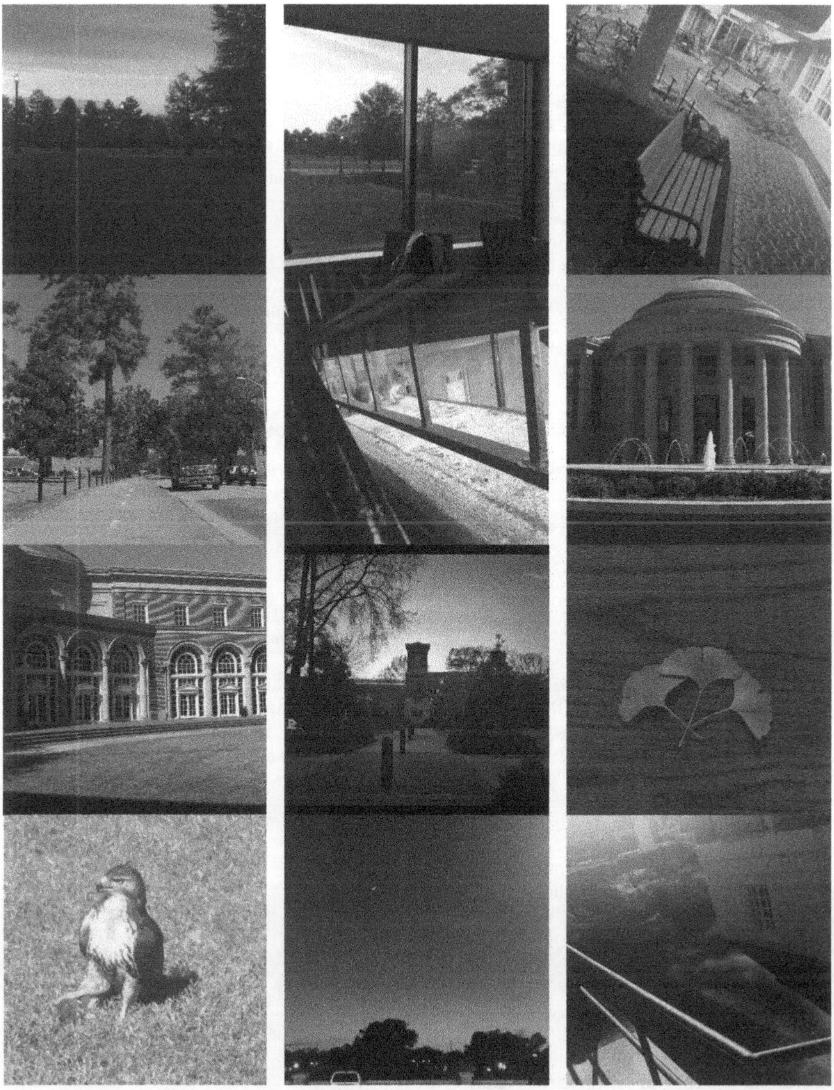

FIGURE 4.2 Ryann's photographs

Listening and reading creates other entry points, other possible imaginings of the future. The listener/viewer is both experiencing along with Morgan, invited into her story as she speaks to them through the audio, at the same time as the disjuncture between audio and text creates moments for other connections. As Campt (2017) writes of infrasounds, 'their affects register at a frequency that is felt rather than heard' (p. 31). Moments of pause, interstices of connection. Viewing the images shared by Ryanne and the text-poem created by Morgan creates opportunities for the viewer to feel co-implicated, to feel brought in and part of the experience. And to wonder how they might be otherwise. Both through the

Oh, man, I really hope not everyone at the University of Alabama is like this

Choose your own adventure book

I decided I wanted to go to a school that was a good school, but would also be fun to go to

I got stuck hanging out with this group of girls

I would like to take part in that, and, like, that and that and that

Being from the South and moving to the North, I am so bothered by the stereotypes people hold for the people in the South – cannot stand it

We're a college campus and, um, everyone's entitled to their own opinions

You have to come to that realization as a freshman, like, oh, I don't need a dessert with every meal

I'm not racist, BUT, I could never room with a *black* person

Ideally, everyone finds their family on campus and their place

And I'm like, wait a minute

Yeah, I'm not a fan of that

I'll take a moment and look around the classroom and everyone in the classroom is white

Me and a group of my friends would go to Burke every single day of the week for dinner

Everyone's gotta grow up and move on eventually

Hey! There are racist people everywhere, and it's not everyone from Alabama

Well, you probably *are* racist

I know that other things happen

I will be so sad to leave that and leave the university

I think now, being older, and being the bolder person that I am, I definitely would have said something back to her

Some evenings we would spend, like, three hours in there just talking

If there's a stupid character on a TV show, 9 out of 10 times they're going to have a Southern Accent somehow

It's just finding where you fit, I guess, and embracing that

Thankful, glad, happy

It's better to miss something when you leave than to not

Knowing that it exists, and not pretending that it doesn't

It's just become home

FIGURE 4.3 Morgan's non-linear transcript

moments of connection—moments where stories line up, and resonate, as well as moments of disruption or unfamiliarity.

At the installation, both Morgan and Ryanne's images were displayed on a wall where a map of campus was projected. On the map, a line continuously circulated,[7] mapping the paths that Maureen had taken with Morgan, Ryanne, and the other participants in the study, undulating across the campus. These other modalities resonated along with the audio, creating additional resonances and opportunities for connection and participation—entry points into a more complicated telling of campus place. At the installation, a viewer would walk up to a podium where a small audio recorder was placed. Upon picking up a pair of headphones they were brought in, aurally to the tellings and experiences of Morgan or Ryanne. Listening at the same time as they read the transcription of Morgan's audio or viewed the pictures taken by Ryanne offered different entry points and readings that are both representative of the place of campus and its contradictions. This engagement moves the act of listening beyond the three of us as co-researchers to provoke moments of resonance and participation for the listener. Audience members were invited to think and experience along *with* us. In the audio, they become caught up in the clack of Maureen's shoes on concrete, the pitch of Morgan's laugh, the tone of Ryanne's voice, the sound of a campus bus stopping in the distance. Through the process of listening during the installation, the viewer/listener was invited into the experiences of both Morgan and Ryanne, in dialogue with and conversation, to participate. The audience was asked, through this close engagement, to consider their own responsibility to place, and imagine their own future real conditional tense of what *must be*.

At the same time, the process of engaging in this project together continues to offer critical questions about ethics and representation and participatory research. Completed as an extension of Maureen's dissertation research, our relationship as co-researchers has always already been marked with hierarchy and power. This includes the individual level—the initiation of the research as an offshoot of Maureen's dissertation and our involvement in Creative Campus which positioned Maureen as the 'research team leader'. In addition, our relationship is also marked by the broader logics of academia and qualitative research which position researcher as expert and knower—indeed, even writing this chapter we have wondered what it means (or how to approach) writing with a participatory ethic. There are many ways in which we could have been more participatory, more disruptive of the traditional hierarchy of researcher/researched, more collaborative. For example, we could have co-produced our research questions or our method of data generation, Morgan and Ryanne could have interviewed Maureen, one another, thought of data generation outside the logic of the 'interview' entirely. Acknowledging these limitations, we hope that our exploration through the process of creating and presenting these sound portraits has offered a unique consideration of what is made possible through sound and sounded methods.

As we have explored in this chapter, participating in this project became a means for the three of us 'to examine echoes across time and contexts, open relationships within and between ecologies … to consider complex interrelations and orientations inside and beyond people' (Daza & Gershon, 2015, p. 641). Creating sounded portraits offered a participatory methodology oriented to futurity, one that, both in their creation, their installation, and their presentation in this chapter, make palpable the circulations of white supremacy and anti-Black racism that flow and function in geography and place. Thinking with Campt's (2017) conception of futurity as 'the power to imagine beyond current fact and to envision that which is not, but must be' (p. 17) we imagine the participatory potential of our work beyond the moment of the installation. They resonated then, they did something, they connected with the audience there and then in particular ways—and, they continue to resonate. We hope that as you listen, you also wonder along with us, that listening to these clips disrupts the taken for granted machineries of whiteness, that you find different tellings, different readings, different entry points to being and becoming in/with place. We offer both making and listening to these portraits as a practice of refusal that implicates both us as co-researchers, community members, students, and participants at the same time as it implicates you as the listener/viewer in imagining a different future. We imagine a politics of responsibility, of futurity, and invite you along with us.

Notes

1 See Weiss (1992) for a dated, but still relevant historical overview of the Machine. In addition, a podcast series released in August 2018 titled 'Greek Gods' explores the Machine through the 2017 student government elections (Yurkanin & Archibald, 2018). Listen at https://www.stitcher.com/podcast/alcom/reckon-radio.
2 This compilation and its theoretical implications is explored more in-depth in Flint (2021b) and can be listened to here: www.maureenflint.com/race-place
3 This was included in the Institutional Review Board consent form for all students in the study as a possibility. In other words, as this project was conceptualised as an arts-based research project, all participants were able to provide consent at varying levels—from having their voice recorded, to consenting to having audio or images from interviews and focus groups used and shared in findings and output from the study. All students whose voices were used in the sounded assemblage were sent the audio and invited to provide feedback or ask to no longer be included.
4 Listen to Ryanne's audio here: https://www.maureenflint.com/ryanne-sound-portrait
5 Here, Ryanne is referring to a series of racist Instagram videos posted by a University of Alabama sorority woman that had received national attention that summer.
6 Listen to Morgan's audio here: https://www.maureenflint.com/morgan-sound-portrait
7 This can be viewed at: https://www.maureenflint.com/walkingpaths

References

Arnfred, M. (2015). Polyphonic sound montages: A new approach to ethnographic representation and qualitative analysis. *Journal of Organizational Ethnography, 4*(3), 356–366. 10.1108/JOE-10-2014-0034.

Baker, A. M., Sonn, C. C., & Meyer, K. (2020). Voices of displacement: A methodology of sound portraits exploring identity and belonging. *Qualitative Research, 20*(6), 892–909. 10.1108/JOE-10-2014-0034

Ballinger, C. (2008). Authority. In L. Givens (Ed.). *The Sage encyclopedia of qualitative research methods*. SAGE.

Bayne, S., Gallagher, M., & Lamb, J. (2014). Being 'at' university: The social topologies of distance students. *Higher Education, 67*(5), 569–583. 10.1007/s10734-013-9662-4.

Bessette, J. (2016). Audio, archives, and the affordance of listening in a pedagogy of 'difference'. *Computers and Composition, 39*, 71–82. 10.1016/j.compcom.2015.11.004

Burke, G., Lasczik Cutcher, A., Peterken, C., & Potts, M. (2017). Moments of (aha!) walking and encounter: Fluid intersections with place. *International Journal of Education Through Art, 13*(1), 111–122. 10.1386/eta.13.1.111_7

Campt, T. M. (2017). *Listening to images*. Duke University Press Books.

Crain, A., & Ford, M. (2013, November 11). The final barrier: 50 years later, segregation still exists. *The Crimson White*. http://www.cw.ua.edu/article/2013/09/the-final-barrier-50-years-later-segregation-still-exists

Daza, S., & Gershon, W. S. (2015). Beyond ocular inquiry: Sound, silence, and sonification. *Qualitative Inquiry, 21*(7), 639–644. 10.1177/1077800414566692

Droumeva, M. (2015). Curating everyday life: Approaches to documenting everyday soundscapes. *M/C Journal, 18*(4). 10.5204/mcj.1009

Duffy, M., Waitt, G., & Harada, T. (2016). Making sense of sound: Visceral sonic mapping as a research tool. *Emotion, Space and Society, 20*, 49–57. 10.1016/j.emospa.2016.06.006

Evans, J., & Jones, P. (2011). The walking interview: Methodology, mobility and place. *Applied Geography, 31*, 849–858. 10.1016/j.apgeog.2010.09.005

Feld, S., & Brennis, D. (2004). Doing anthropology in sound. *American Ethnologist, 31*(4), 461–474. 10.1525/ae.2004.31.4.461

Flint, M. A. (2018). Ruptures and reproductions: A walking encounter with a campus tour and a Confederate monument. *Cultural Studies ↔ Critical Methodologies, 60*(14), 91–104. 10.1177/1532708618809136

Flint, M. A. (2021a). More-than-human methodologies in qualitative research: Listening to the Leafblower. *Qualitative Research*, 1–21. 10.1177/1468794121999028

Flint, M. A. (2021b). Sounded histor-futurit-ies: Imagining posthuman possibilities of race and place in qualitative research. *International Journal of Qualitative Studies in Education*, 1–7. 10.1080/09518398.2021.1885071

Gallagher, M., Kanngieser, A., & Prior, J. (2017). Listening geographies: Landscape, affect and geotechnologies. *Progress in Human Geography, 41*(5), 618–637. 10.1177/0309132516652952

Gershon, W. S. (2013a). Sonic cartography: Mapping space, place, race, and identity in an urban middle school. *Taboo, 13*(1), 21–45. 10.31390/taboo.13.1.04

Gershon, W. S. (2013b). Vibrational affect: Sound theory and practice in qualitative research. *Cultural Studies ↔ Critical Methodologies, 13*(4), 257–292. 10.1177/1532708613488067

Gershon, W. S., & Van Deventer, G. (2013). The story of a poet who beat cancer and became a squeak: a sounded narrative about art, education, and the power of the human spirit. *Journal of Curriculum and Pedagogy, 10*(2), 96–105. 10.1080/15505170.2013.782593

Kuntz, A. M., & Presnall, M. M. (2012). Wandering the tactical: From interview to intraview. *Qualitative Inquiry, 18*(9), 732–744. 10.1177/1077800412453016

Ladson-Billings, G., & Tate, W. (1995). Toward a critical race theory of education. *Teachers College Record, 97*(1), 47–68.

Madison, S. (1993). 'That was my occupation': Oral narrative, performance, and Black Feminist thought. *Text and Performance Quarterly, 13*(3), 213–232. 10.1080/10462939309366051

Mayo, K. (2017). *The Selma Portrait Project*. http://kathrynmayo.com/the-selma-portrait-project1.html

Snowball, J. D., & McKenna, S. (2017). Student-generated content: An approach to harnessing the power of diversity in higher education. *Teaching in Higher Education, 22*(5), 604–618. 10.1080/13562517.2016.1273205

Stevenson, A., & Holloway, J. (2017). Getting participants' voices heard: Using mobile, participant led, sound-based methods to explore place-making. *Area, 49*(1), 85–93. 10.1111/area.12296

Ulmer, J. B. (2017). Writing slow ontology. *Qualitative Inquiry, 23*(3), 201–211. 10.1177/1077800416643994

Waitt, G., Buchanan, I., & Duffy, M. (2020). Lively cities made in sound: A study of the sonic sensibilities of listening and hearing in Wollongong, New South Wales. *Urban Studies*, *57*(10), 2131–2146. 10.1177/0042098019871170

Weiss, P. (1992, April). The most powerful fraternity in America. *Esquire*. http://www.welcometothemachine.info/pdf/esquire.pdf

Wood, D. (2010). *Everything sings: Maps for a narrative atlas*. Los Angeles: Siglio.

Wozolek, B. L. (2015). *The presence of absence: The negotiation of space and place for students of color at a predominantly white suburban high school.* (Publication No. 3726987) [Doctoral dissertation, Kent State University]. ProQuest Dissertations Publishing.

Yurkanin, A., & Archibald, J. (2018, August 27). Greek gods (Nos. 1–3) [Audio podcast episode]. In *Reckon Radio*. Stitcher. https://www.stitcher.com/show/reckon-radio

5

THE RADIO PLAY AS RESTORATIVE JUSTICE EDUCATION: A CREATIVE COLLABORATION BETWEEN A GRASSROOTS ORGANISATION AND ARTISTS

Tanyss Knowles and Frank J. Tester

Introduction

In 2019, The Vancouver Association for Restorative Justice (VARJ) collaborated with Cree playwright, Renae Morriseau, and members of the community, in the writing and publication of a radio play[1] script telling a story about a restorative justice process (Morriseau, 2020). The play, *A Mural of Social Change*, informs the public about how restorative justice works in real-world settings. The collaborative creation, the story, and the experience of listening, all mirror key values of restorative justice practice. These include the importance of the facilitator's preparation, and features such as community engagement, transformation, and active listening.

Restorative justice (RJ) is, by some accounts, not easily defined. As Latimer et al. (2005) note, 'The term has been used interchangeably with such concepts as community justice, transformative justice, peacemaking criminology and relational justice' (p. 218). The definition used in this chapter is consistent with the material and experience we discuss: a perspective on restorative justice that draws its definition and practice from Indigenous traditions found in Canada, New Zealand, Australia, and the United States (Borrows, 2006; Chartrand, 2018; Vieille, 2012). A pre-condition for RJ work is acknowledgement by the offender of the wrong they have done. As noted below, RJ can also play a role in the prevention of behaviours that might often result in conflicts with legal consequences.

RJ is, according to some practitioners, to be distinguished from retributive justice, the latter focusing on the punishment of offenders, rather than on rehabilitation. Others have challenged this distinction, noting that restorative justice fails to recognise 'the complexity of punishment processes outside the formal courtroom' (Roche, 2005, p. 75). Furthermore, what might be considered as healing or punishment is very much dependent on the perspective held by the offender. Punishment can, in extreme circumstances, be part of Indigenous justice systems, as

DOI: 10.4324/9781003164227-10

portrayed by Zacharius Kunuk (2001) in his Cannes winning film, *Atanarjuat (The Fast Runner)*. In the film, family members involved in the murder of Atanarjuat's brother are banished from their camp; a punishment likely with fatal consequences, given the need for group cooperation to survive arctic conditions at the time.

Restorative justice practices typically take three forms: circles, conferences, and victim-offender mediation. The Vancouver Association for Restorative Justice (VARJ), responsible for the work discussed in this paper, is closely aligned with the RJ tradition of circle work. Circle work is a method were victims—including those representing a community or group, where appropriate—may be included with the offender(s) in addressing harm done and its resolution. It is attractive because of its dynamic and the insights and social relations it fosters. It can be used for restorative, as well as educational and preventative purposes. Circles can be used to build and strengthen communities in dealing with issues of justice and their social origins. As Katic et al. (2020) note, and as Gwathney (2021) illustrates, circles used in classroom settings can contribute to the development of values, insights, and identities that address problems like bullying and racism (Davis, 2019). With this in place, restorative circles can be used to deal with issues that subsequently arise in this setting.

Circle work focuses on restoring an offender to their community, family, and culture. It involves work with participants in order to prepare them for the revelations, processes, and feelings involved in circle work. RJ recognises the importance of healing for both the offender and the offended. The offended—an individual, group or community—is an integral part of work that addresses the behaviour of an offender. The dichotomy of 'offender and offended' can be viewed with some scepticism as many conflicts offer more nuance and opportunity for self-reflection for all those involved than these labels suggest.

RJ focuses on the rehabilitation of offenders through reconciliation with victims and the community at large. It is an approach to justice that emphasises healing in victims; accountability, reflection, and personal growth on the part of offenders, and the involvement of citizens in creating healthier, safer communities. Circle work often addresses the behaviours and histories of those involved, and relevant personal and social issues in working towards resolution. What happens in circles can be personal, emotional, and challenging. Issues of confidentiality and trust are critical. Communication skills, grounded in theoretical considerations, are essential to effective circle work.

VARJ is a grassroots organisation promoting the concept and the use of RJ. Its working board includes RJ and community development practitioners and academics, among them, the authors of this chapter.[2] VARJ's mission is to further public understanding of the principles, values, and practice of restorative justice, and to contribute to restorative justice initiatives in the City of Vancouver, in Canada, and internationally. Since its establishment in 2005, VARJ has consistently run up against the challenge of lack of public understanding of restorative justice and how it works. While British Columbia has been at the forefront of the development of RJ practice in Canada (Asadullah & Morrison, 2021), the understanding of politicians and the general public at the provincial and municipal levels remains an obstacle to progress in the use of RJ in dealing with matters of youth, and especially criminal justice.

Therefore, VARJ sought out an engaging way of answering the common question we receive: what does an RJ process actually look like? In producing a radio play, our aim was to reach the general public with educational and informative information that would attract interest and attention. While film could be used to act out and/or record an event and its resolution, issues of confidentiality and finding the funding and other resources needed for such a production were daunting challenges for a non-governmental organisation like VARJ. Restorative justice circles are sensitive and confidential process; therefore, filming would be intrusive and inappropriate.

In the writing and publication of a radio play script, VARJ collaborated with Cree playwright, Renae Morriseau to tell a realistic, but fictional story, intended to leave the reader/listener with some idea of what a restorative justice process 'looks' like without compromising the confidentiality of an actual RJ process (Morriseau, 2020). In the radio play, the listener witnesses a harmful incident between two characters. The story of Derrick and Laura, and Derrick's transgressions, highlight a complexity of issues, histories, and experiences. The play also, through the experience of these characters, brings to life the facilitation of an RJ contract focused on healing multiple relations.

In this chapter, we discuss the collaborative process undertaken to develop the script and to record the radio play in the format of a podcast. The process mirrored key values found in RJ practices and processes. These include community inclusion, collaborative approaches to understanding and appreciation, and the transformation of understanding, listening, reflecting, and acting. We build on literature dealing with arts for social change, popular education, and the history of the radio play as an educational tool. We explore the impact of this project; the result of hosting a listening session and virtual focus group. We note the perspectives and reactions of listeners and their experiences of engaging with restorative justice through a radio play. Finally, we discuss how community organisations can collaborate with artists to develop a significant and impactful piece of art that furthers important conversations in the community.

In this instance, our efforts corresponded to a resurgence of the use of radio plays and podcasts in the presence of the social isolation imposed in Canada during COVID-19 restrictions. The result was a solution to the problems of confidentiality that would have been encountered if an actual RJ process was filmed or recorded. The result allowed for the distribution at a local, regional and international level of a tool for furthering an understanding of what an RJ process actually looks (and sounds) like.

Envisioning the project

On a rainy Vancouver evening, board members of VARJ met to discuss how to develop a project that would help fulfil our organisational mandate of furthering public understanding of restorative justice. The project also aimed to build confidence and capacity in potential RJ facilitators, and to increase support for the process by stakeholders in the criminal justice system. The discussion of these grassroots volunteers revolved around the idea of finding a creative way of teaching the community about restorative justice. We were interested in building awareness of not just what RJ is, but also *how it works* —as a way of dispelling common myths held by the public towards RJ as an alternative to retributive justice. The most common of these is the notion that RJ is easy on offenders; that it offers them a 'slap on the wrist' and a way out of their troubles that is 'soft on crime' and disregards the needs of victims.

While even RJ experts cannot narrow RJ down to one specific process, the public is often even more muddled about how RJ actually works. Even among a public that has some idea of how RJ processes work, there remains a conviction that RJ is best suited to youth, who deserve to be forgiven and re-educated with regard to their misdemeanours. So-called hardened criminals are another matter. The use of RJ with adult offenders typically encounters more resistance. Furthermore, the extent to which facilitators see crime as a personal problem, rather than something with social and cultural origins, has a role to play in what the resolution looks like. The extent to which this is true can be a function of the culture in which the facilitator was raised and the degree to which this socialisation has been critically examined. Looking at the acceptance of RJ in American culture, Gregory D. Paul (2015) makes a similar observation:

> In short, the groups had differing prescriptive justice beliefs. Facilitators tended toward seeing crime as a personal problem, enhancing the victim's role, and treating the parties respectfully. Public members, while also believing sensitive treatment to be important, felt more strongly than facilitators about the use of punishment and unbiased procedures. (p. 283)

Seeing crime as a personal problem and not as a social issue that includes personal developmental histories can play a role in how RJ processes unfold. These matters are worthy of exploration and discussion by anyone who has witnessed an RJ process, yet it is difficult to produce a script that covers all possibilities. VARJ's hope is that, in discussions that might follow listening to the play, the bigger notions of justice can be explored.

RJ processes are an effort to transform the way we engage with and punish wrongful acts. Proponents of RJ believe that wrongful acts, or crimes, create needs and responsibilities for those involved, as well as the larger community in which the acts take place (Zehr, 2002). This is why RJ often involves direct and facilitated communication between the affected person(s) and the responsible person (s), as well as inclusion of community members who can support, offer guidance, and contribute to transformative understanding (Menkel-Meadow, 2007).

VARJ wanted to communicate that no matter which RJ process one chooses; at its heart, RJ is a re-imagining of justice that prioritises transformation. With these considerations being central to VARJ's goals in this project, our objective became the creation of something that was accessible, compelling, and aligned with the principles of RJ. Nevertheless, communicating the *how* of restorative justice is vital to having an informed public that understands RJ as an adjunct or alternative to conventional western retributive forms of justice. The result was a one-hour radio play that told the story of an offensive encounter and showcased a paraphrased, yet accurate version of RJ in practice.

A radio play was ideal for our purposes because it encompassed many of the key values of RJ, as well as communicating a story in a compelling and accessible format for a widely distributed audience.

The project: *A mural of social change radio play*

In the fall of 2019, VARJ secured a grant from the Law Foundation of British Columbia enabling us to commission a playwright to take the lead in creating a script. Written by Vancouver-based Cree playwright, Renae Morriseau, the play focuses on a young Indigenous artist who is accosted while painting a mural for an arts festival, hence the title given to the play. Renae Morriseau is a talented Indigenous playwright, storyteller and singer with experience in community building theatre projects. She skilfully integrated the ideas, objectives, and values of VARJ into a story and corresponding script by engaging in a collaborative process with board members, RJ practitioners, and other community organisations. The consultation process, which is discussed in more detail later, informed the depth and details of the RJ process as a backdrop for the story. The script follows the story of two young people, Laura and Derrick, whose lives end up intersecting in unexpected and surprising ways. The initial stages of their relationship are negative, including the violent assault. Through the RJ process, they are able to start healing and move on with their lives in positive and responsible ways.

The play takes audiences through the RJ process by highlighting how it can have lasting, positive effects on those involved. The play's story also highlights the careful preparation of the facilitator for those listening for training purposes. The Board of Directors at VARJ developed a discussion guide to go with the script as an educational tool. The script and discussion guide were written and published with accompanying images from an illustrator (Morriseau, 2020). VARJ hired local and professional actors and a sound producer to record a reading of the play with music and Foley effects. An engaging and polished podcast of the radio play was produced and uploaded to podcast platforms and is freely accessible to the general public on the organisation's website.[3] As COVID-19 lockdowns and stay-at-home orders were in place at this time, VARJ was in a unique position to be able to move forward on this community engagement project when the usual ways of engagement were not allowed. In lieu of a public launch and reading of the play, VARJ was able to produce and promote a radio play without anyone needing to leave home.

Radio plays as education: Theoretical considerations

Radio plays have been used as education tools since the 1920s (West, 1997). Even as television and a wide range of social media have increasingly dominated the media landscape, radio plays have persisted as a tool for both education and social change (Singhal et al., 2004). There are several reasons for the convenience and effectiveness of radio plays: short rehearsal time, they are cost efficient, the ability to cast against physical type, and the lack of set and staging. As well as these practical considerations, radio theatre promotes 'theatre of the mind' and can expand the audience in terms of geographical reach and demography (Verma, 2012). Unlike a stage presentation, the radio play is highly dependent on sound—both voice and Foley effects to convey contexts and events—presenting its own unique challenges and opportunities. Writing his first radio play for the Kingston, Ontario Short Wave Radio Festival, held in November of 2020, playwright Robert Downes put it this way: 'It definitely got me thinking a lot more about sound and what you can do with it and what you can do with just voices and sound effects,' he said. 'I think anything like this, where you're not limiting yourself but limiting what you can do, makes for a really interesting sort of test and gives you more to work with in a weird way' (Hendra, 2020, p. A2).

In terms of an expanded audience, radio plays allow the content to be broadcast far afield and accessed within the household. Compared with live theatre preformed on stage—often requiring paid admission—the medium is far more accessible. Radio plays can reach large groups of people, especially in places where television and internet are not available in every household. A recent American study shows that radio reaches more people than television or smartphones (Verma, 2017). Radio plays have been used effectively for public health education purposes with, for example, health-focused radio dramas being broadcast through regions of Africa where access to alternatives means for reaching a large audience are limited (Sofowora, 2007; Verma, 2017).

For VARJ, the adoption of the radio play medium was complimented by the scale at which the play could be promoted. Although not aired on any radio station, the creation of a podcast available free, online, allows for a wide audience in terms of age and location. Little did we know, in planning the project and as we were launching the play that the COVID-19 pandemic would restrict people to consuming more media at home as live theatres, movie theatres, music concert halls, and public spaces closed their doors. The medium suited the parameters of social distancing without limiting an opportunity to engage the public and to reach diverse audiences.

Sound challenges the imagination in ways that visual media cannot; sometimes surprising the listener with the emotional potency of that which is heard, but not seen. The radio play script by Renae Morriseau was accompanied by Foley and specialised sound production that brought the scenes to life. Being able to experience the altercation and emotions transpiring between the two main characters through sound, produced a challenging and engaging effect for listeners. With this production, VARJ set out to capitalise on 'theatre of the mind':

> The key to radio drama is sound—is imagination—is what you can do by stirring somebody. You can't do that with television. You can show them the pictures and say 'This is what it is'. But so what? So you sit there like a dummy and accept the car crashes but you can't add anything to it. There's nothing bloodier than the blood you see in your imagination. What are they gonna do? Pour a lot of ketchup on the television screen? It's still bloodier in your own mind.
>
> *(Hand & Traynor, 2011, p. 4)*

The authors of this statement make reference to the power of imagination, highlighting the difference between hearing and the more traditional theatrical experience of hearing and seeing. From this perspective, the audience not only imagines what they are hearing but subsequently, it can be argued, what they see becomes meaningful and impressive by virtue of the fact that it is the listener's own creation. In this regard, it can be argued that the radio play is more democratic and participatory than visual alternatives. It empowers the listener and engages them in the act of creation. Radio plays insist, in important ways, upon an active, as opposed to a passive audience. Into what is heard are drawn personal histories and experiences that may be subject matter for an active process of visualisation. This does not easily happen when the visual accompanies that which is heard.

The script for the radio play, *A Mural of Social Change,* was also published with a discussion guide for further exploring the content of the play and the practice of restorative justice. VARJ uses the play, in written or broadcast form, for educational purposes with youth. Co-author, Frank Tester, in assisting with the conceptualisation of the project brought a working knowledge of Habermas' theory of communicative action to the task (1984). The play illustrates and demonstrates the reciprocal expectations that make meaningful communication possible: validity claims to truth, appropriateness, and sincerity with respect to what is spoken. Working in RJ circles, facilitators benefit from the concept of validity claims by paying attention to facts, logic, shared meaning, and the relationship between speakers, thereby invoking an ethical standard that contributes to undistorted communication in the interest of building comprehension, trust, knowledge and consensus (Cukier et al., 2004). Communication must also be appropriate—'on topic'—and sincere.

While meeting these expectations is important, circle work, in a manner similar to listening to the radio play, further builds the democratic potential of participants who come to 'see' events in relation to their lived experiences. Circle work further democratises the experience, doing away with obvious hierarchies characteristic of western approaches to discussion, where presenters and experts are typically accommodated across the front of a room—perhaps even elevated by a stage—with the recipients of privileged knowledge lined up to listen. Circle work eliminates these obvious relations of power and privilege, thus complimenting and facilitating what Habermas' theory of communicative action sets out to achieve.

To explore these ideas further, we invited a range of people to join us online, listen to the play, and engage in a discussion exploring these and other theoretical ideas. Our focus group included a wide range of ages, backgrounds and experiences. It was ethnically diverse, but with more women than men participating. It included a professional dancer, a retired university professor, a lawyer working primarily with Indigenous peoples, a teacher with experience working internationally, an RJ practitioner, an engineer, a student, and a lawyer and human rights advocate, as well as the authors of this paper. The quotes that follow are taken from the discussion—circle work online—that followed listening to the play.

Listeners reported that the experience was 'immersive'. The combination of a compelling story, skilful actors, and professional sound production, allowed listeners to 'feel like they were in the room' and that these elements 'brought it alive' for them. Listeners explained that their experience of listening was richer than if they had just read the script. One focus group participant explained:

> I definitely found it a richer experience than it would've been had I read [the script]. And, I feel like if I'd seen a video or an actual play that perhaps wouldn't have been as rich either ... [Video and TV] show you what they show you, and you don't have to think about it any further ... In this medium, I had to fill in the spaces.

Others commented that the soundtrack and Foley created a 'felt sense' of the experience, and the voice acting allowed for more complexity and nuance in the characters because it challenged preconceived notions of one-dimensional victim and offender personalities. The soundtrack compelled listeners to use their imagination in a rich soundscape instead of defaulting to preconceived notions of 'victim' and 'offender', as we discuss below.

While VARJ's primary intent was to create public awareness of the process involved in restorative justice, our aim included creating an accessible, compelling and thought-provoking experience aligned with the principles of RJ. The creation of a radio play allowed us, on a limited budget, to meet our goals, as a radio drama lends itself to compelling, imaginative, and cost-effective storytelling. Furthermore, the process that unfolded as VARJ pursued the creation of the restorative justice radio play aligned with key values of RJ: community inclusion, transformation of understanding, and listening.

Restorative justice, reconciliation, and the story

In the story of *A Mural of Social Change,* Laura is a young Indigenous artist commissioned to paint a mural about reconciliation on the side of a building in Vancouver, British Columbia. Derrick, a teenager in the foster care system, engages

Laura about her work. While the encounter starts from genuine curiosity, it quickly becomes politically and emotionally charged and results in a violent incident. Derrick starts spraying paint on the mural Laura is creating. She attempts to stop him and sprays paint in his face. He then assaults her, kicking her while she is on the ground. He breaks her arm. He flees and when caught, resists arrest. During the encounter, the audience gets insight into Derrick's racist and resentful views, his misconceptions about Indigenous communities, and the pain and trauma that he has suffered in the foster care system. It was important to VARJ that the play explored these complex themes, reflected in the character's developmental histories and experiences, the conflict resulting from their interactions, their shortcomings and, in light of history and experience, the very 'human' and unfortunate behaviours to which these give rise.

It was also important the play included an examination of commonly held ideas about RJ processes, including what they mean for both victims and offenders. During the focus group, some participants acknowledged the misconceptions that they held, or things they had heard about RJ. Specifically, the conversation explored how RJ is perceived as 'an easy way out' and 'a less painful alterative to the justice system'. One participant explained that listening to the play allowed her to 'feel the tension', and for her it clearly conveyed the message that the RJ process is not a comfortable experience for those involved in 'this model of justice, [where] people's emotions are heard and acknowledged'.

Another focus group participant admitted that their preconceived image of the offender was challenged by the complexity of the story and the way the actors used their voices to reveal complex personalities. These resulted in the listener being more perceptive and open to the humanity of each character, an insight well aligned with restorative justice values.

> If I had read this, my preconceived notions would have coloured the play a bit more … perhaps I wouldn't have seen the layers and complexity of the individual. Simply put, [I] wouldn't have seen the person as a full person … I think the medium [of a radio play] did provide more layers of information about the characters.

In the story, after the police arrest Derrick, Laura is given the opportunity to pursue a restorative justice process with Derrick instead of resorting to use of the traditional and retributive justice system. Her auntie Rayna, who is included in the restorative justice process as Laura's support person, encourages her to choose the restorative justice path by saying:

> Laura, you go through restorative justice or you go through the court system. You go on the stand … re-live what happened; being cross-examined and all. Yeah, sounds fun. This restorative justice thing is about you, him, and what he did. Let's give this kid some history lessons. Maybe he can do, I don't know, community hours with me in the Downtown

> Eastside. It's up to you, but we can influence what this process is and some community healing along the way. This kid needs to understand more than just Canadian law and punishment.
>
> *(Morriseau, 2020)*

Laura agrees to move forward with the RJ process and the audience is taken through the experience of a facilitated RJ circle process where both Laura and Derrick are in conversation with a trained RJ facilitator about what happened. The experience includes Laura's auntie Rayna, a social worker, Derrick's older cousin, and a few other community members. The story is interwoven with information and perspectives that highlight the need for reconciliation in our communities. The play deals with an incident involving an Indigenous person and a non-Indigenous person, thereby informing the theme of reconciliation in relation to restorative justice practice.

The Indigenous lens is important to the story. Since its inception, VARJ has hosted seminars, workshops, and discussion groups, often with Indigenous organisations in Vancouver, including the Vancouver Aboriginal Transformative Justice Service Society and the Aboriginal Friendship Centre. As a result of these partnerships, VARJ has become aware of the paucity of attention given to the importance and the potential of RJ in contributing to reconciliation between Indigenous and non-Indigenous peoples in Canada.[4] What exists on YouTube is limited to a lot of dated 'talking heads' material. There are no materials that draw both Indigenous peoples and non-Indigenous peoples together, showing how restorative justice can be mutually beneficial and healing. For this reason, the radio play, *A Mural of Social Change: Restorative Justice*, conveys to the listener how an RJ process unfolds in addressing a cross-cultural conflict and the impact on those involved.

Community inclusion

In a meta-analysis of RJ process, Suzuki and Yuan (2021, p. 12) found that 'support networks, embedded in human connections, offered tremendous support to individuals who committed a crime and victims who were trapped in conflicts'. In *A Mural of Social Change,* community supports like Laura's auntie Rayna, Derrick's social worker and Derrick's cousin, illustrate the ways in which restorative justice practice draws on support networks to create an inclusive process that, in the process, further educates a participating community.

Of critical importance in getting from an idea to a play that embodied VARJ's ideas, interests, concerns and objectives, was collaborating with a playwright able to take these elements and express them in a story and corresponding script. In order to do this, we invited many voices from the RJ community to participate by commenting on drafts of the playwright's work so that she could get a sense of the diversity of concerns and approaches that exist in the field of restorative justice practice. The consultation process assisted the playwright in narrowing down an RJ style that she could use for the backdrop of the story.

The first stage of script writing involved interviews and community meetings between the playwright and community practitioners of RJ processes in Vancouver and Nelson, British Columbia. Nelson was chosen as a community that in many ways contrasts sharply with Vancouver. It is a small, rural town set in the southern interior of the province. Furthermore, its RJ process is managed through the municipal police department, a model that is criticised by some RJ practitioners. Consistent with VARJ's commitment to circle work and democratic speech, we are open to different institutional models for RJ practice, recognising the uniqueness of different geographical, historical, and ethnic contexts for organisation of services and for practice. This consultation process contributed to the depth and details of the story because the playwright was able to hear, first-hand, about the practices, stories, and experiences of RJ in very different communities.

The value of community inclusion aligns with a key value in restorative justice practice. Including the RJ community in the exploration phase of the project, where the ideas and information for the script were coming together, mimics the way in which community members are often included in a RJ process. In the story, community supports for both Derrick and Laura are invited into the RJ circle. Their voices are heard and their support for these two people is essential to the transformation and healing journey. They offer advice, support and their own experience of being impacted by the incident, highlighting how the RJ process honours inter-connectedness of community healing, and the complexity of a crime that affects more than the two people involved.

Our focus group participants absorbed this learning about RJ. One participant explained, '[In listening to this radio play] I took away the role of community in this process and the importance of support structures, and personal support systems … how they can help people move onto better places in their lives.' Another stated, 'I agree with the importance of the support people and notion that there really is a hopeful positive alternative to regular justice processes.' While yet another participant reflected on the roles of the facilitator and the people involved, including community members, by identifying that while the facilitator 'is there to ask good questions' and 'hold space', everyone involved needs to figure out what they need from the process.

We are consequently optimistic that we can further the public conversation—through the use of this play—not only around ways to use RJ for minor crimes, but also for situations that involve violence. As an organisation promoting restorative justice, we commonly hear a resistance to RJ as an alternative to justice for violent crimes because it is not 'punitive' enough. After a discussion about preconceived notions of RJ as a practice that 'lets people off easy', one participant reflected that the play 'succeeded in presenting a model that *can* work'.

Transformation of understanding

An overarching theme found in all restorative justice processes is the transformative impact of restorative justice on offenders and often on the offended.

Critical roles for RJ practitioners are to develop and bring forth the best qualities and capabilities of participants—their humanisation (Suzuki & Yuan, 2021). RJ practice, seen in this light, requires that practitioners are oriented to the social history of individuals in unique and particular material and social circumstances, and that they have an orientation to ontological concerns: existence, being, becoming, and different realities. However, different approaches to RJ suggest different emphases when it comes to training facilitators.

Dyck (2005) deals with these, arguing that facilitators need more training on the structural dimension of criminal conflict, that there is too much emphasis on the inter-personal dimensions of conflict, and that too many facilitators focus on the 'ideology of harmony': the 'naïve idea that shared feelings bring about empowerment of parties to the process' (p. 527). A victim/offender mediation (VOM) approach to restorative justice tends to de-emphasise reconciliation, and emphasises victim healing, offender accountability and the restoration of losses: a social work case management approach (McCold, 2005).

The play is an attempt to suggest, if not convey, the conceivable importance of different approaches and skills that might be required for RJ practice. While reference is made to circle work in the play, it is not overly emphasised. The content has parallels with a case management approach, inviting in follow-up discussion, questions about the training, skills, biases, and orientation that RJ practitioners might bring to the work they do, what students in a department of criminology might receive by way of training, and what social workers might bring to the role are different skills, insights, and priorities. Sadly, RJ is a neglected subject in schools of social work in Canada. A radio play that does not visually emphasise circle work allows for an important discussion about the skills and orientation required of RJ practitioners and social workers.

A key way this project showcased the theme of humanising the offender and the offended was through the careful preparation of the facilitator, the complex characters' motivations and developmental histories, and the acknowledgement of the need for an on-going healing journey. Through consultation and feedback from RJ practitioners, the playwright skilfully integrated these concepts into the story and characters. This was reflected in one focus group participant's assertion:

> I feel like the complexity of the characters is essential in this play … it seems like the whole point of restorative justice. And, I think, that what I value most in this play is the showing the people as human beings and not just victim and criminal. There's more complexity than these labels and so people deserve a process that is more complex.

The project drew on the ideas of popular education scholars to consider how this educational tool could also be transformative by embedding the experience in optimism and hope (Hooks, 2003).

The idea that hope, compassion, and joy are central to a transformative practice is found in the writing of bell hooks (2003), and is also common throughout RJ practices and values. Suzuki and Yuan (2021) found that a common theme across RJ processes is coming to a changed outlook. They claim that

> RJ … offered a setting where human beings with normal needs, being individuals who committed a crime or victims, are free to exchange their emotions and views, thereby reaching a mutual understanding, or at least leaving the conflicts and moving on with a positive outlook. (p. 11)

As one focus group participant explained, 'I'm taking away a bit of hope, which is a very positive thing. Because it's like, "Oh this can work, here is an example" … I think this model succeeded in presenting a process that *can* work.'

Listening

Finally, the key element of a radio play is listening. While other forms of theatre may invite multiple senses to experience the visual, audio, and, on occasion, sensory experience of theatre; a radio play is focused solely on the act of listening. The alignment between values and the content of this project is an exercise in respectful listening: a fundamental restorative justice value. 'This is what restorative justice is about: the deadly simple empowerment that comes from creating pacified spaces where we listen to those that we feel have wronged us and those we think we may have wronged' (Braithwaite, 2003, p. 36). Mark Umbreit and Marilyn Amour explain that peace-making and RJ circles are a 'process of dialogue, relationship building, and the communication of moral values in order to promote accountability, healing, and compassion through community participation in resolving conflicts' (as cited in Rhodes, 2021, p. 50).

Listening was central to the experience of the radio play, more so than just the act of hearing an entertaining story. The concept of listening was embedded in the values of the story, just as it is in the values of RJ. What becomes clear in the play is that each participant in the RJ process has a story, and their story has a right to be heard. The story communicates that RJ is a model of justice facilitated in a way that deliberately creates space for emotions, stories, and the 'messy-ness of human experience' to be expressed. A focus group participant explained,

> Something that was emphasised for me was the importance of everyone having the sense that everyone around them has heard and would understand them. Even if the outcome doesn't necessarily change, it was important that the play emphasised that everyone has a story and gets to tell it, and the fact that they have a story is value neutral … everyone has the same right to speak.

This dedication to allowing multiple voices to be heard is coupled with the active listening experience of a radio play. This is further facilitated by a story that explores the core value of RJ as a process of intentional listening. As Rhodes (2021) explains 'the process of learning and integrating the guidelines, along with participating and engaging in the Circle process, helps to reinforce mindfulness and listening from the heart' (p. 61). Another focus group participant reinforced that the immersive experience of active listening to the radio play helped to reveal the emotional landscape and communication styles that would require active listening in a real-life RJ process:

> I could picture myself in the room with the characters as a part of the story, just observing it very closely. Yeah, definitely, it gave me a good glimpse of what was going on and how the people were interacting with each other … in the circle and communicating with each other. I could feel the tension.

Being perceptive to the dynamics between people, considering opportunities to create space for others' stories, and responsively hearing to emotions of those around the circle requires the listener to be present and self-reflective; practices of mindfulness and genuineness that Rhodes (2021) aims to teach his students about RJ practice. A radio play that invites active listening is an opening to practice. While active listening—employing respect and mindfulness in the act of listening—can seem subtle and theoretical, it is a key to the practice of the circle process and a fundamental restorative justice value (Braithwaite, 2003; Rhodes, 2021).

Conclusion

The Vancouver Association for Restorative Justice (VARJ) set out to create an educational product about restorative justice (RJ) that was accessible, compelling, and aligned with the principles of restorative justice. VARJ chose a radio play to illustrate for a general audience, how a restorative justice process works. The intention was to dispel myths about this model of justice and give the listener a better sense of how restorative justice works in counter-distinction to the practice of retributive justice. An additional aim of the project was to build confidence and capacity in potential RJ facilitators, and to increase support for the process by stakeholders in the criminal justice system.

The result is reflected in the comments of nine people participating in a focus group after listening to the radio play. This focus group shed light on the ways in which listeners immersed themselves in the play, gaining an appreciation of RJ—what it is and how it might work—but also achieving a felt-sense of the experience and values important to a restorative justice process. One focus group participant concluded, 'I come away with a global sense of what RJ is. Before I just had fragments.'

Our collaboration with a playwright, Renae Morriseau, was a rich process where VARJ, as a community-based, non-governmental organisation, offered guidance and expertise on restorative justice. The playwright, with community-based input, brought the characters and story to life through a compelling radio play script. VARJ then employed skilled actors and professional sound production to produce a free, one-hour podcast of the radio play. Through meaningful collaboration and educational goals embedded in the story, the radio play achieved our objectives.

This experience has taught us that community organisations can collaborate with artists to develop a lasting piece of art that furthers important conversations in the community. For VARJ, this project highlighted the engaging nature of radio plays, including the ways it activates imagination and encourages critical participation. Further to our goals, we found that a radio play facilitates active and democratic participation in the growth and understanding of a listener and the possibility of their further and informed contribution to the issue, subject, or concern that is the focus of their attention.

Notes

1 We use the term radio play as the genre of this play rather than 'podcast' or 'audio drama' because the project was originally intended to be a published book that would be performed as a live reading, aired on the radio, and recorded and shared as an MP3 file. The project came to fruition during the COVID-19 pandemic and VARJ had to cancel plans for the public launch and reading of this play. While the book was still published, VARJ pivoted to the creation of a podcast to provide a listening experience. Therefore, we use our original terminology of radio play throughout this chapter.
2 The organisation's website is: http://varj.ca
3 An audio recording of the play can be accessed at: http://varj.ca
4 Exceptions are National Film Board of Canada documentaries, *Home Fire* (2014), *Circles* (1997) and *Hollow Water* (2000).

References

Asadullah, M., & Morrison, B. (2021). 'Communities are not at the periphery, rather they are at the centre of Restorative Justice in BC': An inquiry into the praxis of Restorative Justice in British Columbia, Canada. *Contemporary Justice Review, 24*(2), 172–196. 10.1080/10282580.2021.1881893

Borrows, J. (2006). *Indigenous legal traditions in Canada*. Law Commission of Canada.

Braithwaite, J. (2003). Principles of restorative justice. In A. Von Hirst, J. Roberts, & A. Bottoms (Eds.). *Restorative justice and criminal justice: Competing or reconciling paradigms* (pp. 1–20). Hart Publishing.

Chartrand, L. (2018). *Report on the relationship between restorative justice and Indigenous legal traditions in Canada*. Department of Justice, Her Majesty the Queen in Right of Canada.

Cukier, W., Bauer, R., & Middleton, C. (2004). Applying Habermas' validity claims as a standard for critical discourse analysis. In B. Kaplan, D. Truexll, D. Wastell, et al. (Eds.), *Information systems research: Relevant theory and informed practice* (pp. 233–258). Kluwer Academic Publishers (Springer).

Davis, F. E. (2019). *The little book of race and restorative justice*. New York: Good Books.

Dickie, B. (Director). (2000). *Hollow water* [documentary]. The National Film Board of Canada.

Dyck, D. (2005). Reaching toward a structurally responsive training and practice of restorative justice. In D. Sullivan, & L. Tifft (Eds.), *Handbook of restorative justice* (pp. 527–545). Routledge.

Gwathney, A. N. (2021). Offsetting racial divides: Adolescent African American males & restorative justice practices. *Clinical Social Work Journal*, *49*, 346–355. 10.1007/s10615-021-00794-z.

Habermas, J. (1984). *The theory of communicative action. Volume 1. Reason and the rationalization of society*. Beacon Press.

Hand, R. J., & Traynor, M. (2011). *The radio drama handbook: Audio drama in context and practice*. Bloomsbury.

Hendra, P. (2020, October 30). Festival revives the radio play. *The Kingston Wig Standard*. https://www.thewhig.com/entertainment/local-arts/festival-revives-the-radio-play

Hooks, B. (2003). *Teaching community: A pedagogy of hope*. Routledge.

Katic, B., Alba, L. A., & Johnson, A. H. (2020). A systematic evaluation of restorative justice practices: School violence prevention and response. *Journal of School Violence*, *19*(4), 579–593. 10.1080/15388220.2020.1783670

Kunuk, Z. (Director) (2001). *Atanarjuat* (*The Fast Runner*) [Film]. Igloolik: Isuma Productions.

Latimer, J., Dowden, C., & Muise, D. (2005). The effectiveness of restorative justice practices: A meta-analysis. *The Prison Journal*, *85*(2), 127–143. 10.1177/0032885505276969

McCold, P. (2005). The recent history of restorative justice: Mediation, circles and conferences. In D. Sullivan & L. Tifft (Eds.), *Handbook of restorative justice* (pp. 527–545). Routledge.

Menkel-Meadow, C. (2007). Restorative justice: What is it and does it work? *Annual Review of Law and Social Science*, *3*, 161–187. 10.1146/annurev.lawsocsci.2.081805.110005

Miller, G. (Director) (2014). *Home Fire* [documentary]. The National Film Board of Canada.

Morriseau, R. (2020). *A mural of social change: Restorative justice*. Vancouver Association for Restorative Justice.

Paul, G. D. (2015). Justice perceptions and practices of restorative justice facilitators and the public. *Contemporary Justice Review*, *18*(3), 274–295. 10.1080/10282580.2015.1057678

Rhodes, D. (2021). Pedagogy of circles: Teaching restorative justice to social work students. In T. Lewis & C. Stauffer (Eds.), *Listening to the movement: Essays on new growth and new challenges in restorative justice* (pp. 49–66). Cascade Books.

Roche, D. (2005). Retribution and restorative justice. In G. Johnstone, & D. W. Van Ness (Eds.), *Handbook of restorative justice* (pp. 97–112). Routledge.

Singhal, A., Cody, M. J., Rogers, E. M., & Sabido, M. (2004). *Entertainment-education and social change: History, research and practice*. Lawrence Erlbuam Associates, Publishers.

Sofowora, O. A. (2007). An assessment of the effectiveness of radio theatre in promoting good healthy living among rural communities in Osun State. *Research Journal of Social Sciences*, *4*(2), 51–56.

Suzuki, M., & Yuan, X. (2021). How does restorative justice work? A qualitative meta-synthesis. *Criminal Justice and Behaviour*, *48*(10), 1347–1365. 10.1177/0093854821994622

Thakur, S. (Director) (1997). *Circles* [documentary]. The National Film Board of Canada.

Verma, N. (2012). *Theatre of the mind: Imagination, aesthetics, and American radio drama*. University of Chicago Press.

Verma, N. (2017, December 4). *The golden age of radio drama – It never ended*. [Historyradio.org] https://historyradio.org/2017/12/04/remembering-the-golden-age-of-radio-drama/

Vieille, S. (2012). Mäori customary law: A relational approach to justice. *International Indigenous Policy Journal*, *3*(1), 1–18. http://ir.lib.uwo.ca/iipj/vol3/iss1/4

West, K. (1997). Classic radio theatre in contemporary education. Paper presented at the Joint PCA/ACA Meetings. San Antonio. https://files.eric.ed.gov/fulltext/ED407693.pdf

Zehr, H. (2002). *The little book of restorative justice*. Good Books.

PART III

Sonic Assembly: Building Communities and Publics

INTRODUCTION

Sarah Woodland and Wolfgang Vachon

Arguably one of the central concerns of community engaged arts and arts-led research is community building itself. As discussed in the Introduction, the term 'community' has some detractors, and yet we return to it time and time again in our descriptions of how art can strengthen ties between individuals and groups, strengthen a sense of belonging between people and place, and strengthen the formation of communities and publics around certain ideas and beliefs. Communities and socialities may be constituted, not only of people, but also of material and non-material *things* in relation to one another, and we follow several key theorists in foregrounding the relational quality of sound and listening as essential to understanding ethics and aesthetics in community engaged sound arts. Brandon LaBelle (2018, 2020) has contributed greatly to our understanding of the emancipatory and agentic potentials of sound and listening, specifically (though not exclusively) through describing how they facilitate vibrational and relational dialogue, exchange, compassion, and alliance between and among bodies and things that might transcend the 'visual surface' of the world (LaBelle, 2018, p. 3). In a sense, he presents sound as an invisible affecting force—in some cases a glue, in others a lubricant—that can create both bonding and slippage in the constitution of communities and publics.

In this section, we draw together contributions under the theme of sonic assembly, not because this is the sole focus of the works under scrutiny, but because they all seem to us to share a special emphasis on the social, or in the case of Jill Halstead and Brandon LaBelle's contribution, the 'more-than-social'. The works here explore how the 'social' and 'more-than-social' are constituted through sonic relation and participation, and how ethics and aesthetics manifest in and through these assemblies.

We begin with Halstead and LaBelle's chapter, through which the authors aim to develop a theory for socially engaged art that is informed (and formed) by sound

DOI: 10.4324/9781003164227-12

and listening. Through collective, communal projects such as the self-organised concerts taking place in Budapest as a form of cultural resistance (LaBelle), and private experiences of health through the facilitation of intimate improvised performances with elders in residential care in Bergen (Halstead), the authors focus on the operation of sound and listening 'as key to nurturance and repair'. They argue that the two examples illustrate 'the ways in which sound and listening participate within greater artistic and cultural initiatives by lending a palpable and, at times, insurgent force to what it means to engage'. They suggest such practices hold the potential for 'poetic world making' which they describe as 'the creative and critical crafting of more-than-social relations found by way of sound and listening'. The authors conclude that 'sound and listening emerge as key to both activating the social while complicating and elaborating its borders'.

Scholars and sound artists Christos Carras and Eric Lewis present a personal reflection and case study approach to three media walks delivered in Greece, presenting insights into their ethics and aesthetics, with a particular focus on some of the dynamics of participation. Woven throughout the text are questions and reflections about media walks, sound walks, technology, and how these can create, illuminate, and hide community. Of particular interest is how the community members in one Greek village engaged with the artists, inviting them to *come across* from their initial standpoint as outsiders to support the community in its desire for documenting their oral history. Drawing on these experiences, the authors close with a set of principles for those creating media walks in which community members' 'spaces and life-practices are primary contributors to the media walks' content'.

Sound artist and researcher Budhaditya Chattopadhyay develops the idea of an 'acoustic common' through his practice as a sound artist in various cities around the world. His examples include his own workshop praxis, an augmented book (*Nomadic Listener*), and his sound work and mobile exhibition (*Aural Drift*) to develop these concepts into an inclusive artistic and research methodology. His practice focuses on the methodologies of 'hyper-listening' and 'co-listening' as contemplative, meditative processes that integrate ethics and aesthetics in removing judgements about the causal and material nature of sound. He suggests that this is ultimately a relational act, connecting the listener to soundscapes that are untethered from predetermined orders of signification and meaning, including other human and non-human beings, and epistemological relationships with the world at large.

Finally, the contribution from Vadim Keylin explores the aesthetics of sonic experimentation in participatory art and culture. The author describes how 'un-musical' sounds and noises in experimental art practice are more readily accepted and enjoyed by audiences when they are invited to participate in making them. The author deploys his aesthetic theory to analyse sonic memes, as well as the interactive performance *Let Us Sing Your Place* by the Danish artist Katrine Faber and public sound sculpture *Speakers' Arena* by American-German artist Benoît Maubrey. Keylin suggests that these works withdraw the usual frames for aesthetic judgement by engaging participants in the process of creation. He proposes that enabling the public to engage in sonic experimentation 'promotes communality without negating difference' and facilitates empowerment and agency.

References

LaBelle, B. (2018). *Sonic agency: Sound and emergent forms of resistance* (Vol. 1). MIT Press.

LaBelle, B. (2020). *Acoustic justice: Listening, performativity, and the work of reorientation.* Bloomsbury Publishing USA.

6

MORE-THAN-SOCIAL LISTENING: UNDERCOVER ENGAGEMENTS AND UNDOING AUDITORY NORMS

Jill Halstead and Brandon LaBelle

As a medium and material active within a range of creative practices, sound offers an important and unique means for addressing often challenging topics and experiences. From trauma and illness to social and political oppressions and struggles, sound often figures as an agential and healing force. This includes musical, therapeutic, and ethical approaches in which sound and listening enable states of awareness and attunement that readily move across linguistic and more-than-linguistic registers. This is not to overlook the degree to which sound and listening may be mobilised in order to harm others, and how listening may be put to work in the name of security and defence. Sound can be weaponised in the service of any number of liberal and totalitarian agendas.

While these are important perspectives to consider, our chapter focuses on sound and listening as key to nurturance and repair. In particular, we are keen to highlight how sound and listening participate within collective and communal projects, as well as more private experiences of health, and further, how they contribute greatly to fostering human relationalities, especially within scenes or situations of struggle and loss. Importantly, we argue that such an acoustic understanding is grounded in a potentiality for poetic world making activity, that is, the creative and critical crafting of more-than-social relations found by way of sound and listening. We understand poetic world making activity as activity that aims at reworking a given social order, however marginalised or enclosed.

Following Michael Warner (2005), poetic world making is positioned in relation to public discourse, and the ways in which counterpublics work through a dominant ideological and discursive order:

> Public discourse … is poetic. By this I mean not just that it is self-organising, a kind of entity created by its own discourse, or even that this space of circulation is taken to be a social entity, but that in order for this to happen

DOI: 10.4324/9781003164227-13

> all discourse or performance addressed to a public must characterize the world in which it attempts to circulate and it must attempt to realise that world through address. (pp. 113–114)

Warner's proposal helps understand the operations of counterpublics, and how often such publics find traction through crafting specific vocabularies and behaviours. While Warner is concerned with the performative actions of rhetoric and discourse, we are interested to position poetic world making activity within an acoustic framework so as to capture the more affective and more-than-linguistic basis of sound, and how listening places speech and saying as never far from the unspeakable and the unsaid, the felt and the attentionally attuned. Poetic world making, in other words, is put to work in our chapter as a social as well as bodily theory, where sound and listening contribute to worlding counterpublic discourses as well as movements.

The ways in which sound and listening contribute to social arenas, where hearing and being heard are vital, while also often exceeding the grammar of the social, for instance as found within rituals of celebration and festivity, mourning and grief, is deeply suggestive for understanding sound's relational affordances. In this sense, we're concerned with sound and listening as the basis for both social and more-than-social relations, as well as sonic and more-than-sonic practices. For listening must be underscored as not always anchored to sound itself or the social dynamics of language; rather, listening is awareness, attention, gesture, and a profound capacity for understanding and attunement with others. Listening equally moves in and out of social arenas, allowing for human and more-than-human connections.

In order to unpack the question of sonic engagement, and the acoustic arrangements or actions made in support of poetic world making activity, we offer two case studies. First, this includes a reflection by Brandon LaBelle on current and historical gatherings taking place in apartments in Budapest, and how in recent years self-organised concerts have participated in a general movement of cultural resistance in the city. Second, by examining facilitated *performances-in-the-making* with elders in spaces of residential care in Bergen, Jill Halstead accents sonic engagement as a base for critical questions. While emerging within very different contexts, each study speaks to the questions we are aiming to mobilise, and are useful for understanding the ways in which sound and listening participate within greater artistic and cultural initiatives by lending a palpable and, at times, insurgent force to what it means to engage.

1. Noise and love: Budapest events

In reaction to the Fidesz government of Viktor Orbán, and the overall 'illiberal' agenda initiated since his second election as Prime Minister in 2010 (which he continues to hold), a range of resistant cultural initiatives and actions have emerged. Launched in 2015, the OFF-Biennial brought together a range of galleries, small-scale artist organisations, and cultural venues, along with cultural actors across the city of Budapest, in staging a city-wide art exhibition. The Biennial aimed to create an independent initiative that would circumvent national funding councils, and the general circuit of state-sponsored culture, to act as a base for cultural solidarity across a diversity of communities. In short, the OFF-Biennial was launched as a cultural protest against Orbán's nationalistic agenda and policies, which have sought to restrict democratic processes and freedoms.[1]

Since its launch the OFF-Biennial has established itself locally and internationally, with the organising of two further editions (the most recent edition scheduled to take place in April of 2020 under the title INHALE was postponed due to the COVID-19 crisis). With loose ties to the Open Society Foundation, which, along with many of the educational and cultural institutions supported by George Soros, has been under threat by the government, the Biennial signals a larger movement of independent culture in Hungary against the government's overall project of 'national culture'.[2] For Orbán, the only true culture is founded on narratives of an authentic Hungarian ethnicity. This has been articulated through a range of initiatives, including installing into director positions individuals loyal to the government as well as placing public cultural and education institutions under the purview of newly created private foundations.[3]

Within such a cultural environment, a number of musicians, artists, and organisers have taken to holding concerts in their apartments, giving way to an informal circuit of events that hark back to earlier apartment gatherings from the 1980s. As the photographer and film-maker István Jávor notes, house parties were important meeting points that helped sustain the democratic opposition throughout the Soviet period while also affording a general feeling for social possibility and collective joy (Jávor, 2021). His photographs from the 1980s capture the spirit of the parties, where listening and dancing to music, drinking, and festive embraces, fuel an affective politics.

The series of events under the title Noise N' Roses was launched in 2019 by organiser and musician Lucia Udvardyova along with musician Rozi Mákó. The events take place at a range of off-spaces, from industrial courtyards and artist studios, and in particular, people's apartments. For the organisers, the events are geared towards intimate and non-commercial gathering centred on small-scale musical performances and audio-visual happenings. Yet, in opting for spaces outside the established circuit of venues, the events also entail a great deal of cooperative work:

> The most beautiful thing on the whole is how it has turned into a truly communal effort—everyone chips in, offering their home for the night, delivering the speakers, etc. And then, we immerse ourselves in the music and chatter. The last guest leaves and then we clean up. It was a good night.
> *(Udvardyova, 2021, p. 38)*

The communal effort Udvardyova not only highlights but emphatically embraces is positioned as fundamental to the experience. In this regard, the value of the events is not only found in the time and space of performance, but extend well beyond and can be appreciated as lending to collective and communal efforts and bonds vital to maintaining a culture of freedom.

Organising informal gatherings and events are often grounded in more affective and embodied relations, where stepping away from formal staged performances enable more intimate interactions and shared experiences, as well as a sense of cultural excitement. Zoltán Gaszi, a cinematographer active in the Budapest cultural scene in the 1980s and 1990s, offers a similar perspective: 'These flat events happened not because a performer had no way to perform their work on stage, but rather that they preferred to present their work to their friends and for their own pleasure' (Udvardyova, 2021, p. 53). Such conditions of informality though also contribute to what Gaszi highlights as a cultural commingling, where artists and intellectuals, architects, and musicians, and those involved in activities outside established culture, either by choice or by necessity, could meet and support each other. The informal gatherings acted as 'another realm that enabled us to bear it all' (p. 56).

While informality supports movements of cultural freedom, the organising of events over time also establishes key sites and communal rhythms. For instance, the apartment of the artist and designer Gyula Pauer became a regular meeting point in the 1980s for a variety of parties and performances. In fact, Pauer's apartment operated as an 'open house' every weekend and, as Gaszi poignantly recalls, gave way to important associations and memories:

> In this apartment Ferenc Darvas [composer, theatre musician] played such songs on the piano that we all knew, and these were the greatest musical experiences of my life. These melodic fragments well-known to us created a system of associations. These pieces of music were unrepeatable, they were born from the occasion.
>
> *(Udvardyova, 2021, p. 57)*

With the Orbán regime in power, organising informal parties and underground events has re-emerged as part of the cultural attitude and experience of a younger generation. Noise N' Roses' musical events are thus positioned within a larger framework of related actions and movements and contribute important grass-roots articulations of cultural ethos. This finds particular expression through a focus on experimental electronic, techno, noise, and sonic art practices and expressions, and integrate a sensitivity for collective listening experience.

In parallel to the overarching cultivation of informality and communal effort, the events stay close to a culture of the local: local musicians and artists perform and attend, developing craft and culture that explicitly steps out from the 'event economy' (Petrešin-Bachelez, 2017, p. 6). Operating by way of neoliberalism's entrepreneurial and self-precarising dynamics, in which the production and presentation of cultural events are endlessly performed and circulated across a range of venues and social media platforms, the event economy forces a cultural sensibility geared towards a continuous flow of output. Such an economy, as Nataša Petrešin-Bachelez critically notes, undermines our capacity to listen. Here, listening is positioned as a counter-balance to the rapid outpouring of cultural events, and the accumulation of cultural capital on the part of institutions and practitioners, focusing instead on the crafting of more speculative, slow, caring, and resonant processes. Such a view is integral to the OFF-Biennial as well, where 'off' stands against the more fast-paced prepositions of 'post' or 'anti' as Svetlana Boym (2017) suggests, guiding us off the trail of capitalistic enterprise.

The events of Noise N' Roses may be understood along these lines, as informal off-events where listening and the local interweave in support of a *coming-together*. This can be heard in the variety of artists and musicians who do perform, whose electronic and sonic aesthetics tend towards the atmospheric and unprogrammatic, the hypnotic and fluid. From Roland Németh's cascades of cinematic sound to Zuriel Waise's dance-oriented music, or Judit Konopas' synth-driven abstractions, the musical scene around Noise N' Roses gravitates towards an aesthetic listening that gives room for free expression.

While Noise N' Roses doesn't make any overtly political statements, it is clear that their sense of sonic engagement is bound to articulating a politics of possibility: a material and sensate doing that brings people together—or rather, a being-together that stands behind making possible an experience of listening. As Udvardyova again emphasises, 'Without community, the series wouldn't exist and wouldn't make sense' (L. Udvardyova, personal communication, January 10, 2021).

In this regard, sonic engagement takes many forms, and in particular, may equally wield a certain political action by way of disengagement, an intentional exit. In his book, *Hungry Listening*, Indigenous scholar Dylan Robinson (2020) argues that within the context of decolonisation it may be necessary to enact a form of auditory 'blockade' (p. 24). While listening is often theorised as essential for nurturing greater egalitarian and tolerant systems, as an enactment of understanding and ethical regard, Robinson reminds how listening may also participate in forms of colonialism, extraction, and appropriation. In response, blockading breaks from the seemingly inclusive gesture of listening in order to fortify 'sensate sovereignty' and community identity, especially those vulnerable to dominant power relations.

Notions of auditory blockading are suggestive for considering how the musical gatherings in Budapest support community building by enabling a disengagement from the broader public sphere. Rather, the privacy of the home allows for another type of cultural value which, I might add, has nothing to do with capital and the event economy. Returning to Udvardyova's understanding of the communal effort behind the home gatherings, artistic value may certainly circulate by way of more subcultural circuits, but it does so within the framework of what Richard Gilman-Opalsky (2020) terms a 'communism of love'. Mounting gatherings in people's apartments, borrowing and transporting the needed equipment, buying beer and organising the bar, spreading the word and opening one's doors, all without financial support—the events themselves are free—must be appreciated as a labour of love, as a generosity driven by a belief in each other and the possibility of building a different world. This is what sonic engagement is for Noise N' Roses.

I am reminded of the artist Bálint Szombathy, and his infamous work *Talent Test* from 1967. As part of the underground art scene of the 1960s in Hungary, Szombathy worked at surviving in the context of state censorship, creating artworks that had to negotiate both the risk of open avant-garde expression and conforming to a set of institutional parameters that might enable some form of livelihood. Like many of his peers, Szombathy refused to engage in official culture, opting to stay within the confines of the underground, showing his works in back gardens, friends' apartments, and alleyways. *Talent Test* captures this fundamental political position. It consists of an application to an art school in Amsterdam, advertised as a 'Famous Art School', in which the applicant must identify from a selection of examples which artworks are authentic and which are fake. Szombathy displays his successful application which he ultimately turned down as an emblematic statement of critical exit: the refusal to accept the terms of western recognition, and to refuse his own exile. This in turn stands as a statement against becoming official in general, where established norms of artistic worth are cast aside in favour of a more communal commitment to local friends, and to working from within an off-scene of solidarity (Garrera & Triulzi, 2020).

2. **Unruly listening: Bergen**

The COVID pandemic, which started in 2019, thrust issues of ageing into the collective consciousness. As the virus spread across Europe it became clear that age was a determining factor for the severity of the disease. For those aged over seventy and living communally in residential care the risk was greatest. As health and social care systems struggled, and in some countries collapsed, cases of solidarity and abandonment made headlines globally. Stories of carers who lived in care homes alongside elders for months on end intertwined with instances where elders had been abandoned, left dead, and dying in their beds, as staff too were infected and institutions were overwhelmed.

The social fractures which writ large during the subsequent lockdowns were felt most acutely by the old because as a group they were already vulnerable to social disengagement and cultural isolation. In the pre-pandemic UK, for example, of the two million people aged over seventy-five who live alone, more than half regularly went for more than a month without contact with someone they know. Across pre-pandemic Europe similarities can be seen. In Norway some 70,000 of the 850,000 people aged over 65 were classified as experiencing 'extreme social isolation'.[4] Research suggests elders living in residential care were twice as likely to report feelings of 'severe loneliness' than their peers living in the wider community.[5] A large proportion of those in residential care live with some form of dementia, which means their possibilities for engagement are severely disrupted due to loss of memory, language and mobility, resulting in extreme forms of social and relational rupture.

In this way the pandemic served to spotlight existing concerns about how we live as, and with, elders in neoliberal systems, where old age itself has become routinely defined as a societal challenge measured in terms of financial and social burden. In Warner's (2005) terms, the old have emerged as a counterpublic disengaged from productivity and consumerism; situated outside of the mechanisms of public discourse, they are rendered marginal in ever changing and troubling ways.

It is against this background that my own creative sound practice and research engages with issues of ageing, disability, participation, sound, and listening. It is a border-crossing practice purposefully occupying the liminal spaces between creative musical practice, research, social activism, and radical care work, often undertaken in collaboration with colleagues trained in music and dance therapy or community arts methodology. It is a sonic practice rooted in the corporeal, tuning in to the kinetic, tactile kinaesthetic qualities of sound making and listening (Sheets-Johnstone, 2015) and the intra-action of sound, movement, and affect, through what Michel Serres (2008) calls 'mingled bodies' that are knotted with each other and the world.

Through a longstanding collaboration with Beatrice Allegranti, a choreographer, filmmaker, and dance movement psychotherapist, I have developed a range of sound work for screen dance and live dance theatre performance. Our creative process has been built on our joint interest in embodied methodologies where composition meets choreography and the consubstantiality of sound and movement may generate movement from sound, and sound from movement. Our collaborative research and practice has sought to explore how practices of listening animate, compelling us to move, be moved and move with others and what role embodied audition may play in processes of relationality.

This interest in the multisensory nature of listening found an ethical imperative when developing participatory practice with elders living with dementia specifically, as this group are largely considered the most unruly of listeners to engage. The many forms of degenerative brain disease which result in dementia are experienced as disorders of memory, language, vision, hearing, and movement, and reductions of sense of smell, taste, and touch. Grasp of the familiar wax and wane as faces, language, and time itself become ambiguous and slip in and out of understanding. As a result, there is distress, agitation, anger, and aloofness. Changes in hearing can make listening a particular challenge, by impairing which sounds can be perceived and also creating particular sensitives that can supercharge everyday soundscapes as cacophonous and unrecognisable or muffle them into a distance echo. The internal soundscapes can be equally preoccupying with constant ringing, buzzing, or more elaborate auditory hallucinations. Those living with dementia need medical and social facilitation, and are most likely to live institutionally as a result.

Following Erin Manning, we might ask how artistic practices can 'facilitate emergent collectivities without turning to the neurotypical habit of pathologising difference' (Manning, 2016, p. 147). To honour diverse bodily experiences Manning insists we must begin by asking, 'What can a body do?,' rather than assuming we know what it lacks (p. 276). In this way, engaging those living with dementia must address the power of normalcy to define difference and disability. Disablement ruptures widely held assumptions with regards auditory perception and listening, revealing the many ways normalcy frames how sound practices develop, where they develop and for whom resulting forms of art exist. Honouring differences gives momentum to the collaborative and the participatory in my own creative work, pivoting on practices of facilitation and gestures of accommodation that can mobilise sonic engagement rather than consolidating sonic exclusions.

A moment waiting to happen

The elders enter the windowless, utilitarian room: some are with staff members, others accompanied by a relative, they are welcomed by the six people involved in the workshop—five dancers and one musician. The room is organised to create as much open space as possible, with furniture pushed to the side and chairs arranged in a circle around the edge. The elders are encouraged to sit or stand, whatever they can or wish to do, such choices give space for action and some disorder. As is usual for such workshops we have been given little information about individual circumstances of those attending beyond their age and diagnosis of dementia, along with a general reminder that many tire easily and have decreased mobility and disrupted sensory perception.

The sound and movement begin immediately. I play a prearranged soundscape developed through previous work with people with dementia, the sounds are soft wave like textures, filtered and phased mid-register acoustic string sounds and vocal samples which ebb and flow on a loop. This is a piece the dancers recognise and have moved with before. The timbre and rhythmic flow are designed so as not to agitate and the dynamic level is kept only slightly above the volume level of normal speaking. The group of about 15 people fragment into several smaller groups and fill the open space prepared in the middle of the room. There is informality as the five dancers move amongst the elders, sitting or standing with them. There is much talking and some movement, the energy level in the room increases.

I am outside the well of interaction, set up in a corner of the room with musical equipment of various sorts. We planned that I would use the soundscore as a drone to activate the acoustic space, and then use the same palette of sounds to improvise to the unfolding interactions between elders and dancers. But as I am faced with a sea of bodies and voices moving in different ways, it feels noisy to the eyes and to the ears. I gradually increase the volume and intensity of my playing, searching for the sweet spot where the elders may be alert to the sound but it does not overwhelm or trigger.

My attention settles on the interaction unfolding directly in front of me between two men, an elder with dementia and one of the dancers. As the dancer begins to move to the ebb and flow of the soundscore I improvise short melodic fragments on the piano, mirroring the gestural quality and rhythmic pattern of his movement. He makes a strong gesture, arms outstretched at shoulder height, front leg bent, a kind of warrior pose, he is smiling, and they are making eye contact. The elder begins to mirror the warrior pose, gradually sliding his legs further apart,

I sound this gesture with a slow descending line. The dancer responds exaggerating the pose further by deepening and widening his stance. The elder indicates that he cannot follow this move and instead pulls his legs together and takes his arms above this head, in what reminds me of a classic stance of a flamenco dancer. My impression is confirmed when he suddenly stamps his feet and enunciates a rhythmic 'Olé!' The dancer replies with an exclamatory 'Olé!' I play two notes, c up to f, a melodic 'O-lé!'

The elder then breaks into song, 'So this year I'm off to sunny Spain.' His voice is loud and clear. He smiles broadly at the dancer who continues to follow his lead with elegant flamenco style moves, but he doesn't sing along (he tells me after the session that he didn't recognise the song). As the elder repeats the wordless curve of the melody, I instinctively sing the end of the phrase, 'Hey, Viva España!' It wasn't clear if he had heard me, but as we both continue to sing our voices meet in the air between us, and together we become audible around the room. Together we sing a version of the song's chorus—albeit with missing or improvised words. The elder is now clapping and stamping, making exaggerated flamenco style movements and all three of us are now fully committed to this moment of singing and dancing.

The gestures of *Y Viva España* spread through the group, within a couple of minutes at least half of the elders are singing the refrain 'O, Viva España!' There are vocal sounds which are melodic and exclamatory, laughter, synchronous stamping and clapping: movements are echoed as arms are extended over heads, backs are straightened and chins raised, imaginary castanets are played and what appear to be bull fighting scenes are played out. I am acutely aware that through the physio-sonic gestures a shared dynamic is crystallising, a mutuality. I feel an unexpected emotional transposition as my own childhood memories of this song—long forgotten—emerge and mingle with this shared moment of affective community.

The moment is spontaneous, unrestrained and rowdy, it seemed it was a moment that had been waiting to happen, which found a release through our interaction. We are now singing and dancing together in a way totally unforeseen, in an unlikely atmosphere of playfulness and flirtation, we generate an energy that reverberates far outside the windowless room. The sound catches people outside and staff members not involved in the session come to see what is happening, to check that the situation is under control. They express surprise and some bewilderment at the unfolding scene.

Animating engagement through more-than-sonic practices

This vignette, drawn from a one-off workshop undertaken with colleagues from the Beatrice Allegranti Dance Theatre Company illustrates a moment of affective community through sonic engagement, but in doing so points to the ways we may usefully rethink the terms sonic and engagement.

The workshop was planned to specifically engage with elders living with dementia through participation in a creative *physio-sonic* practice, the institutional space was reconfigured so that expressive movement and sound making were jointly facilitated. For one man, living with disintegrating memory, and hearing loss, this space of physio-sonic possibilities offered up an extra-ordinary expressive interaction which released a moment of remembering and affective attunement with others in the room. Moved to move, the elder found the 'flamenco' stance, the kinaesthetic recognition of this posture being consubstantial with a sonic memory of Spain, which at this moment was recalled as a melodic fragment from a pop song—a 1970s summer holiday anthem. The sounding of *Y Viva España* was a mingled-modality (vocal, linguistic, auditory, visual, kinetic and kinaesthetic) and as such experienced as something more-than-sonic. *Y Viva España* as physio-sonic hook caught hold of many elders in the room, unfolding a shared moment of sonic recognition *bodying* a poetic world of personal significance.

In nurturing the entanglement of sonic and bodily motion, the creative possibilities of being moved affectively remain accessible for the group. 'Motion and emotion are … intrinsically connected: one is *moved by movement* (perception; impression; affection) and *moved to move* (action; expression; e-motion)' (Fuchs & Koch, 2014, p. 1). In the same way movement and sound are also intrinsically connected. Sound is animate matter which *animates*—it moves us to move and be moved, whilst resisting conceptions of separateness. Its liquidity fills spaces and permeates bodies in patterns of resonance which reverberate and evoke. As such matters of more-than-sonic listening and resonance form important foundations for understanding the 'inter' of our inter-actions, inter-subjectivities, and inter-affectivities.

Y Viva España was a moment of mutuality, grounded in a shared recognition and action, facilitated in the commingling of those living with dementia, and those invested in caring for dementia. Facilitation in this instance is a mode of artistic working, not a means to an end but an end in itself. Through compositional and 'choreographic kinship' (Allegranti, 2019) such practices are not tethered to engagement as such but to 'the relational force of facilitation' as 'a necessary and honored aspect of social life' (Manning, 2016, p. 276). Such processes stand in spite of neoliberalist systems which reify individuality by naming those who need facilitation as always 'lacking', cementing their stigmatisation and rendering relation peripheral rather than central to the world. Here, as an artistic practice, facilitation stands as a calling out of the normative regimes that delegitimise *coming-together.*

3. **Sonic engagement and more-than-social art**

Sonic engagement is positioned here as what may contribute to forms of socially minded artistic actions and the ways artistic practice may activate the field of relation and 'affect the complex ecologies of which it is part' (Manning, 2016, p. 50). Whether through collective musical gatherings in situations of oppression, or the complex scenes of illness and care which demand a sensitivity for hearing as well as hearing loss, sonic engagement emphasises sound and listening as being adept at fostering transformation.

From our perspective, sonic engagement operates on both social and corporeal levels, moving from an engagement with questions of language, narrative, and the symbolic ordering of particular collective agendas, to experiences and challenges around corporeality. We are concerned to capture how these two aspects may also foster a sonic theory that can impact on how we conceive of social engagement in the arts. In short, what becomes of social engagement from a sonic perspective? Are there specific perspectives or opportunities that sound and listening offer to the ways in which being engaged may occur?

In the publication, *Artistic Citizenship: Artistry, Social Responsibility, and Ethical Praxis* (Elliott et al., 2016), the editors argue for greater appreciation of art as a form of social and civic engagement. This includes a consideration of the rights and responsibilities of artists and their practices, accenting how art or 'artistry' contributes to human flourishing by lending to the manifestation of a more egalitarian world. In making such arguments, the editors also recognise how notions of rights and responsibilities may rub against traditional understandings of the artist as an autonomous figure, and artmaking as something that should not be anchored to forms of moral reasoning or social instrumentalisation. What happens to the artist, and artistry, when they are positioned within a framework of citizenship? The editors propose that such a move helps in capturing the ways in which artists are always working within specific communities, even a community of fellow artists, and in this sense, as an artist, one operates through societal structures. The artwork or artistic gesture as well is positioned as a public act or object, situated within a communal framework deriving much of its power through its circulation as a social good. That is to say, artists are enabled through the rights of free expression, and through a range of communal and institutional structures, and therefore have a responsibility to those communities and institutions. In this regard, artistic citizenship is already at play, and as such, requires greater discursive and critical questioning.

In raising the questions of responsibility in relation to engagement, artists can be reminded of their particular place within communities, and can be encouraged to more overtly direct their practices towards public issues, or to understand already existing social practices and engagements as expressions of rights and duties. At the same time, it feels essential to question what may go missing when entering into artistic citizenship: in accepting these as guiding terms for how artists come to practice, what do we lose? In what ways does 'social responsibility' undermine the rather unreasonableness of artistic action and the importance of the 'irresponsibility' of the artist and the ability to act outside a framework of rights? As the artist Luis Guerra (2020) suggests, irresponsibility in fact becomes a responsibility within the framework of art, as a way of enacting the creative force of an artistic position or sensibility. While we may strive for social engagement, or a position of human flourishing, particularly in today's polarised environments, it seems necessary to consider the rather anti-social or disobedient nature of artistic actions—that is, the strange articulations, unwieldy meanings, affective resonances, and obsessive antics which art may also deliver. In this way, questions of responsibility within the arts may be usefully reframed in terms of the *response-abilities* and *irresponse-abilities* of artistic practice.

Returning to understandings of sonic engagement we have been pursuing, sound and listening emerge as key to both activating the social while complicating and elaborating its borders. As seen and heard in the context of Budapest, small parties in private apartments become resistant not so much by way of the articulation of political views or that of a public discourse; rather, sound and music fuel forms of communalisation that are mostly driven by the passions of collective sounding and listening, the musical crafting and sharing that come to support or drive communal effort. In addition, as the performative workshop in the care home reveals, social engagement may take expression through other ways of hearing, where sound and music enable unruly forms of listening and acting. We take both these instances as suggestive for a 'sonic sociality' and a theory of sonic art practice that engage others in such a way as to demand a rethinking of the social as what may emerge in resonant and also disengaged ways.

In thinking through these questions and views, we are led to think less about reinforcing practices of socially engaged art, and rather, to consider a more-than-social approach. Sonic engagement may lead us to a more-than-social art position, helping capture how artistic actions and gestures attend to public issues, social worlds, and human challenges by staying close to their excessive, unruly, ambient and uncanny manifestations and imaginaries. These 'minor gestures', to borrow Erin Manning's term, are forces that open 'experience to its potential variation' whilst destabilising normative ideals (Manning, 2016, p. i). Locating the minor in forms of insurgent perception across a register of pathologised and marginalised states—the autistic, the schizophrenic, the psychotic, the depressed, the institutionalised, and out of sight—she observes:

> The minor isn't known in advance. It never reproduces itself in its own image. Each minor gesture is singularly connected to the event at hand, immanent to the in-act ... but the minor gesture also exceeds the bound of the event, touching on the ineffable quality of the more-than. (p. 201)

The more-than-social art position we unfold here follows Manning by generating 'a field of resonance for the minor' (Manning, 2016, n.p.) sourcing productive ways of reworking social arrangements to establish new points of meeting and being-in-concert or being-in-common.

As Giorgio Agamben reminds in his work on forms-of-life, conceptualisations of what counts as 'community' often fail to recognise how being-in-common is founded on a profound unrecognisability (the limits of language): that if we are to approach community as the creative undertaking that it is, it is essential to stay close to the fundamental inertia of the common as what exceeds identity and identification. In this way, a 'true community' is that which never names itself as such (Agamben, 1999, p. 47).

Agamben's arguments provide useful input into reflecting upon artistic citizenship, and approaches to sonic engagement we are pursuing. As the editors of *Artistic Citizenship* argue, to be a citizen is to act as a political subject within a specific community; it is to be a figure of the city, a participant within a given polity and set of discourses—and importantly, it is to be a legal subject. Yet, from an artistic perspective, community and citizenship also often find their expression through aesthetic, material, and affective experiences, as processes of encounter and invention that find traction by tensing what counts as legal and right. The editors do suggest as much by underscoring artistry as an 'ethical praxis' and how artists bring forward a profound sensitivity and engagement with the situatedness of people and places. Yet, ethical praxis could be said to require an ongoing invention of approaches that also lend to the worlding of other ontologies, other knowledges, and other relationalities that are less about social engagement and more about the 'self-creating and self-organising' drive of poetic world making activity (Warner, pp. 68–69). As our two cases demonstrate, such poetic world making hinges on possibilities for participation and action by the creation of space which opens up 'the existing fields of relation toward new forms of perception, accountability, experience, and collectivity' (Manning, 2016, p. 31). The liquid, relational qualities of the more-than-sonic/more-than social troubles the traditional partitioning of the senses and the field of experience as bodily/private or communal/public, reconfiguring these as radically open, porous, and entangled modalities. In places of care or political oppression this more-than gives life to moments of disruptive mutuality that may be prohibited or otherwise inaccessible because it is a 'force that makes felt how a process is never about an externally situated individual, but about the ecology it calls forth' (p. 31).

The artist as a figure, and artistic actions in general, may work through understandings of political subjectivity by showing the limits of the political, captured not only by participating in an informed public sphere, but rather, by way of a range of gestures, including: the experimental reworking of states of matter and affective relationalities, the expression of diverse forms of embodiment and enactment, the construction and deconstruction of institutional discourses, the erotic engendering of collective practices, from sonic commingling to unruly pedagogies. These forms of expression or action may further echo Agamben's (2016) description of 'the swindler', as that figure who not so much breaks the rules, but 'pretends to keep playing when in reality he has already left the game' (p. 242).

Agamben's notion of swindling is suggestive for appreciating the ways in which art may not only civically and socially engage, but does so by often worlding another game altogether, another communal scene and therefore, another polity—not so much to 'redistribute the sensible' following Jacques Rancière (2013), but to create a different sensible altogether. From ethical praxis to swindling trickster, from conscientious aesthetics to rituals of musical escape and recovery, a sonic social art is posed as what not only relates us to the social or civic sphere (as an identifiable scene), but additionally to the currents of life always underpinning and exceeding the social in general. This includes amplifying the differential ways in which hearing may take place, to allow for incursions onto what counts as sound and listening as well.

Notes

1 There are numerous instances where restrictions on social and cultural freedoms have been moved into place. For example, in 2019 a government policy was initiated so as to gain more control on theatres; in order to receive cultural funding municipalities must agree to have the government play a role in its management and the appointment of directors. More broadly, the government has directed cultural funding in support of new cultural institutions that celebrate and reinforce 'national culture' and heritage to the detriment of more liberally-oriented institutions and cultural scenes. In addition, organisations must label themselves, in publications and websites, as having received support from a foreign entity, which acts as a way of marking 'foreign influence' and anti-nationalistic organisations.

2 As an example, a law passed in 2017 (known as the 'Soros law') freezes funding to organisations of over 22,000 euro from outside Hungary. While justification of the law was aimed at thwarting money being used for laundering or for the financing of terrorism, it was understood as an attack on the Open University and the Soros Foundations in particular, which have been instrumental in supporting cultures of democracy and policies in favor of human rights. (Following the law, the EU requested further information on the necessity of the law on the part of the Hungarian government; in 2020, the courts of the EU finally declared Hungary is breaking EU law by restricting the financing of NGOs).

3 One such example is the National Theatre as well as the University of Theatre and Film Arts Budapest which, in July 2020 was given over by the government to a private foundation led by Attila Vidnyánszky and former army officer Gábor Szarka who immediately replaced the University's governing board. This led to a dramatic occupation of the University building by students objecting to the 'illegal' take over. https://witness-europe.eu/2020/11/22/hungarys-secret-university-szfe-students-in-budapest-fight-for-autonomy/. Accessed March 2021.

4 Source: Statistisk Sentralbyrå Norge: https://www.ssb.no/en/statbank/list/soskon. Accessed March 2021.

5 Sources: NHS UK (www.nhs.uk) and Age UK (ageuk.org.uk) *Later Life in the UK* (2019). Accessed February 2021.

References

Agamben, G. (1999). *Potentialities: Collected essays in philosophy*. Stanford University Press.

Agamben, G. (2016). *The use of bodies*. Stanford University Press.

Allegranti, B. (2020). Moving kinship: Between choreography, performance, and the more-than-human. In Thomas, H. & Prickett, S. (Eds.), *The Routledge companion to dance studies* (pp. 88–108). Routledge.

Boym, S. (2017). *The off-modern*. Bloomsbury Academic.

Elliott, D., Silverman, M., & Bowman, W. (2016). *Artistic citizenship: Artistry, social responsibility, and ethical praxis*. Oxford University Press.

Fuchs, T., & Koch, S. C. (2014). Embodied affectivity: On moving and being moved. *Frontiers in Psychology*, *5*(508), 1–12. 10.3389/fpsyg.2014.00508

Garrera, G., & Triulzi, S. (2020). *Evasion techniques: Strategies for the subversion and derision of power in 1960s and 70s Hungarian avant-garde art*. Azienda Speciale Palaexpo.

Gilman-Opalsky, R. (2020). *The communism of love: An inquiry into the poverty of exchange value*. AK Press.

Guerra, L. (2020). *In conversation with the author*. For more on the artist see: https://www.luisguerra.org/

Jávor, I. (2021). Once upon a party. In V. Aguado, R. Del Buey, & B. LaBelle (Eds.), *Party studies* (Vol. 1) (pp. 58–73). Errant Bodies Press. http://communitiesinmovement.net/wordpress/wp-content/uploads/2021/06/Party-Studies-Vol1.pdf

Manning, E. (2016). *The minor gesture*. Duke University Press. 10.1515/9780822374411

Petrešin-Bachelez, N. (2017). For slow institutions. *e-flux: Journal, 85*, 1–13. http://worker01.e-flux.com/pdf/article_155520.pdf

Rancière, J. (2013). *The politics of aesthetics*. Bloomsbury.

Robinson, D. (2020). *Hungry listening: Resonant theory for indigenous sound studies*. University of Minnesota Press.

Serres, M. (2008). *The five senses: A philosophy of mingled bodies*. Bloomsbury Publishing.

Sheets-Johnstone, M. (2015). *The corporeal turn: An interdisciplinary reader*. Andrews UK Limited.

Udvardyova, L. (2021). Budapest flat concerts – Past and present. In V. Aguado, R. Del Buey, & B. LaBelle (Eds.), *Party studies* (Vol. 1) (pp. 36–57). Errant Bodies Press. http://communitiesinmovement.net/wordpress/wp-content/uploads/2021/06/Party-Studies-Vol1.pdf

Warner, M. (2005). *Publics and counterpublics*. Zone Books.

7

REINSCRIBING THE NOISE: NEW MEDIA WALK TECHNOLOGIES AND THE POLITICS OF COMMUNITY ENGAGEMENT

Christos Carras and Eric Lewis

Digital media walks constitute a new art practice enabled by mobile devices, GPS and digital cartography. Media walks involve a digital map which one views on a mobile device, which remains centered on one's actual location in physical space. Embedded in the map are zones where assorted media (audio, video, text, image) are triggered for viewing and hearing on the mobile device (usually a cellphone). Various functionalities may be incorporated into the precise media walk platform employed, allowing for multiple medias to be triggered at the same location, cross-fades between audio files, differing distances centred upon a precise location that trigger the opening of a particular media file, and so on. While media walks are in and of themselves neutral as to what sort of media is triggered based on one's physical location in a space, audio files tend to be favored, for the practical reason that they are easily incorporated into media walk software, but more importantly due to the fact that media walks are historically and conceptually the direct descendants of 'analogue' sound walks, and draw upon the theory and practice that has evolved around sound walking.[1] In what follows we will use digital media walk when discussing the technology in its broadest terms, and digital sound walk when discussing a sub-set of digital media walks which only include audio files.

More specifically we discuss a series of digital media walks, all created in Greece, which the present authors were intimately involved in the planning, creation, implementation, and subsequent use, of. This series of digital media walks, collectively named *Rhizomes I-III* (more are in the planning stage) were all realised as a formal partnership between Media Electronique (an Athens' based new media art collective, which author Eric Lewis is a member of) and the Onassis Cultural Centre, of which author Christos Carras is the Executive Director.

DOI: 10.4324/9781003164227-14

Background and theory: *Rhizome 1*

Dourgouti, in the Athenian quarter of Neos Kosmos, is an urban area that reflects much of Athens' twentieth century history. Originally a refugee settlement founded after 1915 and populated by Armenians fleeing persecution, it was also a centre for the resistance to Nazi occupation during World War II. Since the war it has seen arrivals of migrants from the former Soviet Socialist Republics and then the Middle East, as well as substantial communities from Bangladesh and North Africa, alongside mainly working-class or lower middle-class. Over the past decades, it has seen a rapid development of corporate buildings, cultural, health, and education infrastructure and large hotels, as well as many strip clubs.[2]

One of these new buildings on the avenue is the Onassis Stegi (OS),[3] a contemporary cultural centre with an interest in producing new, site-specific, intermedia work. The curatorial team of the OS was well aware of being housed in a building on the sleek façade of the avenue, with little contact with or understanding of urban life only a couple of blocks away. In collaboration with new media art collective *Medea Electronique,*[4] the OS conceived and produced what would be the first in a series of three interactive media walks, *Soundscapes-Landscapes: Rhizome 1* (henceforth 'SLR1') (see Figure 7.1).[5] During the planning and creative phases, the project was articulated around three fundamental concepts: 'Rhizome' taken from the work of Gilles Deleuze and Felix Guattari (Deleuze & Guattari, 1987), the 'flâneur' whose theoretical origins lie in the work of Walter Benjamin (Benjamin, 1997, 2002) and that of 'walking' in its broader

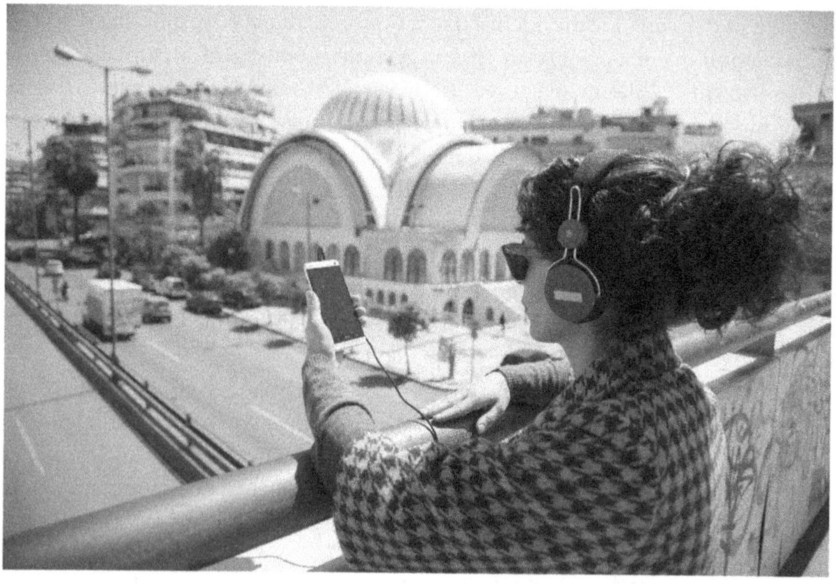

FIGURE 7.1 Neos Kosmos in Athens, within the area covered by SLR1 ©Stavros Petropoulos/Onassis Stegi

theoretical context as well as more specifically related to soundwalks.[6] We shall return to these below.

The end result of the project was an app, developed using proprietary technology, but freely available for both Android and iOS systems, which was launched in 2014.[7] The app's content is closed; in other words, it is not possible for users to add anything to the original media developed by the artistic and production teams. The app enables access to a range of media and contents: field recordings made in the Dourgouti area as well as short electroacoustic compositions (usually based on field recordings), short video pieces again based on videos shot in the area, texts (poems, prose, and historical texts) that either appear on the screen or are presented as audio, interviews with residents of the neighbourhood and of course a map of Dourgouti. The media are activated as the app user walks through the area covered by the geo-referencing coordinates. One interesting (and unplanned) result of the app's closure and of the fact that it has not been updated is that it now functions in part as an archive of the area, since several buildings no longer exist or have changed, meaning that the media now represent 'phantom' images and impressions of urban settings.[8]

While SLR1 shares characteristics with earlier media walks, such as *The Missing Voice—(Case Study B)* by Janet Cardiff (1999) and Toby Butler's *Memoryscape* (2009), it differs from these (and many other media walks) in so far as SLR1 is an open walk in which the participant decides her own route, a decision that in turn alters the ways and sequences in which media contents are accessed. For most media walks are linear, in other words they invite the participant to follow a predetermined course through space. The open nature of SLR1 suggests two further levels of understanding of this media walk's engagement with its site. The first, which was significant in the planning phase of the project, touches upon the concept of 'rhizome' as developed by Gilles Deleuze and Félix Guattari (1980). As they write in *A Thousand Plateaus*, 'Principles of connection and heterogeneity: any point of a rhizome can be connected to any other and must be. It's very different from a tree or a root which fix a point, an order' (Deleuze & Guattari, 1980, p. 13). The indeterminacy of the route taken allows for a multiplicity of sequences of contents and locations, though of course limited by the physical continuity of movement though space. Thus the app design itself suggests the user connect any locations in any order, which can be seen as revealing that these connections, as in a rhizome, were implicitly present all along. [9] The second brings together Deleuze and Guattari's development of the rhizome concept with the significance of walking as a performative act elaborated by Michel de Certeau (1984).[10] The former propose that a model of language should aim to 'reach the *abstract machine* that operates at the intersection of a language with semantic and pragmatic contents of enunciations, with collective assemblages of enunciation, with a whole micro-politics of the social field' (Deleuze & Guattari, 1987, p. 14). de Certeau in a similar[11] way highlights the manners in which 'Beneath the discourses that ideologise the city, the ruses and combinations of powers that have no readable identity proliferate' (de Certeau, 1984, p. 95) and invites us to elaborate further on the capacity of the 'practice of space' that is walking to dissolve functional barriers that instrumentally fix the urban space. de Certeau suggests that 'a migrational, or metaphorical, city [thus] slips into the clear text of the planned and readable city' (p. 93). Walking appears in his writing as one of the surprisingly powerful ways in which people manage to create minor spaces of emancipation within the structures of power that tend to homogenise and control, leading him to propose that 'it thus seems possible to give a preliminary definition of walking as a space of enunciation' (p. 98). The experience of embodied, located presence; the autonomy of movement and direction, the slowness and closeness to the environment and the zooming in on detail that it permits; the liberation of associations and images in the thought process as the body moves through an ever-changing environment; the implications of being exposed to that environment for one's understanding of humans' effects on it; the realisation that space is not smooth and that apart from geographical and physical obstacles, the distribution of access and freedom of passage are conditioned by powerful forces of ownership, segregation and repression; the feeling of 'liberation' in relation to work and consumption that a walk can inspire, all these and other aspects indicate why the practice of walking (that most basic human practice) has been imbued with new meaning in our mechanistic and now also digital world. As Frédéric Gros (2015) writes, 'the walker considers it to be a liberation to be disentangled from the web of exchanges, no longer reduced to a junction in the network redistributing information, images and goods' (p. 4). It is in this space of liberation that new constellations of meaning can appear.[12]

We mention these theoretical approaches because they are very much part of the framework from within which the SLR1 project emerged and as such shed light on its political significance. The decision to create a media walk that requires the participant to effectively perform it without a predefined trajectory, that aims to bring out multiple possible layers of meaning from points in the urban fabric, that does so in a part of the city that is 'minor' in terms of its position in standard depictions of Athens, but 'major' in its political significance, and to do so by making audible and visible the words and worlds of local communities strongly indicates an anti-hierarchical stance. Likewise, the insistence on a multi-modal approach suggests that the kind of semantic space that SLR1 aims to create—an 'extended field of relations (between people, media, socio-cultural reality and the environment)' (Bubaris, 2017, p. 27)[13]—is one that perhaps enables the emergence of minor spaces of emancipation for the walker-participants, such as those suggested by De Certeau. SLR1 is a (successful in our opinion) example of an intermedia artwork, that actively engages the participant and references traditions of critical thought. However, what is its relation to the communities of Dourgouti and what are their relations to it? How different are the interactions engendered by such intermedia works from those one experiences from the back of a tour bus, or via a random unscripted stroll? The degree of engagement that such media walks and sound walks are likely to produce is a product both of the technical design employed, and the content embedded in the walk, and neither separately or together can guarantee the emergence of such emancipatory spaces, but can to varied degrees catalyse such emergence.

To achieve these goals the designers of both the content and architecture of the app need to themselves develop intimate knowledge of the spaces and communities at play. Media and sound walks can only engender potentially politically active engagement if their creators have themselves been so engaged as part of the walks' creation.

We would contend that the process of creating SLR1 is in many ways close to the practice of a sensory ethnography, even though none of the creative team were academic ethnographers and the result is presented and promoted as an intermedia participatory artwork. Sarah Pink proposes that sensory ethnography 'involves the researcher self-consciously and reflexively attending to the senses throughout the research process, that is during planning, reviewing, fieldwork, analysis and representational process of a project' (Pink, 2009, p. 10). The creative process of producing SLR1 did involve extensive interviewing and fieldwork and it does purport to create knowledge and enable a multi-modal form of knowing, even if this knowledge is not delivered in an academic format.[14] As Pink notes, it is possible that 'ethnographic and artistic representations might create a sense of intimacy sufficiently powerful to invite empathetic understanding and communicate experiential knowing to audiences' (p. 134). The crucial term here is 're-presentation' and the directionality of the knowledge that this implies. The interviews, fieldwork, and recordings can be construed as part of a research process and the composition of the media walk itself as a process of representation of Dourgouti and its communities by the creative team.[15] This is rendered even more complex by the deliberate blurring of boundaries between documentation and interpretation: as was mentioned above, the media in SLR1 include interviews, texts selected by the creative team, and audio-visual compositions whose aesthetics are those of the artists involved but are layered onto the sensory experience that the participant has of the space and often frame the more 'documentary' elements. Such projects are inevitably part of 'political and ideological agendas and power relations integral to the contexts and circumstances of ethnographic processes' (Pink, 2009, p. 23), and, we would add, to artistic processes as well. The theoretical framework of SLR1 analysed above brings with it a specific 'political and ideological agenda', which is not in itself reprehensible of course, but is not necessarily one that informs the attitudes and beliefs of any of the individuals and communities represented in the work. In so far as it contains elements of a sensory ethnography it is part of a politics of research methodologies that has been rightly critiqued for its neo-colonial residues—it is about creating empathy, yet without engagement, it is in keeping with the logic of 'outreach' without reciprocity.[16] However, it also starts to transcend these well-studied limitations by placing in dialogue the artistic representations as found in the fixed-media of the media walk, with the actual sensations one receives from the spaces themselves when taking the walk, in an uncontrolled and unpredictable way. To this degree noise enters into the experience, both literally as cars honk, drowning out the media you hear through your headphones, and metaphorically as the neighbourhood itself alters your experience and cuts against the representational veracity of the fixed media. And so SLR1 enriches the concept of a sensory ethnography, and, in effect, allows the spaces themselves to partially control the form this ethnography takes, which starts to resemble more of a collective improvisation. As we will soon go on to discuss, recognition of this loosening of creative agency to now involve the spaces themselves and those who occupy them, suggested to us both the possibility, and ultimately the political necessity, of even further loosening our control over the narratives, the representations, that such ethnographic methods produce.

The power relations also include the institutional setting of the project and the technical choices made even before the actual fieldwork starts, aspects that are always present in any project but are likewise never without significance. We have already alluded to the technical choices that determine the project's form. The use of a closed app that does not allow anyone outside the creative team to add or modify content is a strong barrier to deeper participation on the part of local communities. The very choice of engaging with an urban area and those who live there through the interface of mobile devices brings into the project a slurry of issues that are important but too complex to analyse here, related to the myriad ways in which mobile technologies organise and structure spaces as objects of financial gain, as fields of political confrontation or as grids of surveillance.[17]

Composition/Improvisation: *Rhizome* II

Following on from SLR1, Onassis Stegi and Medea Electronique developed a second version, *Soundscapes Landscapes Rhizome II (henceforth SLR2)* this time in the areas of Gazi, Kerameikos, and Metaxourgeio, once again in Athens (see Figure 7.2).

Though this media walk used a different platform (about which more will be said below) the approach was similar, with even thicker layers of ethnographic, audio, video and textual content.[18] When planning SLR2 the creative team had envisaged correlating the points on the territory covered with two levels of media content: on the one hand content related to what the walker could experience, on the other, content related to underground elements that were not available to their perception; underground streams, catacombs, metro lines, archaeological remains, and so on. This initial plan was only partially realised, but it did emphasise the significance of the map: if elements of content are not available for perception, then only their placement on the map connects them to the walking experience. In a strange way, maps in media walks (perhaps digital maps in general) have a status in between that of a normal fixed-scale map and a 1:1 scale map such as the one imagined by Borges in *On Exactitude in Science* (1998) both because they expand and contract and because the content that they index aims to achieve a high level of veracity. Indeed, by layering content that goes far beyond what is available to the senses, including sounds and images that are derived from sensory inputs through processes of creative interpretation, they gesture towards a theoretically infinite level of detail. In any case, the importance of the map for SLR2

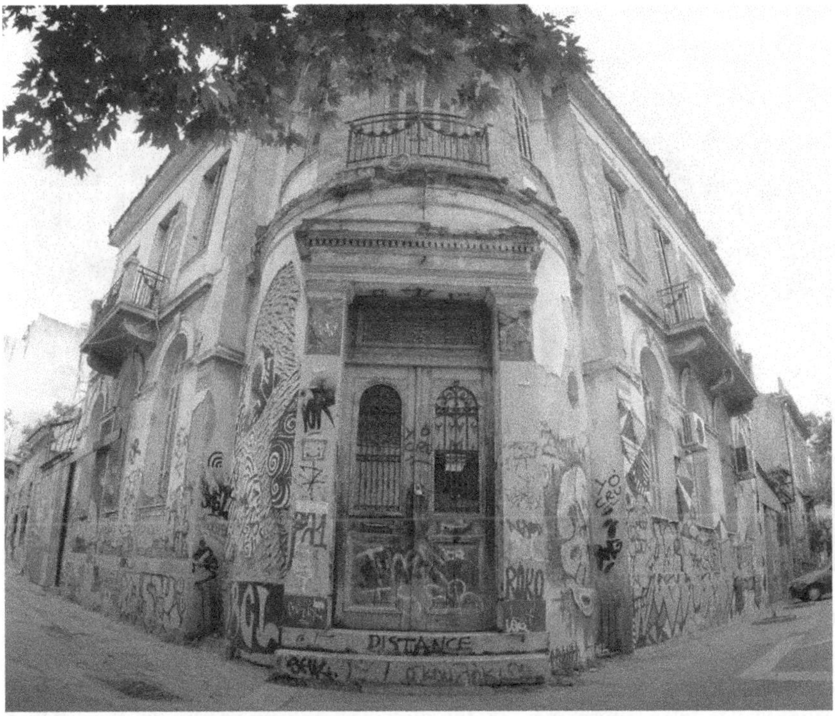

FIGURE 7.2 A street in Metaxourgeio, an area included in SLR2, ©Medea Electronique

led to a parallel project, an extended reality tour in the Onassis Stegi gallery space, entitled *City Above—Life Below* (see Figure 7.3).[19] Here, the visitor could walk around the space encountering content derived (and in many cases expanded upon) from SLR2 through a variety of media: highly directional speakers, headphones, video projections, VR installations. The media were distributed in the gallery in a way that approximated their spatial relationships in the urban space. This offshoot of SLR2 highlights, indeed exacerbates, issues of representation. It suggests that the map, an integral part of these media walks, does not merely represent space but also acts itself as a representational space.[20] This iteration of SLR2, which emerged organically from the initial one *via* the process of mapping, suggests that the mediating technologies, practices, and concepts that come into force cannot be separated from a politics of space, because they further deterritorialise the content which has been 'sampled' from the communities that populate the urban space and in the process. This sampling process, as is the case with all selections, filters out the 'noise' that undermines the process of representation because it does not neatly fit the categories established by those performing the representation. The mediating technologies also potentially act as barriers for those who do not have access to them or for whom they are unfamiliar.[21]

SLR1 & 2 are part of an ongoing and varied corpus of site-specific work commissioned and/or produced by the Onassis Stegi in relation to the city of Athens. For

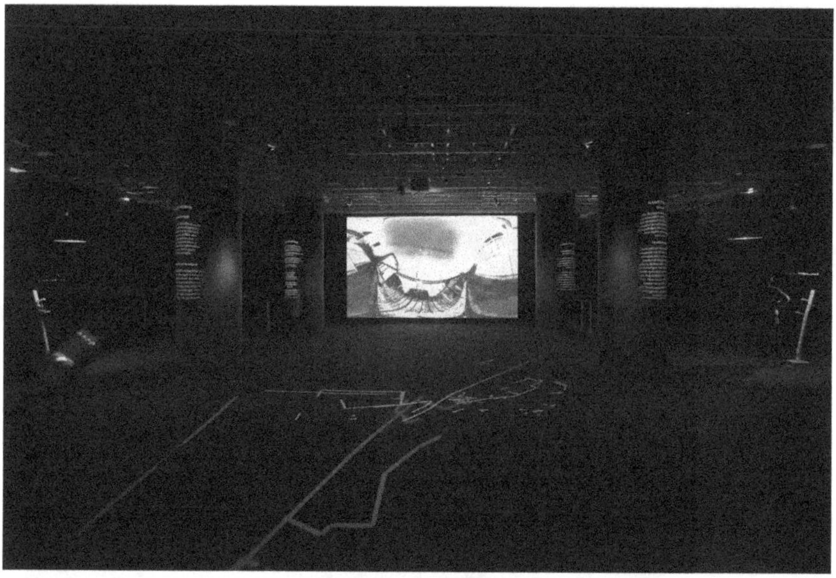

FIGURE 7.3 City Above—Life Below © Kiki Papadopoulou/Onassis Stegi

the most part, these works aim to enable a heightened sensitivity, often with explicitly 'progressive' political overtones, in the audience or participants (in the case of participatory works) in relation to the city, its structures of exclusion, its commercialisation, and the threats posed to civic engagement in public space. Many (if not most) of these projects to varying degrees involve non-artist residents of the city and sometimes communities. An important element to keep in mind is that the residents and communities are invited to engage in formal performative contexts (the site-specific media walks themselves) that emerge from contemporary artistic practices (from land art to performance-installations) that may well not inform their conception of art or their relationship to their environment. Nonetheless, both their, and the audience's experience is mediated by the presuppositions of these practices.[22]

Let us return to the idea of composition as a process of representation by the artist. An alternative conception of composition was proposed by Jacques Attali and has been influential in discourses of anti-hierarchical artistic production. Attali wrote:

> Beyond the rupture of the economic conditions of music, composition is revealed as the demand for a truly different system of organisation, a network within which a different kind of music and different social relations can arise. A music produced by each individual for himself, for pleasure outside of meaning, usage and exchange.
>
> *(1985, p. 137)*

This kind of composition might conceivably characterise a self-organised community media walk and bring forth a politics of empowerment. As we have seen,

SLR1 adopts a different approach to politics, one that promotes the transformative force of knowing in non-conventional ways, but which nonetheless preserves the separation of the knowing and the known.

With this background of what was successful in SLR1 & 2 and what was lacking, we thought about ways of opening up the architecture of the app itself, and the process of content creation, to make the whole process, and the final product, more participatory, inclusive and itself an act of community building and identity formation. One way of conceptualising the evolution of our thinking is to consider the ambiguities inherent in Attali's notion of 'composition'. Attali states that what is needed is:

> Not a new music, but a new way of making music … . It is the advent of a radically new form of the insertion of music into communication, one that is overturning all of the concepts of political economy and giving new meaning to the political project. The only radically different course open for knowledge and social reality. The only dimension permitting the escape from ritual dictatorship, the illusion of representation, and the silence of repetition. Music, the ultimate form of production, gives voice to this new emergence, suggesting that we designate it *composition* … . Composition thus appears as a negation of the division of roles and labour as constructed by the old codes … .to listen to music in the network of composition is to rewrite it … . The listener is the operator … . Composition, then, beyond the realm of music, calls into question the distinction between worker and consumer, between doing and destroying … . Is Composition future or past? Is there a noise that can organise the transition toward it from the grey world of repetition?
>
> *(1985, pp. 134–135)*

For Attali composition is not the polar opposite of improvisation, but is a species of improvisation, which drives the production of music, or really any collective artistic endeavour, away from repetition. And once one realises this meaning of composition for Attali, one can see how the passages just quoted can serve as a theoretical model for a new conception of the media walk, one which SLR1 pointed towards the need of. Referencing Attali quoted above, what was needed was a new way of making a media walk, one which would insert sounds into new modes of communication, which would over-turn the political economy of media walk (including the traditional sound walk) and SLR1, which would undermine repetition, itself a product of the fixed media internal to SLR1, which would undermine the labour roles which produced SLR1 by blurring the distinctions between app designer and app user, that would allow the listener to be an operator, that would allow 'noise' to organise our media walks away from repetition and the hierarchical model of their creation and use. SLR1 allowed the noise of spaces to enter into the experience, but not the noise—the opinions, narratives, politics, and lived experiences of those who occupy these spaces, who call them home. Like perhaps the first generation of sound walks, we had prioritised the natural/built sonosphere over the personal and political sonosphere. We needed an improvisational (that is compositional in Attali's sense) model of media walks, and that is what we set out to develop.[23]

A three-way partnership was established with the Onassis Cultural Centre serving as the overall project organiser and supervisor, Medea Electronique as in charge of project realisation and implementation, and the Zentrum für Kunst und Medien (ZKM) in Karlsruhe, Germany as housing the design team for creating the new app. This app was limited to sound files, and so was exclusively for the creation of digital sound walks. This new project, SLR2, required a custom-built app with the following features:

1. Non-site specific—a sound walk[24] could be created anywhere in the world.
2. User-created sound walks—anyone who downloads the app can then create their own sound walks, with regards to location, content, and where and how this content is triggered by those taking the sound walk.
3. Free—there should be no paywall for use of the app (above and beyond the need for access to a computer and/or mobile device).
4. Modifiable—anyone should be able to add content to a given sound walk, use the sound walk, or create a new sound walk covering the same territory.
5. Simplicity—the process of creating, downloading and using a sound walk should not be complex, it should not require technical expertise.

These features allow for the creation of sound walks that are, in a sense, more democratic, for they can be employed and created by individuals and communities without outside technical support or assistance. They undermine simple repetition and representation, creating a new political economy for sound walking. The noise that accomplishes this can be varied—it is noise because it is user-determined, it can itself be unruly, discordant, a composition that decomposes established and accepted structures, hierarchies, and dogmas.

The app was called MyCitiesMySounds[25] and operated on both iOS and Android devices. Yet while the app's functionalities allow for the creation of more participatory sound walks, thus enabling an ethical engagement with those who would otherwise in all likelihood remain passive subjects, the process by which the app was introduced to the public, via its initial use by Medea Electronique, needed itself to model the potentially liberatory and communal features built into the app. This was something Medea took to heart, initially with its use of the app to create a series of sound walks in, with and for the small village of Sellasia in southern Greece, which is where Medea holds its annual improvised new media arts residency program, Koumaria.

The app allows anyone to record sounds on any device, and then pin them to a global map using the GPS meta-data associated with the recording (or one can pin the sound file to whatever location one wants). One can also create circular proximity zones around the pinned location that will determine the physical area within which the file will be triggered, and its volume relative to the centre of the circle. These 'trigger circles' can overlap, allowing multiple sound files to play simultaneously with varied and changing relative volumes based on how you move through the physical space. On the mobile device you use while engaging in the sound walk, the map of where you are and your surrounding area appears (re-centring itself as you move), displaying the sound-triggering circles so you know where they are. No route is predetermined, and so the sound files may be triggered in any order, or not at all. The non-linear nature of the sound walks the app presents, the ability of anyone to add sound-files to a walk, either a pre-existing one with permission, or a new walk they create, coupled with the fact that no walk leader is needed (anyone can participate in any walk at any time), opened up possibilities for a new ethics and politics related to such walks. In addition, the app downloads all its data onto one's mobile device, and so no continuous internet connection is required, lowering considerably the cost of participating in any walk.

Community agency and autonomy: *Rhizome* III

How can these features of the app itself which offer more agency and autonomy to the users, turning them also into potential creators of walks, be reflected and magnified by the precise process of creating a walk? The mere existence of an interactive and flexible app might mean nothing to a community unaware of its existence, lacking experience in the creation and employment of mobile apps, and unfamiliar with the creative, social, and cultural potentials of media walks in general, and those which SLR2 facilitate in particular. Digital apps of this sort do not follow the logic of 'if you build it, they will come', and so Medea thought hard about how this app could be made to be a catalyst for greater community engagement and ownership of the sound walks to be created.

The Koumaria residency takes place at an organic olive oil farm on the outskirts of the village of Sellasia. This village has a year-round population of under 300, which swells in the summer when many former residents of the village and their families return, many from Chicago. After a number of years creating site-specific and performance art at assorted locations in the village, we started to rethink our relationship with the village and its inhabitants. Our residency was self-consciously about community formation configured around collective, highly improvised new media art practices. Yet the community we focused on forming was composed of Medea members and our invited residents. Should we not extend this to include our relationship with the village, and so the villagers themselves? We began to conduct interviews with villagers and create an annual performance just for them which takes place at one of the cafes in the village (see Figure 7.4). We formed a close relationship with the unofficial village historian and asked him each year to give a tour of the village and surrounding areas. We sought village teens who might be interested in spending time at the residency and accompanying us on our trips to assorted locations near the village where we would develop and create site-specific works.[26]

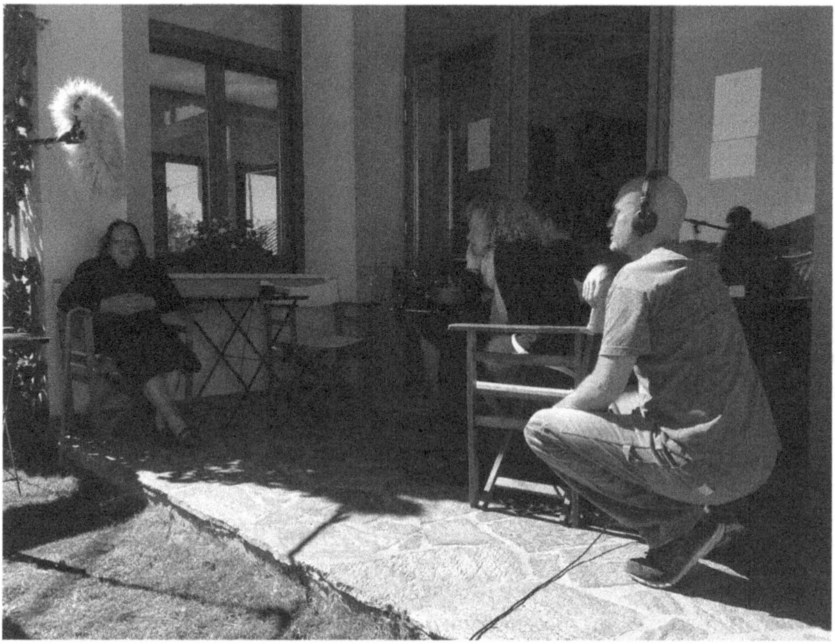

FIGURE 7.4 Interviewing residents of Sellasia for SLR3, ©Medea Electronique

We came to realise that the villagers were quite interested in what we were doing, and looked forward to our annual residency. We would be beckoned into the cafes, questioned about our equipment and its uses. Residents would be asked where they were from, and what they thought of the village. Spaces we had used in previous years and for which we had to seek permission to use, or find the person with the key, would be open the next year without our asking. We transitioned from going to the village to pick up supplies, viewing it as space merely for us to engage in minor commerce, to engaging with the village in a more organic way. We would linger in the cafes, and make sure we frequented each of them in turn. If there was an event in the village that was important to the villagers, we made sure we attended—we were expected to attend. We would place large orders for honey from the local bee-keeper or seek out high school-aged teens who might want to spend the day with us, observing, assisting, and socialising. More important still, they saw the possibility of leveraging our technical expertise in audio and video to document village narratives and histories. They would request copies of photos we took (some proudly then displayed on café walls), audio recordings of interviews we conducted, video of local homes and other village locations, and soon started to make specific requests of us to, say, conduct a series of interviews with a village elder concerning village history. Slowly we felt as if we were being allowed access to the collective memory of the village, which equally had us realise that this memory, like all memories, was highly perspectival. We also noted that the new doors in the village that were opening up to us corresponded with the increased integration of villagers into our activities, that there was, and continues to be, reciprocity in our relationships. While we viewed this as being granted access to the village's sense of identity and history, it was really more a process of our becoming part of the village's ecosystem—we became in a small but important way, part of the village itself, and this only happened when we also granted villagers access to *our* spaces, activities, and memories. This reciprocality resulted in the activities of our residency themselves entering into the village's memories, in effect a powerful form of epistemic validation of our partnership with the village.

It also became clear that many of the village stories we were collecting were highly site-specific—'here is where my grandmother's home once was', 'that house was the Nazi headquarters during the war', 'this is where a fresh source of water once flowed', 'here is the old village threshing floor'. The village had undergone many radical changes to its design over the past centuries, as a result of wars, bandits, need for fresh water, migration, and changing agricultural methods and patterns. A good portion of the village is in assorted states of ruin, with bits resembling an archaeological site, others with partially crumbling, yet still occupied homes, yet other parts of the village having recent renovations and modernisations. This added up to an ideal situation for employing the new app in a way not just respectful of the villagers but incorporating their voices and agency. The challenge became, how can we incorporate village narratives into the walks, encourage villagers to generate content for it and assist us in its design and overall function? And, in turn, how might the app and the content created for it perhaps assist them in any distinct community goals they might have? And how could this all be accomplished while still creating media walks with creative and aesthetic content reflective of our goals and identity as a new media art collective?

We decided on a flexible methodology, but one with certain rather fixed characteristics. First, we asked the villagers themselves what locations were important to them. Some were obvious, like the three small cafes which are the locus of village daily life. Others were far from obvious, including locations like the old village threshing floor, or the home of a village elder off of a labyrinth of back lanes and paths. We did not then simply 'collect data' from these locations but worked closely with the villagers and owners of the land or buildings identified to learn what was important *to them* about these locations. It may be the location of an important event in the history of the village, or the home of a villager held in high esteem, or something else altogether. We then invited the villagers themselves to tell us these stories at the very locations they picked out and documented these sessions with both audio and video. Only then, after coming to an understanding, however thin it might be, of the significance of a place to the village, would we craft a performative piece at and about the location and invite villagers to witness the performances.[27] These performances were normally multi-media, involving music, sound, movement, and often video projections, the complexity of each performance often a function of the accessibility of electricity at the chosen site.

We would then take all this collected 'data' and in effect parse it two ways. First, we would create documents, videos, audio recordings and photos that the villager's themselves requested, full recordings of an interview with a village elder say, or photos of daily life in the cafes. We would treat these materials both respectfully and professionally, creating photo albums and the like. As a result of this collective work the village came to desire to create a small museum of their history, using the materials we collected. We realised we had, and continue to have, a moral responsibility to work closely with the village on this, even if it means using collected materials in ways we did not originally envisage, taking much time and labour to create appropriate exhibits and collect 'data' that is unlikely to serve purely creative purposes.

We realised that not only was proceeding in this way the only ethical way to work, but that only once the village was aware of, and crucially took ownership of, the media we were collecting would they then endorse and appreciate the sound walks we created. They would know why certain files were pinned to the locations that they were, they would understand that our sound art and musical accompaniments were inspired by their own stories and memories, and they would consider themselves not objects of study but collaborators.[28] We soon found ourselves granted access to intimate spaces in the village, with the tacit understanding that we would treat such visits and the media collected respectfully, and often with the overt understanding that we would return this media in forms that the villagers themselves wanted it in. As a result of this now 12-year partnership with the village, Medea Electronique has undertaken the development of a highly complex and innovative archive of all the materials collected, spurred on by the moral compact we in effect have with the village to make these materials accessible in a professional and highly aesthetic way (see Figure 7.5).

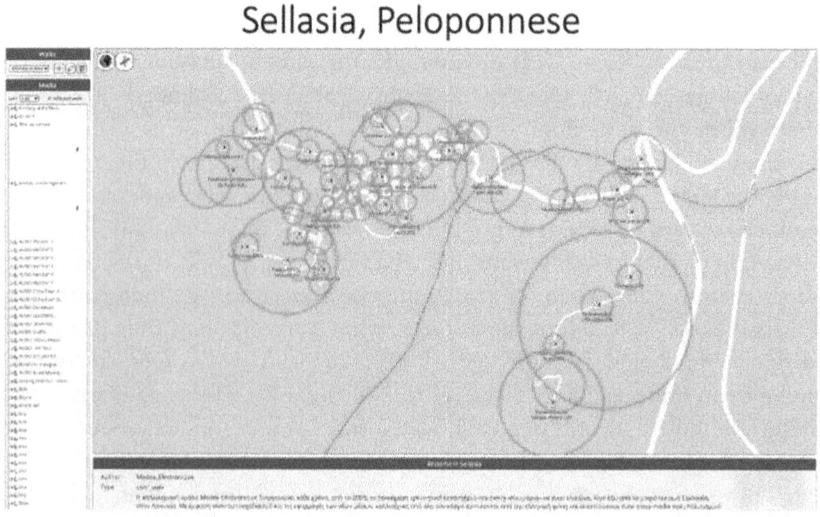

FIGURE 7.5 Screenshot of the Sellasia area in the app back-end interface

And so, the Koumaria residency, which has as a goal the creation of a micro-community amongst its participants, now looks to continue to form community with the village as well. The annual presentation of our work, much of it sound-centred and site-specific, draws and attracts a large percentage of the village. Politicians from the surrounding region attend, plans for civil works are discussed, we have been brought into the greater ecosystem of the village, and are called upon to help with the planning and implementation of cultural and heritage projects. Where once our sonic interventions in the village would produce either quizzical stares by the villagers, or be seemingly ignored, now such things are a source of pride, and often can be heard not simply as noise, but as a cultural product of the present vitality of the village and its past. Or perhaps they are still heard as 'noise' but noise in Attali's sense, noise which is a harbinger of new forms of civility and action, new forms of interpersonal relationship as this village now comes to a sense of its own identity as including annually an explosion of new media art, sound walks, sound-centric installations, and the return of artists who are now familiar and friends.

Suggestions for principled engagement

We want to end with some suggested principles to guide artist-community sound-based projects, which emerge out of our experience engaging in such work. While we recognise that contextual details make generalisations both difficult, and open to revision, we suspect many of these are transferable, perhaps with revision, to other places and contexts.

1. Engage with the publics/communities you will work with/in from the beginning. It is tempting to conduct community-based sound art projects with a fully formed project already in mind, and simply modify details in the light of interactions with the community. Proceeding in this way is unlikely to result in a project reflective of the community's needs and interests, is going to face hurdles in attracting community buy-in and does not take advantage of community knowledge and alternative epistemologies. In addition, it sets an artist/community dynamic that is likely to create annoyance on the part of the artists when their preconceived project faces questions and modifications from the community. It is far better to approach a community with a technology and/or method, but without its implementation or content already thought out.
2. Realise that your artistic goals, and the community's goals, while overlapping, may well be distinct. What this often requires is multiple modes of project implementation. Your community may be most interested in the documentation, archiving, and distribution of narratives created by community members. As an artist you may be primarily interested in using such materials as content in a sound-art piece. Create multiple implementations of your project—in this case an easily accessible archive of narratives, and the sound-art piece. Use your resources and expertise to do both, and do both well! You

will also garner more interest and buy-in concerning the art component if the community sees that you have taken their primary interests to heart, and they will be better positioned to see the value and meaning of the art component under such conditions.

3. Whenever possible, hire locals to work on your project. There is no better way of demonstrating your actual care and desire to work with a community than by offering members of the community paid employment. This is particularly true with university-based projects, where relationships between universities and surrounding communities are often fraught, and grounded in vast differences in power, access to resources and histories of oppression and inequity. Spend time and resources in training community members in the technical and artist aspects of sound-art, as well as possibly on basic issues related to rights and on how to use the project to bring attention to their community, if this is desired.
4. Leave the community you work with/in tangible marks of the project. This might be equipment, in situ installations, an exhibit. And do what you can to maintain these.
5. By empowering the community you work with to have the means and knowledge for creating their own sound-art projects, you should be prepared to take on a role of technical/creative advisor for their initiatives. Just as you once came to them, they may come to call on you, and you should be prepared to play a role in *their* projects in the future.
6. Encourage participants/users of your technologies/art that have them physically enter a given occupied space not to simply be new media art tourists. Request that they eat in the local restaurants, shop at local stores, support local businesses. Do this actively, supply information with your art-work, place installations, and needed technological devices in local businesses.
7. Distribute authorship/ownership of works created. Credit your community and/or named partners in it as equal authors and owners of any rights that might accrue.
8. If you have been involved in the development of the app, have a plan for how to maintain it and update it to meet technological advances. Making the code open-source is essential in order to mobilise people to assist in this process.

For the authors, as well as for the other members of the creative teams involved in the *Soundscapes-Landscapes* projects, each iteration was a step in a non-linear learning process. Technologies, their use and implications, conceptual framing and its impact on inclusion and representation, aesthetic choices regarding both content and format, and the politics of the relationship between artists, producers, communities, and audiences needed to be articulated each time anew. In this process, tacit and/or explicit presuppositions and motivations play a determining role. Affirming the primary significance of community engagement at the outset of a project guides the elaboration of all the above elements in directions that are rich in potential for *all* participants.

Notes

1 The artistic practice of 'soundwalking' emerged as such in the 1970s but has multiple antecedents that contributed to its emergence, for example the work of John Cage, fluxus artists such as George Brecht, and most significantly R. Murray Schafer's *World Soundscape Project.* At its most essential, it is, in the words of soundwalk pioneer Hildegard Westerkamp, 'any excursion whose main purpose is listening to the environment. It is exposing our ears to every sound around us no matter where we are' (2001, para. 1). Useful overviews of the history and practice of soundwalking can be found in Drever (2009) and McCartney (2014).

2 See Dourgouti Island Hotel | urbandig-project [WWW Document], n.d. URL https://www.urbandigproject.org/dourgouti-island-hotel-project (accessed 1.10.21).

3 Stegi—Onassis Foundation [WWW Document], n.d. URL https://www.onassis.org/onassis-stegi (accessed 1.10.21). Christos Carras is the Executive Director of the OS.

4 Portfolio [WWW Document], n.d. Medea Electronique. URL http://medeaelectronique.com/ (accessed 1.10.21).

5 Soundscapes/Landscapes—Medea Electronique—Onassis Foundation [WWW Document], n.d. URL https://www.onassis.org/whats-on/soundscapes-landscapes (accessed 1.2.21).

6 See the relevant section on 'Walking' in Paquette and McCartney (2012) and in Carras (2019)

7 SoundscapesLandscapes Player—Apps on Google Play [WWW Document], n.d. URL https://play.google.com/store/apps/details?id=com.calvium.soundland&hl=en&gl=GR (accessed 1.10.21).

8 For documentation on *Rhizomes I* see: http://medeaelectronique.com/portfolio/soundscapes-landscapes-rhizome-1/ For a documentary on the complete Rhizomes project see: https://vimeo.com/295663942

9 The concept of the 'rhizome' was important during the formulation of the project. One could argue however that it is atemporal, a structural feature which allows any two points to be connected. This potential is, however, materialised through movement and in time. In a sense, a promising analogy might be open scores, such as those by John Cage where sequence is left open to the performer.

10 On performativity in relation to media walks, see Wagner (2020).

11 But not identical: de Certeau proposes that 'the act of walking is to the urban system what the speech act is to language or to the statements uttered' (1984, p. 97). This relationship is not the same as that which Deleuze and Guattari (1980) posit between linguistic and rhizomatic conceptions of enunciations, but it shares a conviction that myriad forces and relations operate in ways that subvert the authority of grand schemata.

12 The bibliography on walking is extensive: historical surveys such as Joseph Amato's *On Foot* (2004), the compendium on *Walking and the Aesthetics of Modernity* (Benesch & Specq, 2016) or Rebecca Solnit's *Wanderlust* (2001); literary works such as W.G. Sebald's *Rings of Saturn* (2007) or more popular texts such as Will Self's *Psychogeography* (\2007) for example.

13 This article takes SLR1 as one of its case studies.

14 Something that is not rare in sensory ethnography. See for example, Feld (2005).

15 There are asymmetries between how this process is conceived of by the content creators, the users, and members of the communities the walks suggest engagement with. As we will discuss below, further iterations of the Rhizome project realized the importance of allowing greater control over modes of representation by members of the community whose spaces were being represented, which itself can be seen as allowing for more bi-directional, and so more equalitarian, transfer of knowledge.

16 Critiques of modes of engagement by colonial and majority cultures of colonized and minority peoples form a central tenant of post-colonial studies and assorted anti-oppression movements and intellectual traditions including critical race and feminist theories. Perhaps the most pointed critiques emerge out of whiteness studies, and the models developed of the stages of anti-racist thinking that white-folk pass through,

where empathy without full commitment to anti-racist thought and action, is viewed as a far too common form of racism disguised as anti-racism.

17 The tools we use to create a media work are not neutral; they are based on the same kind of interfaces and functionalities that are familiar to us from a whole range of applications that enable the pervasiveness of social media and other systems that mediate our interaction with the contemporary world. That this pervasiveness has become a kind of 'second nature' akin to what Antonio Gramsci (1992) called 'common sense' makes it even more insidious and dangerous. For an excellent analysis of the predictive, extractive, and behaviour-modelling functions of these interfaces, see Zuboff (2019).

18 https://www.onassis.org/onassis-stegi/onassis-stegi/medea-electronique

19 https://www.onassis.org/whats-on/medea-electronique-extended-reality-tour

20 This would be another essay, but it would no doubt be worth thinking about media walks in relation to Henri Lefebvre's conceptual triad of spatial practice, representations of space, and representational spaces (1991, p. 33).

21 For further documentation on *Rhizomes II* see: http://medeaelectronique.com/portfolio/soundscapes-landscapes-rhizome2/

22 On the significance of canons and their institutionalisation in the field of soundwalks, see Flügge (2020).

23 While we draw upon Attali here, since his language and context is very close to that employed by the designers of the *Rhizomes* project, it is important to recognize that theorists and practitioners of Black diasporic sonic arts have long made very similar connections between composition, improvisation and political, economic, cultural, and social networks that both serve to perpetuate and reinscribe white supremacy, and the potentials for improvisation to serve both as a model and site for fighting against such systems of oppression.

24 Note, SLR2 allowed only for embedded sound files, and so was more of a digital sound walk than a media walk.

25 http://mycity-mysounds.zkm.de/

26 For documentation on the Koumaria Residency see: http://medeaelectronique.com/koumaria/

27 An archive of projects realized at Sellasia and its surroundings can be found here: http://medeaelectronique.com/koumaria/works/

28 Indications of this increased degree of involvement and sense of ownership by the villagers in our projects is difficult to measure precisely, but is surely reflected in facts such as the increased attendance by villagers in our annual presentation of our residents projects, the clear fact that the village would plan certain of their activities to take place during our residency, the increased requests we would receive from villagers to use their home and land for a new media project, or to interview and document a family member, or even a comment we might receive about a name we used picking out a local landmark in one of our posters, asking for a modification.

References

Amato, J. A. (2004). *On foot: A history of walking*. New York University Press.

Attali, J. (1985). *Noise: The political economy of music* (B. Massumi, Trans.). University of Minnesota Press.

Benesch, K., & Specq, F. (Eds.) (2016). *Walking and the aesthetics of modernity: Pedestrian mobility in literature and the arts*. Palgrave Macmillan.

Benjamin, W. (1997). *Charles Baudelaire: A lyric poet in the era of high capitalism* (H. Zohn, Trans.). Verso.

Benjamin, W. (2002). *Paris, capitale du XIXe siècle: Le livre des passages* (J. Lacoste, Ed.). Passages. 3e éd. Les Éd. du Cerf.

Borges, J. L. (1998). *Collected fictions* (A. Hurley, Trans.). Penguin Books.

Bubaris, N. (2017). Media and walking: A multimodal approach. *Punctum. International Journal of Semiotics*, *3*(1), 27–44. 10.18680/hss.2017.0004

Butler, T. (2009). "Memoryscape": Integrating oral history, memory and landscape on the River Thames. In P. Ashton & H. Kean (Eds.), *People and their pasts* (pp. 223–239). Palgrave Macmillan. http://link.springer.com/10.1057/9780230234468_13

Cardiff, J. (1999). *The missing voice (Case Study B)* [Artangel]. https://www.artangel.org.uk/project/the-missing-voice-case-study-b/

Carras, C. (2019). Soundwalks: An experiential path to new sonic art. *Organised Sound*, *24*(3), 261–273. doi:10.1017/S1355771819000335

de Certeau, M. (1984). *The practice of everyday life* (S. Rendell, Trans.). University of California Press.

Deleuze, G., & Guattari, F. (1980). *Mille plateaux*. Collection 'Critique' t. 2. Éditions de minuit.

Drever, J. L. (2009). Soundwalking: Aural excursions into the everyday. In J. Saunders (Ed.). *The Ashgate research companion to experimental music* (pp. 181–210). Routledge. 10.4324/9781315172347

Feld, S. (2005). Places sensed, senses placed: Towards a sensuous epistemology of space. In D. Howes (Ed.), *Empire of the senses: The sensual culture reader* (pp. 179–191). Routledge.

Flügge, E. (2020). Sounding in paths, hearing through cracks: Sonic arts practices and urban institutions. In S. K. Groth & H. Schulze (Eds.), *The Bloomsbury handbook of sound art* (pp. 295–316). Bloomsbury Academic.

Gramsci, A. (1992). *Prison notebooks* (J. A. Buttigieg, Trans.). Columbia University. Press.

Gros, F. (2015). *A philosophy of walking* (J. Howe, Trans.). Verso.

Lefebvre, H. (1991). *The production of space* (D. Nicholson-Smith, Trans.). Blackwell.

McCartney, A. (2014). Soundwalking: Creating moving environmental sound narratives. In S. Gopinath & J. Stanyek (Eds.), *The Oxford handbook of mobile music studies* (Vol. 2) (pp. 212–238). 10.1093/oxfordhb/9780199913657.013.008

Paquette, D., & McCartney, A. (2012). Soundwalking and the bodily exploration of places. *Canadian Journal of Communication*, *37*(1), 135–145. 10.22230/cjc.2012v37n1a2543

Pink, S. (2009). *Doing sensory ethnography*. Sage Publications Inc. 10.4135/9781446249383

Sebald, W. G. (2007). *The rings of Saturn* (M. Hulse, Trans.). Vintage Books.

Self, W., & Steadman, R. (2007). *Psychogeography: Disentangling the modern conundrum of psyche and place*. Bloomsbury.

Solnit, R. (2001). *Wanderlust: A history of walking*. Penguin Books.

Wagner, K. (2020). The city as stage: Performativity and urban space. In C. Seiffarth (Ed.), *Bonn hoeren urban sound art/Stadtklangkunst 2010–2019* (pp. 23–27). Beethovenstiftung für Kunst und Kultur.

Westerkamp, H. (2001). Soundwalking. [Writings by Hildegard Westerkamp]. https://www.hildegardwesterkamp.ca/writings/writingsby/ (Accessed 10 December 2021).

Zuboff, S. (2019). *The age of surveillance capitalism: The fight for a human future at the new frontier of power*. Public Affairs, Hachette Book Group.

8

HYPER-LISTENING AND CO-LISTENING: REFLECTIONS ON SOUND, SELFHOOD, AND SOLIDARITY

Budhaditya Chattopadhyay

Prelude: Listening to our time

During a visit to Egypt a couple of years ago, I came across a scripture in the Cairo National Museum that read 'maa-kheru'—an expression in Egyptian, translated into English: 'true voice'—as provided by the Museum. The term first appears in the court case between Horus (in some texts Osiris) and Seth, which ended in the verdict that Horus was the legitimate heir to the throne. The judges decided that his statement (voice) was 'true', present and legitimate. Some scholars choose to translate 'maa-kheru' as 'blessed' (Anthes, 1954; Grenfell, 1906)—drawing on the idea that one who speaks in a self-aware voice and listens with an attitude of mindful reciprocity, survives the test of time. The scripture struck me as a prompt for contemplating truth and essence in sound perception and cognition through social and cultural engagements. The phrase *true voice* resonated with a statement from *Manusmriti*, a religious text from Vedic India: 'Utterance (vāk), brought forth all the universe' (Daniélou, 1995). Ancient expressions such as these help us understand the complex relationship between voicing, sound, spatiotemporal contexts, listening and the world. This led me to examine the emancipatory role of listening in contemporary life, which is marked by intensified social, religious, and political issues and a global environmental crisis—centred on the issue of disconnection. This includes estrangements from nature, from the self, from the community, and others. How can we practice listening to build a bridge between these fragmented parts of the universe we share together as humans and non-humans through an acoustic solidarity? As a sound practitioner, who endeavours to speak earnestly and listen with an honest attention while engaging with the many shades and splendours of the time and its felt realities, the idea of a true voice resonates with me. In my view, the *artist* speaks in this way, voicing personal realisations consisting of, among other things, the reflections, illuminations,

DOI: 10.4324/9781003164227-15

streams of thoughts, perceptions, and poetic contemplations upon listening in-depth. Indeed, I would suggest that to be true to oneself, and becoming truly present, is the very foundation on which art stands. This premise encourages reading personal experiences and their comprehensive interpretations through art's lens. As a practitioner, I see that art affords a sensitive listening ear to engage with human conditions. This truthful perspective makes artists indispensable to society for sustaining its sanity and cognisance, but at the same time their probing works and anti-establishment stance are often deemed problematic for the mechanisms of a nation state, which relies on governance based on control by surveillance and policing. There are many instances of artists incarcerated or exiled for voicing dissent. Following Plato's suggestion, artists' true presence antagonises the Republic and the city (Denham, 2012; Murdoch, 1977) forcing it to often take an oppressive and violent position by a closure of their sensate and empathetic listening ears through surveillance and control.

In this visit to Egypt in 2019, I faced this antagonism while strolling around Tahrir Square in search of site-specific ambient sounds. As I passed by a series of closed down buildings, I came across a desolate wall and an adjacent closed door with a large drone sound emanating from inside. My curious ears were trying to locate the source of the drone. My practitioner's hands were busy taking the sound recorder out of my pocket and putting the microphone on. My embodied presence at the entrance of the closed door was so engulfed by the blinding sound that I didn't realise a couple of military personnel were approaching me. As I pushed the recording button to keep a trace of the drone sound, I heard someone called me from behind. When I looked back, I saw two military personnel in full uniform and with rifles in hand were scrutinising me. One man's hand was on the rifle pointed towards me. He asked me what I was doing there. I replied that I was recording sound. He asked again, keeping the rifle pointed towards me, what was the purpose of recording sound. I understood that seeing a South Asian person in this disputed locality with a machine in hand disturbed the ordered sense of normality of this military officer. He seemed not being able to connect the dots. Of course, I got the point: I was experiencing the intensified surveillance mechanism of post-revolution in Egypt and any imagined or potential anomaly in this system would need to be dealt with. Comprehending my position, I started talking about how sound art is defined, what is the role of recorded sound, and how urban noise is a great source of artistic inquiry into everyday practice. My voice, resonating with its typical South Asian accent coupled with my non-white skin, however, put these two officers in a mode of rapid racial profiling. My deliberately vociferous approach to make a human connection with them was immediately disparaged.

In *The Parasite*, philosopher Michel Serres (2013) notes how noise can pose a hiding mechanism in adverse social situations for both non-humans and humans: 'He (*the parasite*) becomes invisible by making, on the contrary, a lot of noise. One can hide by being too visible or too perceptible. The parasite hides behind the noise' (p. 218). The social noise I impersonated to both withhold and justify my presence at the post-revolution Tahrir Square was considered an aberration in the ears of the soldiers and appeared to be reduced ultimately to an antagonistic and marginal voice in regard to the security and sovereignty of contemporary Egypt. The officers first took away my recorder thinking it to be a technological device to transmit data internationally. And then they grabbed hold of me—a perpetual traveller and a sonic drifter—for interrogation. It is another story how I came back from their clutch and how some of these recordings survived. By narrating this personal experience, I hope to illustrate the general antagonism with which artists are treated in an authoritarian regime obsessed with control and surveillance. They are afraid of the power of context-aware listening in exposing this mode of protectionism. I wonder how artists might deal with the contemporary militarisation of public spaces and regulation of public sounds by employing listening as an emancipatory practice.

The role of the artists as interlocutors for social engagement can be referred back to the situationists, who were active in Europe in the 1950s and 1960s and whose ideas have attracted the deep interest of urban theorists and art historians. They were social revolutionaries made up of avant-garde artists, intellectuals, and political thinkers, who employed the concept of 'psychogeography' to describe the self-aware, subversive, and experimental practice of subjectively and mindfully exploring urban terrain by means of playfully walking or 'drifting' (English translation of the widely used French word *dérive*) across the city (Coverley, 2010). This practice was designed to unsettle and disrupt the systems of consumerist homogeneity as well as the state control, surveillance, and policing of everyday practice in the city. Some differences notwithstanding, the *dérive* or drifting is generally considered to be a chosen method for the psychogeographic exploration of a place, a site, or a city in a manner 'in which the contemporary world warps the relationship between psyche and place' (Self, 2007). In this same practice of *dérive*, I met the military personnel in Cairo like a ghostly manifestation of its current authoritarian regime morphed into a performative political confrontation. This experience, however, was nothing new to me. I have had similar encounters on the immigration borders in the West, for example in Heathrow Airport, Schiphol, or in Copenhagen Airport on a number of occasions. The nomadic drift I embody as a listener-artist, appearing free from a specific racial profile, doesn't fit into the security protocol of the European fortress.

By drifting in such a self- as well as context-aware navigation through a city and listening to the various sites with an inclusive and contemplative perspective, an artist indulges in a subjective or mental journey through the sound environment in 'an attempt to transform the urban experience for aesthetic purposes' (Coverley, 2010, p. 10)—as a psychogeographic exploration. A listener can excavate an acoustic geography by 'nomadic listening',[1] drifting through the realm of locative sound, altering the perspectives of a place into something private, meditative, or imaginative, which Guy Debord suggests, is a playful-constructive behaviour (1958) and a constructed 'performance' (Sadler, 1999, p. 105). The search for playfulness, freedom, democracy, and equity is always fertile in the mind of the artists who are, as Raymond Williams (1963) suggests in his seminal work *Culture and Society*, arguably the true voices of society. As Arnold Berleant (1978) argues in *Artists and Morality: Toward an Ethics of Art*, this revolutionary position of artists in the social formation grows out of a moral responsibility. In my case, it is an inner urge for social engagement and community-building against state violence that constantly informs my artistic practice and creative thought processes. I believe that the practice of non-judgemental, democratic, and emancipatory listening gives birth to an imagined acoustic common and a sense of solidarity based on collaborative listening, in which a social situation encourages an equal sound, shared resonance and listening to all the voices present. My artistic practice with listening suggests that this capacity for equity, inclusion, and interconnectivity emerges from a personal action of self-attunement that works against the over-arching judgemental modes of listening prevalent in today's times.

Against judgemental listening: Going inward to build capacity for engaging outward

At the time of writing, many societies the world over are being impacted by intensified conflicts born out of rising nationalism, and racial tension. Listening—as our curatorial note for *The Listening Biennial* (Berlin, London, and other venues across the globe, 2021) suggests[2]—may enact an intervention in this conundrum to offer a reconciliatory and peace-building platform. At the core of this statement is the belief that listening practice, when employed as a mode of engagement with the self, and the sensate subjective formation that this entails, enables us to engage with the community in an inclusive manner. In the implementation of conflict resolution, acceptance and reciprocity are essential, and above all, the ability to listen, as I attempt to show here. These values are, I argue, useful for us to navigate the challenging times ahead, which will certainly be marked by scarcity of natural resources and unprecedented human-made environmental decay (Morton, 2013), leading to climate calamities. I discuss here how two complementary practices of listening that I respectively term *hyper-listening* and *co-listening,* can be useful when performed together to counteract these crises.

In this essay, I expose the methodology and research process of a listening-driven workshop series, as well as some of my artistic projects, that employ an inclusive and contemplative listening practice as an approach to community building, co-resonating, and conflict resolution. Philosopher Gemma Corradi Fiumara urges 'a propensity to listen' (1990, p. 73). This intentionality in the listening act may help in paying heed to others, without making immediate judgements and may potentially lead to bridge the troubled water of difference. Similarly, the UNICEF programming guide on peace-building argues that the root of all conflicts is essentially embedded in a lack of ability to listen to the other.[3] In this context, *hyper-listening* and *co-listening* operate as a set of individual and collective exercises and experiments in sonic medi(t)ation that intend to encourage listening to self and others without making immediate judgement. The idea revolves around mindful listening as a conscious and sentient act to compassionately engage with the self and the surrounding environment, and the bodies that inhabit this environment, namely humans, plants, animals, and other sentient beings. This departs from the western modernist idea of listening as a humanistic practice for navigating everyday environments—a form of sensory orientation from a cartographic and utilitarian perspective focused on immediate meaning-making in order to arrive at specific conclusions. I discuss in this essay how hyper-listening and co-listening involve the mindful engagement with the body of a situated self, as well as the people and non-human others inhabiting the site, by generating a sensation of sonic inter-connectedness, social collectivity, and co-dependence that may lead to a sense of solidarity and wellbeing in a volatile time. The interest in mindfulness and sonic meditation foreground an affective mediation of the self and others as parts of a common sonority informed and shaped by an attentive and context-aware listening.

In my recent book *The Auditory Setting* (Chattopadhyay, 2021a), I hint at how a process of sensitive self-attunement to the site and the surrounding environment may help imagine and enact new kinds of acoustic cohabitation and communication between human and non-human life-forms present in a setting. In the research conducted in the book, I show how the compulsion to achieve anthropogenically liveable and urbanised settings often leads to technological rendering and phenomenological reduction of the site-specific sounds. Such rendering excludes the non-human entities cohabiting in such settings under the pressure of the Anthropocene. This leads to the psycho-acoustic alienation of sensitive listeners from an embedded listening state. Drawing from this research, I hope to show in this essay adaptive perception, in which a listener may attune to situated noises without judgement, grounding the listener in their place of listening through a natural bridging between the self, the other, and the environment. Going inwards through listening to self builds the capacity for engaging outwardly. This idea of selfhood draws from rasa theory—a contemplative abstraction of aesthetic experience, in which the inwardness of human feelings suffuses the surrounding world of embodied forms that can be suggested but not described (Barlingay, 2007). This is a conceptual premise of my artistic practice with listening, and the listening-driven workshop series that I have been conducting over the last six years across various settings.

Self-attunement: Three artistic positions and a workshop proposal

In my work, hyper-listening and co-listening are central methodologies to engage with the self and the inter-connected others in my environment, as illuminated through examples of my own work.

My workshop series *Hyper-listening: Praxis*[4] has been conducted in various arts, humanities, and cultural institutions across Europe, and Asia.[5] Each of these workshops develops into a series of site-specific sonic meditation experiments that employ the concepts and methodologies of mindful listening described above.

I begin the workshop series with a set of exercises and experiments involving sonic meditations that encourage an engaged and 'expanded' mode of listening (LaBelle, 2006, p. 4). In these experiments, I invite people to listen to the seemingly mundane auditory situations surrounding the self at a specific personal corner of a city (e.g. around the gallery or art organisation's urban location) and follow the unfolding of the situation. The listener closes their eyes and engages with auditory associations that are memory-driven and might encourage personal recollection. The listener is asked to (con)textualise these associations by chronicling them in scribbles and spontaneous notes. They then present these generated texts and other materials as the outcomes of the workshop as performative readings in a collective setting with other co-listeners. These notes register a myriad of thought-streams triggered within the psychogeographic evocation of the situated listening experience. The texts and writings hint at a trajectory of listening-driven thoughts and memory associations that the listener engages with. Often these notes appear poetic and contemplative. Given the emotive responses of the participants, this exercise appears to be empowering the listener to broaden their sonic worlds expanding the immediate moment towards the context-aware subjective associations, reveries, desires, and deeply embedded assumption. For example, one participant wrote an emotionally moving account about the passing of his father. Another recounted a memory of a loved one long lost. Thus, the listener is able to relate better to their personal narrative, memory, trauma, and longings through a therapeutic process of self-listening and consequently, listening to their environment more self-consciously (through the self). I would suggest that this personal exposure allows them to relate better not only to themselves but also, as a consequence of this self-attunement and sonic sensitisation to their surroundings and others as an inclusive community of fellow listeners.

As a method, the participants are asked to walk alone or 'drift' through the city following listening to the urban sounds and locate certain sites that trigger a multitude of associative thoughts, imaginations, and/or personal memories. They then focus on this site, being attentive to its environment. They are guided by me, the workshop facilitator, to reconnect with these auditory associations embedded in everyday navigation, aiming to produce self-awareness and subjective formations marked by an elevated and emancipated selfhood. In essence, I conduct this series of workshops as an exploratory exercise to generate and nurture an awareness of self and others as interconnected beings of an acoustic community that is surviving by a sense of solidarity. Hyper- and co-listening help recognise the potential of listening to others in our environment not as bodies of conflict, but extension of the self as a larger common. While hyper-listening explores the self for awareness and empowerment manifested in a series of automatic writings among other self-expressions, co-listening engages with others in the shared environment, through performing the writings and other generative materials. There is a sense of ceremony or sacredness in the sharing, and an underscoring of careful and attentive involvement with the others performing their personal texts.

In my listening-driven artistic practice, I am curious to discern how a sensitive itinerant listener might become absent-minded or fall into a reverie when listening to certain sounds in their everyday mobility, navigation, and engagement with various sites. These sounds might be mundane—sounds to which we do not normally attend such as a distant car horn out the office window, a flushing toilet or the early morning drone of the street cleaning car. These sonic phenomena are nothing special; they appear and evaporate in our immediate environment during a working day without leaving any trace. However, some of these sounds may induce a listener to elevate onto some other perceptual planes perhaps not directly related to the source or place of occurrence of the sonic phenomena. These sounds open doors to an alternate sonic world beyond their immediate meaning or materiality. I try to understand how they might unsettle a sensitive and context-aware listener in such a way that an elevated and poetic state of contemplation is experienced.

The subtle formation of the auditory situation is the point of curiosity in my current artistic practice, particularly in my listening-driven exploratory and creative writings. For example, *The Nomadic Listener* (Chattopadhyay, 2020) is an augmented artist book project based on the notion of hyper-listening, considered here as a case study for discussion. Based on my artistic research on migration, contemporary urban experience, and sonic alienation, the book is composed of a series of texts (60 in total) stemming from psychogeographic explorations (namely walking, listening, and site-specific writing) of a number of contemporary global cities developed between 2012 and 2019 spanning three continents. Each of these 60 pieces is an act of attentive listening, leading to the recording of the specific urban location in sound and text, attempting to attune to the sonic fluctuations of movement and the passing of events. Spread over 60 pages of the book, these pieces conjure meditations on the minutiae of life, interwoven with my own memories, associations, desires, and reflections on being in the city. Sensate readers and listeners are brought inside a tender map of contemporary urban experiences and the autonomous assemblage of the often-lonely, surprising, and random interactions found in the act of listening through a city without any fixed direction but following its sounds, like a sonic drifter. In this mode of sonic navigation, hyper-listening is a useful concept and methodological entry point:

> Water splashes on the jetty since medieval times; the sound of this primordial contact between nature and culture cannot make me forget that some people didn't keep their words to me. The unkempt city is slowly drowning—the water level is rising with each passing moment. The words that weren't kept, the faces that turned away, the bird that never returned to the house garden—all these worldly imperfections are weighing heavily on the decks of the motorboats. The misgivings of the city-dwellers have piled up to make the dustbins at the street corners spilling over. Water is flowing through the darker corners of the city bringing up skeletons of failed marriages, unborn babies, love letters thrown away, unmade beds and pillows—all floating in the knee

> deep water, reborn for further speculation. People who didn't keep their word to the city, now rush to jump into motorboats to escape the city's rebirth as a heartbroken wasteland.
>
> *(Chattopadhyay, 2020, p. 56)*

Written in Venice while listening to the overflowing lagoon from a jetty, this urban sonic meditation imagines the future of this Italian city in the aftermath of global warming and climate change. As widely speculated, Venice, like many other cities close to the sea, will be submerged in the rising level of the sea water in a few decades. The context-aware listening act triggers such imaginaries, scribbled in a poetic form, where the real meets the surreal, and gives the passages of thought-stream a narrative flow. In a mode of auto-ethnographic narrative, the field recording at the Venetian jetty traces the setting of the sensory interaction with the past, present, and future of the city with many unfolding stories to be engulfed by the rising water level in darker times to come. The auditory situation of this moment unfolds through the context-aware process of hyper-listening.

The above-mentioned project, *The Nomadic Listener* (2020), focuses on contextualising the stream of thoughts or the personal contemplation that sounds trigger, particularly for a migratory listener. More often than not, these thoughts, when registered or written down as scribbles, appear as lyrical, touching upon certain reflective, abstract, and introspective states of the mind. Expanding the ideas of the Situationists (Coverley, 2010; Sadler, 1999; Self, 2007) into sound studies, we can examine how acoustic geography engages mental construction of space through hyper-listening. As I have argued elsewhere (Chattopadhyay, 2013, 2017, 2021a), such a psychogeographic approach may have led my practice in the project to self-attune my wandering ears for surviving in contemporary European urban environments marred by anti-migration policies, racial differences, pandemic, climate catastrophe, and rising nationalism—all marked by state policing and surveillance. My artistic-subjective position as a sensitive migratory entity, seeking urban safety, freedom of movement, autonomy, and wellbeing, is informed and supported by this sonic intervention in a city through recognising a personal narrative.

My second artistic project *Aural Drift* (2021)[6] (Figure 8.1) is also invested in the emancipatory practice of hyper-listening and co-listening in contemporary urban environments. In European cities, gentrification is a matter of concern, as the process of noise abatement wipes out diverse voices and sounds from the urban atmospheres. Therefore, such settings may alienate sensitive listeners. The acoustic architecture of a city is a flux of sonic textures, signals, functional and navigational sounds, language, and sonified human-machine assemblages, subject to constant change over time. In the form of an auto-ethnographic field research, I examined the transformative auditory situations of Offenbach. The poetic archaeology gave a hearing to unheard of nooks and crannies in public spaces, such as a public park, around a bridge, a busy railway station, or inside a basement car shade. This

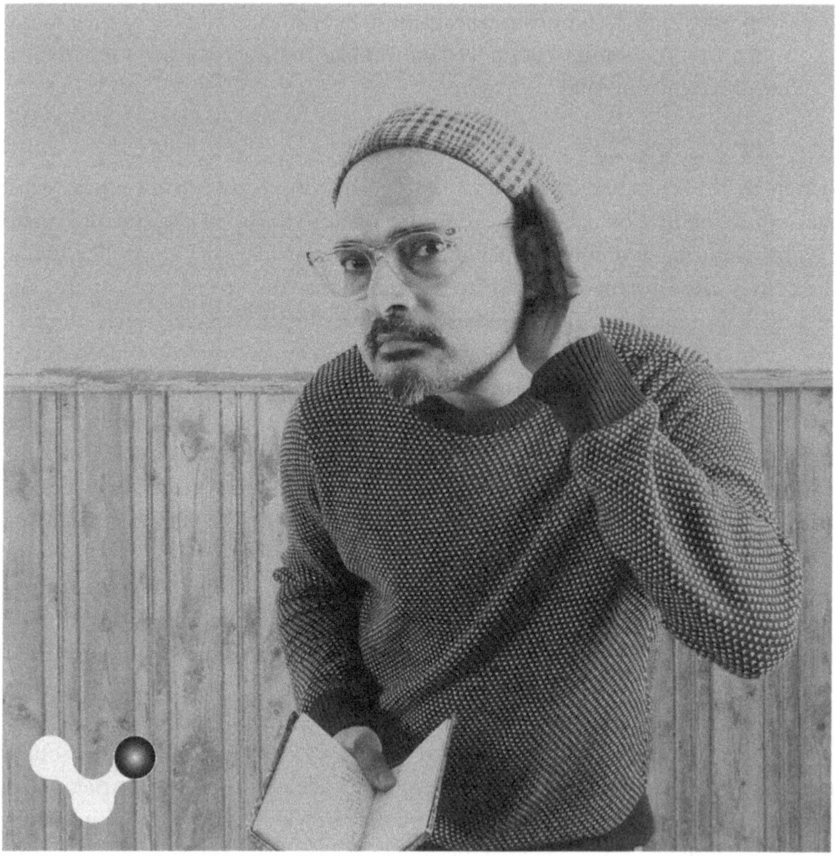

FIGURE 8.1 Poster from *Aural Drift*, by Budhaditya Chattopadhyay (May–June 2021). Photo: © Künstlerhaus Mousonturm

provided a way of engaging with the city as a nomadic entity; free, at least momentarily, from any state-imposed identity, keeping the listening movement itinerant and outside any surveillance radar.

Aural Drift manifested as a mobile exhibition of listening traces, retrievable and audible through QR codes placed at specific locations in the city (Figure 8.2). The work was meant for augmented modes of hyper- and co-listening and reading as reflections on methods of urban engagement. The public installation opened across the city of Offenbach for the passers-by who could access the work on their mobile devices and personal headphones while traversing the exhibition sites. In its psychogeographic approach of free drifting, poetic contemplation and de-territorialised listening, the project questioned the model of cartography that quantifies land into (visual/digital) objects. Instead, it suggested an alternative, sound-based navigation, in which intersubjective engagement with place by a poetic mode of self-attunement is central.[7]

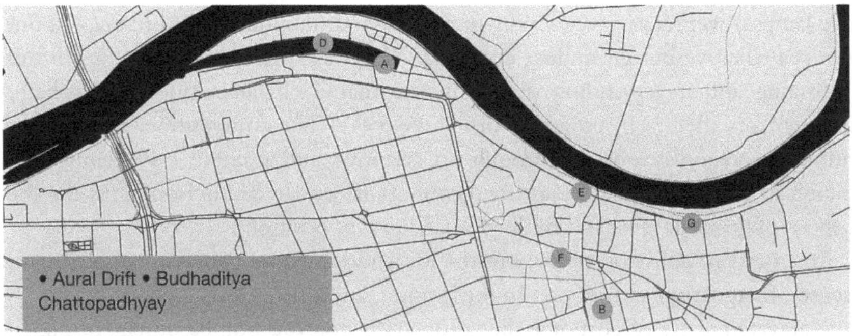

FIGURE 8.2 Interactive map of *Aural Drift* exhibition in public spaces by Budhaditya Chattopadhyay (May–June 2021). Photo: © Künstlerhaus Mousonturm

A number of my artworks aim to shed light on the perception of a rapidly developing and expanding city by engaging with the multilayered sounds of a temporally coalescing past and present. Many of today's cities carry remnants of a rural past in the contemporary hyper-modernity—take for example Bangalore. In my articles (2013, 2014), I have shown how a high-tech city like Bangalore's contemporary urban environment is an archive of rural and pre-modern sounds such as roadside cane juice made in frugally made machines, rural hawkers selling on the streets homemade sweets with archaic voices, or offering their services to sharpen kitchen utensils using indigenous technologies. This merging of time reflects in the cumulative urban ambience, in which sounds from the past, such as the pre-modern sounds coalesce with sounds from current urbanisation and digital convergence. I aim to devise an unfolding situation of the vibrant but clamorous cities from the Global South, such as Mumbai, Colombo, Kolkata, Casablanca, or Cairo, in corresponding acts of psychogeographic drifting, deterritorialized listening and field recording. A number of my artworks examine the processes of sensing an apparently chaotic and disorganised city with its multisensory complexity to develop an intuitive and personalised urban design.

My third artistic project that incorporates the ideas of hyper- and co-listening manifests is *Indecent Whispers* (2021)—a sound art composition derived from attentive listening to the intimate surroundings and various everyday objects in conversation with the self during a self-quarantine. The period between 2020 and 2021 saw sonic environments around the globe change beyond measure as normal life came to a halt, streets emptied, and the world slowed down, sensitising our ears to the subtlety of life on earth and our own existence as a species. For many, this situation also opened the auditory setting for new sounds to occupy the silence and define the sonic environment for an extended isolation. For me, staying indoors resulted in greater appreciation of or reconnection with the immediately surrounding sonic environments, coupled with a deeper awareness of the human effects on the environment as climate crises intensify. The sounds of everyday objects in the private surroundings, such as the refrigerators, toilet flushes, fans, air-coolers, and the table lamps, reverberate more loudly to our ears as if sounding their presence in our lives as a talkative but grounding entourage. They are the true companions in this continuing and never-ending silence in the many self-quarantines, and solitary confinements after each sporadic journey we make. The composition is a dialogue with these everyday objects through an attentive and mindful co-listening and voicing in earnest, with a desire to reconnect to our lived environments for the purpose of grounding us and finding emancipatory positions.

An emergent consciousness about the local and the quotidian was coupled with a sense of stagnation and repose. In these times, a thriving social media exacerbates an aggravated sense of corporal alienation with nonstop bombardment of digital presences. As I have shown in my book *The Auditory Setting* (2021), often this vociferous onslaught of vibrating data seems confusing and misleading to one who is looking for a presence of mind in a moment of global disorientation. On the one hand, there is a serious physical and psychic disconnection and divide between people and different spheres of the society due to prolonged lockdowns and travel restrictions; on the other, there is a need felt on a global scale for more social exchanges and solidarity that are unrealisable at the moment. As a result of this conflict, individually we are growing an inner sanctum. This is a moment of solidarity with ourselves and our immediate environment, namely the home. The world got an unanticipated chance to slow down and listen more closely to itself. While this deeply solitary inner listening is practiced, one may ponder upon the idea of true presence. For me, solitude is always embedded with a desire to connect with myself more contemplatively and also to rediscover the objects around that constitute my immediate environment of the home. Sounds of everyday objects at home draw my attention to them in this disconcerting silence and never-ending stagnation. I give more attention to these everyday objects as if they are part of my intimate world. In solitary confinement I seek refuge in this sense of belonging as a way to build my cognitive universe.

In my perpetual isolation during multiple mandatory self-quarantines in 2020–2021, I nurtured a deep longing for inter-personal connection—particularly when I returned to India in September 2020 from Europe after an arduous trip and I was isolated for 15 days in my parent's apartment. I was forced to be alone, as my mother was not allowed to be around. In that traumatised housing complex, coming back from COVID-infected Europe was scary for the inhabitants, and the local regulations made me go under scrutiny and distancing. In this two-week's self-isolation, digital platforms hardly provided any solace. Time was still and I had to innovate ways to keep a sense of sanity. Two specific methods that I employed were to write my thoughts down in the form of poetic texts, and focus on the immediate sounds in my room to ground me. *Indecent Whispers* was the result of this contemplative engagement with my domestic environment through hyper-listening to and co-listening with everyday objects in the form of contemplative listening traces as scribbles while conversing with the sounds of incarceration. The work aimed to explore the notion of human-machine assemblages in sound practice, and produce new knowledge in post-pandemic, post-humanist social formations, contributing to

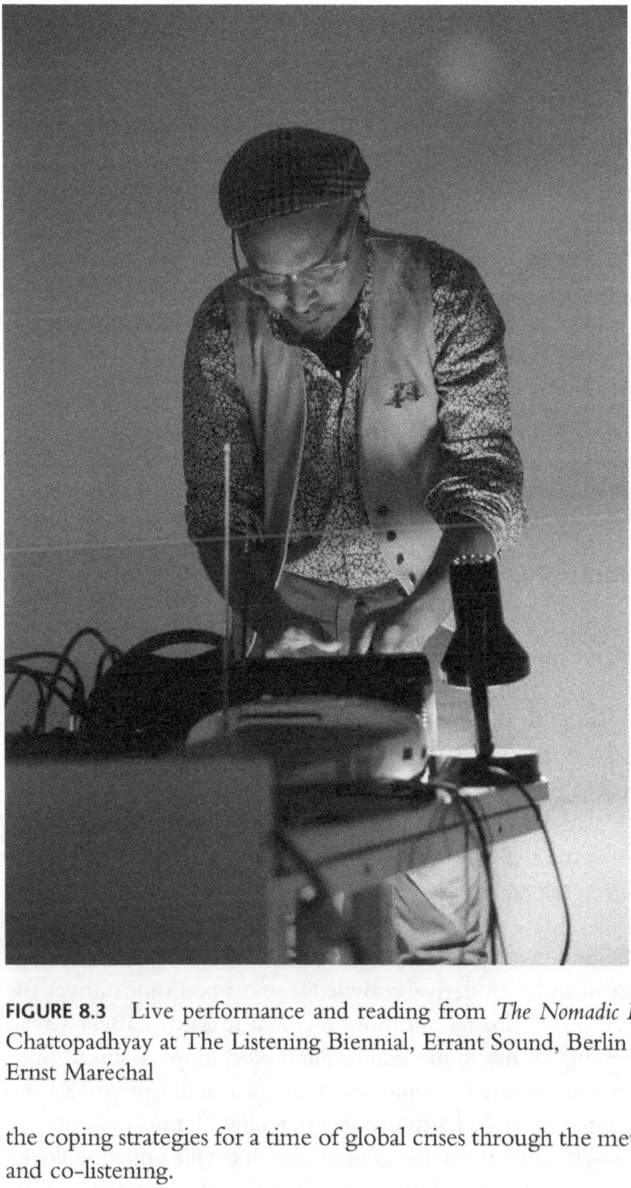

FIGURE 8.3 Live performance and reading from *The Nomadic Listener* by Budhaditya Chattopadhyay at The Listening Biennial, Errant Sound, Berlin (July 2021). Photo: © Ernst Maréchal

the coping strategies for a time of global crises through the methodologies of hyper- and co-listening.

I employed these ideas in the form of a listening workshop in the last edition of *The Listening Biennial* (2021)[8] (Figure 8.3) in which, alongside co-curating the main program, I led the workshop in The Listening Academy. The Academy focuses on listening as a philosophical, artistic, social, and somatic issue. This entails a relation to sonic, performative and ecological practices, sound studies, and experimental pedagogy centred on listening. The Academy offers a generative and nurturing framework for sound researchers, artists, and practitioners to engage in collaborative exchange and the sharing of knowledge to find new directions in sound studies and sonic practices. The curatorial position that *The Listening Biennial* takes is to consider how listening enacts an intervention in today's conflict-ridden and polarised environment. My entry into this debate is with the proposal for hyper-listening and co-listening practices as key to facilitating dialogue, understanding, solidarity, and social transformation. To listen in this way is to extend the boundaries of the familiar, the recognised, and the known to be more inclusive about different personal and political contexts. In addition, such modes of listening afford more egalitarian, decolonised, and ecologically attuned relations, challenging exclusionary systems and human exceptionalism by way of empathic, attentive, and more-than-human orientations: to hear beyond the often-fixed schema of self and other. In my proposed methodology, the very act of listening is a powerful position to take, as it may open and hold, support, and attend, and afford escape and deep friendships. And yet such listening is greatly undervalued and neglected across contemporary societies, especially those where nationalism, authoritarianism, and state policing prevail.

Hyper- and co-listening: Critical engagement

Hinting at how the listening subject and the surrounding world may connect in an everyday context, Steven Connor suggests, 'The auditory self discovers itself in the midst of the world' (1997, p. 219). This connection between sound, listening, and subjectivity is a field of inquiry for some of the leading scholars of sound studies (Connor, 1997; LaBelle, 2010; Voegelin, 2010). While Connor focuses on the capacity for ontological attention to the self and the surrounding environment enhanced by listening, LaBelle, more explicitly suggests that listening, as a relational project, 'Reroutes the making of identity by creating a greater and more suggestive weave between self and surrounding' (2010, p. xxi). For these scholars, the act of listening enables listening subjects to produce and position themselves in relation to everyday sound worlds. This mode of self-attunement to the surrounding environment in order to derive knowledge and impressions drives my artistic practice and pedagogy, albeit I am more inclined towards LaBelle's relational approach to listening. In my sound practice and workshops, I aim to open up the ears for the intimate reveries, memories, and other auditory associations surrounding relational listening, in order to connect the self with the environment.

Having this focus on self-attunement and going through detailed methodologies of the workshop and my general approach to sound practice in the above discussion, I have proposed two new modes of listening methodologies, termed hyper-listening and 'co-listening'. Influenced by Fiumara's (1990) ideas around a propensity to listen—and inspired by her intentionality on the act of listening—hyper-listening operates as a set of inward listening exercises in the form of meditative engagement with the self. Michel Chion (1994), in his seminal work *Audio-Vision*, delineated three primary listening modes, drawing on the phenomenological perspectives of composer Pierre Schaeffer (1966). These are: causal listening, semantic listening, and reduced listening, respectively. Causal listening consists of listening to a sound in order to gather information about its cause or source, or (sounding) object. Semantic listening refers to an interpretation of messages from speech and other vocal sounds through a spoken language or a predetermined code (e.g. Morse code). Reduced listening, on the other hand, focuses on the volume, pitch, tone, textures, and other characteristics of the sound itself, independent of the cause or the source, or the sounding object, and of its meaning. According to Chion, 'Reduced listening takes the sound—verbal, played on an instrument, noises, or whatever—as itself the object to be observed instead of as a vehicle for something else' (1994, p. 29). In musical practice, reduced listening is useful for composing and performing sounds based on their character.

Challenging the epistemic object of sound embedded in the causal listening mode—and taking a clear departure from reduced listening approach of concentrating on the sound itself for ontological and compositional exploration—hyper-listening instead meditates on the inner resonances a listener feels when listening to a sound. This listening approach is mindful about the subjectivity of the listener (rather than the sonic process of production, transmission, and reception). It is also based on a sensitive exploration of the trajectory of affective and emotional impact a particular auditory situation (rather than a specific sound and its characteristics) has on a listener. Hyper-listening revolves around listening as a conscious act to compassionately and holistically engage with the self and to meditate on the inner process of vibrating, more in a metaphorical and mindful sense than corporeally. In this mode I learn to stay present, unlearn meanings attached to the sound source, as the practice of listening requires an openness towards what is being listened to. Not seeking one particular answer, this practice develops from my continued engagement with myself as a reservoir of experiences and knowledge, and an antenna for absorbing emotional cues from my immediate surrounding for a careful inner resonance of cognising and imagining. In my articles 'Beyond Matter: Object-Disoriented Sound Art' (Chattopadhyay, 2017) and 'Auditory (Con)texts: Writing on Sound' (Chattopadhyay, 2015), I have delineated how hyper-listening to the self and the environment can make a listener sensitive and vulnerable to the context, and help recollect personal memories for re-listening. Exposing these narrative threads for re-listening can be therapeutic in the sense that they are acknowledged and accepted. Here listening acts as a psychosomatic de-cluttering and detox process. Departing from the idea of a functional mode of listening as immediate meaning-making for the purpose of everyday cognition and navigation, I suggest the practice of hyper-listening, which explores the aesthetic and mindful aspects of the listening act. This approach may generate processes of affective engagement with lived environments by going inward through a production of context-aware subjectivity. Such self-awareness through sound may nurture sensitivity towards social inclusion and acceptance of others in the context of today's turbulent times.

Philosopher Jean-Luc Nancy has argued that a philosopher is one who hears but does not listen, 'Or who, more precisely, neutralizes listening within himself, so that he can philosophize' (Nancy, 2007, p. 1). His argument challenges the epistemic discourse in the field of sound studies that predominantly equates listening with understanding, audibility with intelligibility, or the sonic with the logical (Nancy, 2007, p. 3). Indeed, a sensitive listener may ask why is there a constant judgemental thrust in creating a causal connection between the sounds and the listening process? Apparently, by keeping the epistemological and the causal as the frameworks to study sound, a sensitive listener may miss out on the resonating quality of subjective listening. It would therefore be a worthy project to explore the contemplative potential of sonic phenomena at the listener's end. Taking as my point of departure the phenomenological premises of sound (Ihde, 2007; Nancy, 2007), subjective and personal listening experiences are made the basis of hyper-listening, which encourages the framing of hyper-spatial and beyond-temporal auditory situations through their transcendental potential. Bearing this theoretical context in mind, the idea of hyper-listening employs the methodologies of mindful listening whereby seemingly mundane auditory situations are listened to by means of their fluctuating and spatiotemporal developments beyond the here and now. In the *Hyper-listening: Praxis* workshops, as discussed earlier,[9] these evolving situations are (con)textualised by chronicling the emerging associative thoughts, imaginings and/or personal memory into listening traces, helping the participants to engage inclusively with themselves situated in their environment, producing subjectivities marked by an elevated and emancipated selfhood.

In my artistic practice, for example, in the augmented book project *The Nomadic Listener* (2020) (Figure 8.4), I have experienced that listening is a private act, and there are subjective positions taken at each moment when there is a process of sensing and cognition unfolding in an auditory situation. Nancy suggests

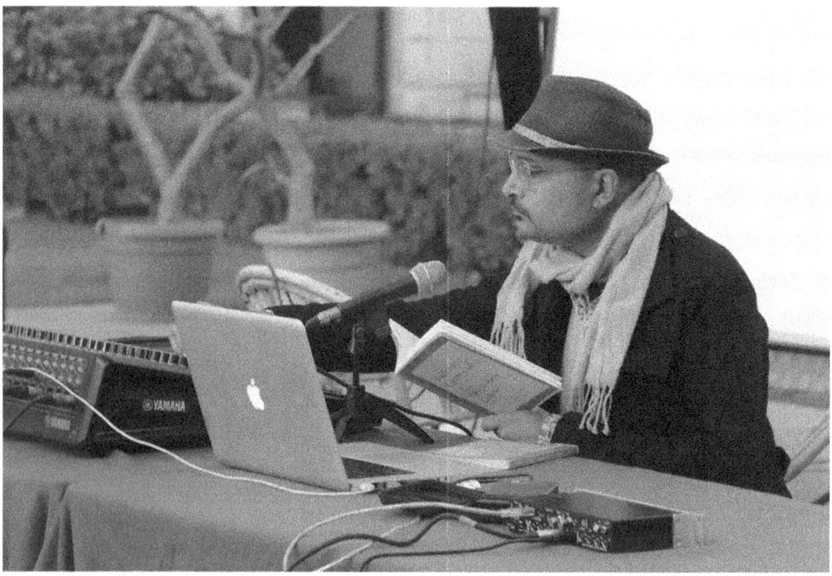

FIGURE 8.4 Live performance from *The Nomadic Listener* by Budhaditya Chattopadhyay at Tahrir Cultural Center Cairo (April 2021) Photo: © Hagar Masoud

that listening suspends between sensing and understanding (2007, p. 2). This in-between space is where contemplation opens up every sounding experience for multiple interpretations. Once inward listening is practised, it sensitises the listening body to its surroundings through this moment of multiplicity. The sonic meditation embedded in going inward through the listening process makes a listener accommodative enough to listen to others, not through othering ears, but as part of a common listening body. Martin Heidegger (1962) recognised the primacy of listening in creating meaning and in developing our relationships with one another. Indeed, as a number of wellness and meditation practitioners suggests,[10] if one is a better practitioner of inner listening, the capacity to connect outside of the self enhances. As in a sonic meditation, this outside does not remain outside anymore. It is accommodated within.

This is the ground on which the idea of *co-listening* grows. It is an evolving process of connecting aurally with the co-habitants of the environment in which the individual listener is also situated. The inward listening capacity empowers the listener to listen in depth, with acceptance and reciprocity, in order to reach outside and accommodate the world. Hyper-listening in this proposal is a personal capacity-building exercise to become inclusive and resonant towards other listening bodies in a mode of co-listening. The term 'co-listening' is seldom used and has previously been defined in organisational communication as having equal time limits for talking and listening between conversational partners.[11] Michael C. Stewart used the term 'CoListen' (2018) in his PhD dissertation, where he investigated young adults' habits of technologically assisted listening to music together on social media to develop a streaming music player. Music researchers Kirk et al. (2016) applied the term to understand mobile music sharing, and digitally mediated mobile music consumption.

I would suggest, however, that co-listening could be expanded from this stiffer definition limited to music listening in social media, towards a socially and community-driven listening mode, in which listening together both in situ and online, reaches a whole new affective dimension of social cohesion nurturing solidarity. While equity between and respect for the listening partners are important, the non-judgemental and compassionate approach in co-listening is central, and therefore cannot be restricted to professional contexts. What happens when there is a fertile pause in speaking, thinking, looking out the window, and coming back to the conversation with a new perspective? The immeasurable and ineffable moments in listening are as important as making meaning by following a steady pattern. The focus of co-listening is on this sense of emergence, indeterminacy and flux while listening together. This 'togetherness' involves the surrounding environment, and the bodies that inhabit this environment. Co-listening entails a compassionate engagement with the other listeners in a shared situation or context. Co-listening requires love and care between humans, between human and non-humans, animals, and other sentient beings, such as plants.

In the co-listening workshop series, participants explore social formation in shared sounds—by connecting listening bodies divided by geography, social and political conflicts or issues of security. I hope that the participants across borders and cultures will explore how to listen together, to each other, and develop care and kinship by connecting voices across diverse backgrounds, for example from Global South and Global North, East and West, with site-specific sounds and voicing of situated realities. The practice of community listening and sharing of sounds as social connection suggest a sonic engagement that includes a more worldly attuned environmental focus and proactive socio-political commitment, community building, and collective action. The outcomes of the workshop will be presented as an online archive of accessible sounds contributed by local participants, currently using social network applications such as WhatsApp, Zoom, and Skype, and by innovatively using these apps to bypass their technological barriers for unmediated listening. This includes listening to each other's personal narratives, monologues, and utterances with patience. The workshop will also give birth to the social platform *Sound Across* that explores the mediatory role of listening in conflict resolution.

These workshops aim to mobilise the idea of hyper-listening and co-listening practiced both as a personal development and community art practice in collaboration with other participants. I intend to promote the practice of engaged listening as a methodology apropos of spreading an awareness of inclusion and contemplative acceptance of the co-habitants at a listening site. The scholarly unboxing of this artistic method helps recognise the potential of listening to others in our environment not from a position of difference, but as extension of ourselves as a part of an acoustic common. The practice underscores that there is no 'other' as sound tends to socially connect everything surrounding us as we inclusively engage with ourselves as listeners.

Acoustic solidarity: Concluding remarks

Through my artistic research- and practice-oriented approach in sound, selfhood and community, I have come to realise, as mentioned earlier, sonic phenomena often trigger poetic contemplation if we open our ears both to the environment and to others with an inclusive and empathic approach. This is a methodology for engaging more compassionately, accommodating others' views and perspectives towards building a community of humans and non-humans for the future with a sense of solidarity and kinship. Hyper-listening and co-listening are concepts and experimental forms of listening that operate as a set of exercises and collaborative acts intending to encourage listening without judgement. A propensity to listen without making immediate judgements may potentially lead to affective engagement with each other. My sound practice aims to challenge the utilitarian idea of listening as immediate meaning-making to arrive at conclusions in order to navigate contemporary times. I underscore in my practice how the exploratory modes of mindful, context-aware, and temporally nonlinear aspects of listening transcend ontological and epistemological constraints of everyday sound towards including the poetic-contemplative states of a listener. Being in this state generates social connections and affective engagements with others by a production of socially inclusive sonic subjectivity. This sense of subjectivity draws on memory and reflections that arise from situated listening, connecting with each other by nurturing an acoustic solidarity.

Notes

1 My own term used in my article 'Auditory situations: Notes from nowhere.' *Journal of Sonic Studies* 4, and later incorporated in my book *The Nomadic Listener* (Chattopadhyay, 2020).
2 A global event co-curated by me; see: https://listeningbiennial.net/about
3 See the full text here: https://www.unicef.org/media/96576/file/Programming-Guide-Conflict-Sensitivity-and-Peacebuilding.pdf
4 This is an ongoing series of workshops and collaborative experiments. Further information is available on my website: https://budhaditya.org/projects/doors-of-nothingness/hyper-listening-praxis/ and a documentation is accessible on a publication: https://www.degruyter.com/document/doi/10.1515/9783110528329-028/html
5 Including among others, at VOX-POP, Amsterdam (Café Chercher, April 2018); ZKM Karlsruhe (inSonic2017 festival, December 2017); Goethe University Frankfurt (May 2017); University of Applied Arts Vienna (September 2016); Q-02, Brussels (April 2016); Overgaden, Institute of Contemporary Art, Copenhagen (February 2014).
6 See: https://www.mousonturm.de/events/aural-drift/
7 The exhibition was accessed at the following sites in Offenbach opening on 4 June 2021: 1. Hafenplatz/Ufer; 2. Elisabeth-Selbert-Steg (harbour); 3. Main bank (Waagon am Kulturgleis) Mainstraße 1, 63065, Offenbach am Main; 4. Mainstraße 127; 5. University of Design (main campus), Schloßstraße 31; 6. Wilhelmsplatz; 7. Offenbach Central Station. See for further information: https://www.mousonturm.de/events/aural-drift/
8 See: https://listeningbiennial.net/
9 See: https://budhaditya.org/projects/doors-of-nothingness/hyper-listening-praxis/
10 There are experiential accounts of Yoga and wellness trainers, in which the meditative practice of inner listening is endowed with a capacity to transcend the body. See: https://kripalu.org/resources/inner-listening—such accounts are based purely on lived experience and personal wisdom, but hardly any scientific evidence is available.
11 See: https://www.dinaglouberman.com/approach/co-listening/

References

Anthes, R. (1954), The original meaning of M' ḫrw. *Journal of Near Eastern Studies, 13*(1), 50.

Barlingay, S. S. (2007). *A modern introduction to Indian aesthetic theory: The development from Bharata to Jagannåatha*. D. K. Print World.

Berleant, A. (1978). Artists and morality: Toward an ethics of art article. Leonardo October.

Chattopadhyay, B. (2013). Auditory situations: Notes from nowhere. *Journal of Sonic Studies, 4*(1). https://www.researchcatalogue.net/view/269178/269179

Chattopadhyay, B. (2015). Auditory (Con)texts: Writing on Sound. *Ear | Wave | Event*, 2.

Chattopadhyay, B. (2017). Hyper-listening: Praxis. In R. Mateus-Berr & L. Reitstätter (Eds.), *Art & design education in times of change: Conversations across culture* (pp. 171–175). de Gruyter.

Chattopadhyay, B. (2020). *The nomadic listener*. Errant Bodies Press.

Chattopadhyay, B. (2021a). *The auditory setting: Environmental sounds in film and media arts*. Edinburgh University Press.

Chattopadhyay, B. (2021b). *Sound production practices in the Global South: Co-listening to re-sounding plurilogues*. Palgrave Macmillan (in press).
Chion, M. (1994). *Audio-vision: Sound on screen* (C. Gorbman, Ed.). Columbia University Press.
Connor, S. (1997). The Modern Auditory I. In R. Porter (Ed.), *Rewriting the self: Histories from the renaissance to the present* (pp. 203—223). Routledge.
Coverley, M. (2010). *Psychogeography*. Pocket Essentials.
Daniélou, A. (1995). *Music and the power of sound: The influence of tuning and interval on consciousness*. Inner Traditions.
Debord, G. (1958). Theory of the Dérive. In K. Knabb (Ed., & Trans.). *Situationist International Anthology* (2006). Bureau of Public Secrets.
Denham, A. (2012). *Plato on art and beauty*. Springer.
Fiumara, G. C. (1990). *The other side of language: A philosophy of listening*. Routledge.
Grenfell, A. (1906). Egyptian Mythology and the Bible. *The Monist, 16*(2), 169–200.
Heidegger, M. (1962). *Being and time* (J. Macquarrie & E. Robinson, Trans.). Blackwell.
Ihde, D. (2007). *Listening and voice: Phenomenologies of sound*. Sunny Press.
Kirk, D. S., Durrant, A., Wood, G., Leong, T. W., & Wright, P. (2016). Understanding the sociality of experience in mobile music listening with Pocketsong. *Proceedings of the 2016 ACM conference on designing interactive systems* (pp. 50–61). 10.1145/2901790.2901874
LaBelle, B. (2010). *Acoustic territories: Sound culture and everyday life*. Continuum.
LaBelle, B. (2006). *Background noise: Perspectives on sound art*. Bloomsbury Academic.
Morton, T. (2013). *Hyperobjects: Philosophy and ecology after the end of the world*. University of Minnesota Press.
Murdoch, I. (1977). *The fire & the sun: Why Plato banished the artists*. Clarendon Press.
Nancy, J-L. (2007). *Listening* (C. Mandell, Trans.). Fordham University Press.
Sadler, S. (1999). *The situationist city*. The MIT press.
Schafer, R. M. (1994). *The soundscape: Our sonic environment and the tuning of the world*. Destiny Books.
Schaeffer, P. (1966). *Traité des objets musicaux*. Seuil.
Self, W. (2007). *Psychogeography*. Bloomsbury.
Serres, M. (2013). *The parasite*. University of Minnesota Press.
Stewart, M. C. (2018). *CoListen*. [Doctoral dissertation, Virginia Polytechnic Institute and State University]. http://hdl.handle.net/10919/85052
Voegelin, S. (2010). *Listening to noise and silence: Towards a philosophy of sound art*. Bloomsbury Academic.
Williams, R. (1963). *Culture and society*. Columbia University Press.

9

I MAKE NOISE THEREFORE I AM: AESTHETICS OF SONIC EXPERIMENTATION IN PARTICIPATORY ART AND CULTURE

Vadim Keylin

Introduction

A bald middle-aged man in grey overalls is attaching carved yellow rubber chicken to the exhaust pipe of a dusty black Mercedes. When the car's ignition starts, exhaust gasses pass through the chickens' sound mechanisms, creating an uncanny chorus. From the man's car mechanic garb and his manner of speech, we can assume that he is probably working-class—in other words, as common wisdom has it, not someone likely to be seen at a sound art exhibition in the audience of an experimental music concert. Yet what he is making in this video—which has over two million views on YouTube at the moment of writing[1]—is essentially a sound sculpture, and a pretty inventive one at that. How is that possible? Is sound art not supposed to only be of interest to highbrow intellectuals educated at art academies?

This is a classist stereotype, of course. On the one hand, the belief that modern art and experimental music are of no interest to 'common people'—whether meant as a criticism of art and music or common people—is easy to disprove simply by looking at the composition of Facebook groups dedicated to these forms. On the other hand, as, for example, the reviews of the 2013 *Soundings* exhibition at MoMA show, even the audiences and critics well-versed in modern art at times display a staggering failure to comprehend the more unusual-sounding works.[2]

Nevertheless, the point remains: even a century after the first sonic experiments of the avant-garde, negative reactions to it, whether mocking or outraged, remain quite common. The impulse for this essay was an observation—made through my research into participatory sound art and my everyday engagement with participatory sound culture—that the audiences are surprisingly more open and appreciative towards weird noises when those are the noises of their own making. The essay's goal then is to explore the aesthetic underpinnings and the social mediations of this appreciation. For the former, I draw on John Dewey's (1934) pragmatist

DOI: 10.4324/9781003164227-16

aesthetics as a theory that emphasises aesthetics experience over aesthetic judgement and addresses its agentic aspects. For the latter, I turn to Erving Goffman's (1986) frame analysis and mediation theories of Bruno Latour (1994) and Georgina Born (2005) in order to attend to the social-cultural-technological assemblages that both facilitate participatory soundmaking and emerge from it.

Pragmatist aesthetics and autonomy of art

The emphasis that both participatory art and participatory culture put on creative making as part of the aesthetic experience make pragmatist aesthetics an obvious choice of theoretical framework (Keylin, 2020a). However, there is another equally relevant aspect to it that I want to discuss first. Both phenomena problematise the notion of autonomy of art by being either fully or partially embedded in the order of the everyday and oftentimes rejecting the disinterestedness of artmaking.

In contrast to much of western aesthetic thought from Kant to Ranciere, Dewey (1934) treats the autonomy of art not as art's fundamental quality but as a highly problematic consequence of modernity. He claims that the origins of art lie in the aesthetic experience of the everyday, citing the way pre-modern cultures do not distinguish between form and function, art objects, and useful objects.

The notion of art as an autonomous sphere of human activity fully separated from the everyday life emerges according to Dewey as a result of several processes of modernity. First is the formation of art collections as symbols of personal or national prestige, which leads to the perception of art objects as objects of extraordinary value, made with extraordinary skill. Second, the institution of taste is established as an expression and a means of reinforcing a class hierarchy, excluding persons of lower social status from art experience. Connected to that is the idea of aesthetic judgement that Kant defines as striving to universality and disinterestedness—instituting aesthetic as its own realm, separated from and elevated over the everyday and the bodily. Finally, industrialisation separates the production of art things—useless but aesthetic—and everyday things—useful but unaesthetic, denying the everyday activities an aesthetic dimension.

Essentially anticipating the work of Erving Goffman (1986), Dewey describes the autonomous art of modernity as a frame—a socially agreed upon definition of reality—characterised by the values of virtuosity and exclusivity assessed through a disinterested aesthetic judgement. And at the same time, the same processes redefine the frame of the everyday as one devoid of aesthetic value and aesthetic experience. In fact, as Dewey (1934) writes insightfully,

> So extensive and subtly pervasive are the ideas that set Art upon a remote pedestal, that many a person would be repelled rather than pleased if told that he enjoyed his casual recreations, in part at least, because of their esthetic [*sic*] quality. (p. 5)

It is largely the same repulsion, I would suggest, that drives the public hostility towards avantgarde and experimental music. The avantgarde famously aspired to blur the lines between art and the everyday and do away with the notion of virtuosity—yet did so from a position of autonomous art, leading to the clashing of two frames.[3] As one of the meme-like panels of Craig Damrauer's webcomic *New maths* (2008) succinctly puts it, 'Modern art = I could do it + Yeah, but you didn't'. However, there are also deeper issues at play when it comes to sound in particular.

In his article *Ableism and the Reception of Improvised Soundsinging*, Chris Tonelli (2016) explores the negative audience reactions to experimental singing practices. The vocabulary of such practices includes distinctly 'unmusical' vocal sounds—from screaming and grunts to whispers and croaks—that are often perceived as a failure conform to the 'dominant forms of singing wherein melody and the production of 'pure' pitch content predominate' (para. 7). Tonelli notes how this reaction is steeped in ableist rhetoric: the artists are being mocked for their perceived lack of musical ability. The public's discomfort and outrage in the presence of non-normative voices mirrors that evoked by non-normative bodies. There is a slight difference, however: while everyday ableism operates on the expectation of normative ability, a musician is expected to be exceptionally abled—an expectation against which the ostensible mundanity of extended vocal sounds clashes.

In a different configuration, the clash between the frames of art and the everyday has long been explored in public and site-specific art. However, as Christabel Stirling (2016) discusses in her article *Sound Art/Street Life*, public sound art often provokes hostile reactions even when it is not intended to be provocative. She explores, among other case studies, the controversy surrounding Esther Ainsworth's sound installation *Bridge Links* (2013). The work was installed over a towpath under a bridge in Camden Town and consisted of remixed field recording made in the same area. It was however met with protests from the local residents,[4] whose issue was not with the noise level—as Stirling notes, they had no problem with buskers playing around the bridge—but with how they perceived it as an incursion into their life spaces. To Stirling, this shows that the transgressive character of sound makes sound art illuminate and exacerbate conflicting claims to urban public space—in contrast with how it is often treated in literature as a means of facilitating social encounters and building social connections.

Frames of participation

I am bringing up Tonelli's and Stirling's studies not only because they explicate the political undercurrents of the negative reactions towards experimental sounds. The kinds of artworks they discuss mirror the two artworks I used as case studies to conduct ethnographic observations and audience interviews for my project on creative agencies in participatory sound art[5]—with the distinction of them being participatory.

The first is the interactive performance *Let Us Sing Your Place* by the Danish artist Katrine Faber that I have attended three times between 2018 and 2019 in Aarhus and Copenhagen. The work proceeds in cycles wherein one audience member describes a place that is important to them, after which the rest of the audience form an impromptu choir and try to recreate the soundscape of that place with their voices. The performance is fully improvisational, without any rehearsals or workshops to prepare the participants except for a couple of vocal exercises Faber leads at the beginning, and the audience composition is largely incidental—for example, one of the performances I attended occurred as part of an anthropological film festival. Yet in contrast with the experiences that Tonelli describes, the participants in *Let Us Sing Your Place* readily embrace the unusual vocal palette of singing a soundscape. While the participants' reactions to the performance were not uniformly positive, the resistance that a small minority of them expressed in my interviews with them was not due to unconventional singing. Rather, they found the performance's theme of universal connection, and Faber's way of relaying it in her introductory speech, too reminiscent of *new age* rhetoric—in a way, too conventional—for their liking.

The second is the public sound sculpture *Speakers' Arena* by American-German artist Benoît Maubrey. The work was installed from July to November 2019 in Berlin, on a small square framed on two sides by the Pallasseum housing complex. The sculpture was built of some 300 active loudspeakers that the public could speak or sing through via a microphone or by calling a special phone number, connect their smartphones through Bluetooth and play their own sounds and music, or make Twitter posts with #speakersarena hashtag and have the sculpture speak it in a synthesised voice. Even though the work was installed right next to a huge housing complex—right in the middle of its inhabitants' life space—I did not encounter in my observations the kind of hostile reaction that Stirling writes about. Neither did the project's curator Stephan Kruhl. When I asked him later whether they had any complaints throughout the time that *Speakers' Arena* was installed, he only mentioned that a few older residents were disapproving of public money being spent on modern art. On the contrary, the locals (very few of whom were the arty types that Stirling writes about) were enthusiastically using the sculpture for impromptu concerts or just to hang out with friends and strangers alike.

What these two examples show, in contrast to Tonelli's and Stirling's observations, is that participatory sound art invokes a different frame of reception than that of established formats of art presentation—one that suspends judgement and promotes openness. Applying frame analysis to musical practices, ethnomusicologist Thomas Turino (2008) suggests, that the term music in fact encompasses four quite different kinds of cultural practices characterised by different values and aesthetic goals. These four practices are presentational performance (music performance in the traditional sense), participatory performance, hi fidelity recording (i.e. one meant to represent a live performance), and studio audio art (fully composed electronically).

Turino's distinction between participatory and presentational performances mirrors Dewey's perspective on pre-modern and modern art. Presentational performances in Turino's description are autonomous art forms that maintain strict artist-audience division (reinforced, among other things, through the organisation of performance spaces) and foreground virtuosity and individual skill. Conversely, participatory music performances—religious ceremonies, folk celebrations, dance parties—are embedded in the everyday. Soundmaking in such performances is a social activity, more so than purely aesthetic one, resulting in more horizontal structures, where participation is encouraged and differences in musical skill, downplayed. But probably most importantly, they are judged first and foremost on the quality of participation—the heightened sense of togetherness and the level of social bonding achieved—rather than on the quality of music performed.

It would be no stretch to assume that a similar prioritisation of the social aspects of experience over the musical ones occurs in participatory sound art as well. There is a fundamental difference, however, in the material—sonic—mediations of this sociality. Traditional participatory music practices rely on familiarity and predictability, both on the micro and macro levels. The list of typical musical traits of a participatory performance that Turino (2008) compiles stresses repetition, including such features as short repeated forms, constant meter and rhythm, and lack of dramatic contrasts. Furthermore, he claims that 'musical synchronicity [and] the deep feelings of identification' that participatory performance induces are necessarily facilitated by 'similar habits of style' developed through a prolonged exposure to a musical tradition (p. 19).

Conversely, participatory sound art relies on sounds and sonic structures that are unfamiliar and unpredictable. For example, in my interviews with the participants in *Let Us Sing Your Place*, I noticed that even though the title of the performance has the word 'sing' in it, they were often not referring to what they did as singing, but rather as making sound. One respondent even felt the need to correct me when I worded my question in terms of singing. What this restatement of their actions as 'making sound' meant was that they rejected the frame of virtuosic performance. They stressed the values of communality and altruism in their soundmaking: not trying to show off their singing skills, aiming to be as faithful as possible to the image of place conjured by one of their peers, and not judging or being judged by others.

In other words, it is not just that the social aspects of participatory sound art override the audiences' aesthetic concerns—rather, the art form's reliance on unfamiliar sounds and structures enhances the sense of togetherness and social bonding. By pulling out the frame of reference for aesthetic judgement, experimental sounds level the playing field with regard to the participants' musical skills and cultural backgrounds, inviting horizontal interactions, and shaping emergent communities across social divides.

Sonic memes and emergent communities

Another limitation of the participatory musical performance theory is that it only concerns practices that bring people together in the same time and space. It cannot account for the spatially and temporally distributed experimental sonic practices of participatory culture. However, such practices are arguably of even more interest to the current discussion, as unlike sound art, they are not framed by art institutions and exist as part of popular culture.

Nevertheless, research on online music practices shows that similar mechanisms of social bonding are at play here. In their article *From Microsound to Vaporwave: Internet-Mediated Musics, Online Methods, and Genre*, Georgina Born and Christopher Haworth (2017) trace the networks of mediations through which internet-based music genres unfold. They argue that what gives these genres their particular character is the way they are mediated by the internet in three specific ways: material, social, and discursive. Materially, online platforms, whether legal or pirate, disintermediate the distribution of music, simplifying both the artists' access to audiences and the audiences' access to music. Discursively, 'The web is used incessantly in many contemporary music genres to cultivate, publicise, and disseminate both knowledge and criticism of music, as well as to facilitate discussion', stimulating 'democratization of this discursivity … inciting participation, and speeding up its production and circulation' (p. 611). Finally, social mediations refer to the ease with which online platforms afford the formation of communities around specific music genres and practices. Most importantly, however, the internet stipulates the convergence of all three types of mediations, with the same platforms used for distribution and discussion of music, which become inseparable from the social processes surrounding it.

Of the six case studies of internet-based music genres that Born and Haworth (2017) conduct in their article, the microsound one is particularly salient to my argument here. Microsound—an electronic music technique of combining sonic 'grains' of very short duration into stochastic or regular 'clouds'—has originally emerged within western art music and has long remained a highly specialised academic avantgarde practice. However, as Born and Haworth show, the establishment of the microsound discussion mailing list in 1999 and its affiliated file transfer protocol (FTP) server has led to the emergence of an online subculture comprising academic, non-academic, and amateur musicians and critics. Thus, through the mediations of participatory culture, microsound—while not possessing any stable sonic or structural features that characterise a musical genre—became an online genre in practice.

The example of microsound demonstrates two mechanisms by which experimental music and sound art may give the impression of being popular on the internet: filter bubbles and mentorship networks. The term 'filter bubbles' was coined by the internet activist Eli Pariser (2011) as a critique of the way social media algorithms create 'echo chambers' for their users by reinforcing their connections with people with similar views while separating them from opposing viewpoints. However, even before algorithmic regulation of social media feeds became a thing, similar filtering was performed by internet users themselves, building distributed communities of likeminded individuals. Politically, filter bubbles have been considered a net negative as they polarise public discourse. However, with regard to aesthetic phenomena and practices, they may have a positive effect as well, creating extensive communities around experimental music and sound art genres, access to which is unhindered by physical circumstances (e.g. a lack of local experimental music scene).

Because of such bubbles, a piece of experimental music or sound art posted online is simply more likely to reach people who appreciate this kind of aesthetics and would pass it on further, creating a snowball effect. This happened, for example, with performance videos of Czech noise musician Petr Válek, which became viral in certain experimental music and art-adjacent segments of Facebook. However, when reaching beyond these segments through a repost in a group dedicated to unusual (in practice, mostly just non-western) musical instruments, they were met with a typical 'how is it art?' reaction by some of the commenters.

At the same time, through the expansion of such networks, new people become exposed to experimental music and sound art and inducted into their aesthetics—both as listeners and producers. Henry Jenkins and his co-authors on the influential report *Confronting the Challenges of Participatory Culture* (Jenkins et al., 2006) consider this process a fundamental quality of participatory cultures. They characterise it as necessarily having 'some type of informal mentorship whereby what is known by the most experienced is passed along to novices' and 'strong support for creating and sharing one's creations with others' (pp. 5–6). In the example of microsound, Born and Haworth (2017) show how the affordances of online discussion platforms facilitated informal education of amateur musicians, expanding the knowledge and appeal of the genre beyond its initial status of a highly specialised academic music practice.

Constructing socio-cultural bubbles around themselves and expanding them through informal mentorship, experimental music and sound art have carved an extensive niche for themselves online. However, these two mechanisms have little explanatory power when it comes to forms of sonic experimentation like the rubber chicken video discussed in the introduction. Such practices do not have recognisable sonic identities, nor do they possess 'a community whose agreement forms the basis for the definition of [the] genre' (Fabbri, 1981, p. 54, as cited in Born and Haworth, 2017). Neither their producers, nor their audiences are part of experimental music and sound art communities. For example, the 'Rubber chicken exhaust mod' video was posted to the YouTube channel *Garage 54* dedicated to experiments with cars rather than sound. A cursory search of the platform for similar rubber chicken videos reveals the same results: they are posted by people whose professional identities and channels are connected to car and bike mods and hacks. However, at the same time, this search also shows that such sonic experiments are not just shared and reposted, but repeated and recreated with some variations (using a motorbike instead of a car, for example), retaining some of the social and material aspects of a genre, while eschewing discursive ones. In other words, rather than experimental music genres, these videos are closer to being sonic versions of internet memes.

In her book *Art Platforms and Cultural Production on the Internet*, Olga Goriunova (2012) describes the way such online phenomena come into being through the notion of 'autocreativity'. This term refers to the process by which emergent creative communities online both shape and are shaped by a certain aesthetics through repeating, developing and refining its key aspects in their cultural production. The fundamental quality of this process is reciprocity. On the one hand, it is the shared aesthetic orientation that brings the necessarily open-ended community together, expanding it and recruiting new members. But at the same time, it is the development of the community and interactions between its members that (re)define the aesthetics that lies at its core.

Another way of looking at sonic experimentation online is through Gilbert Simondon's (1992) concept of individuation—the process of becoming into being that constantly oscillates between the individual and pre-individual. According to Goriunova (2014), individuation is central to the aesthetic functioning of internet memes, which, she claims, are 'aesthetic objects that mediate individuation', both individual and collective (p. 54). Moreover, 'Such becoming that has not yet reached or might never reach a completion, or be aesthetically valorised, is core to many digital aesthetic processes', where 'creative expression is produced in a manner of a "grey zone" between the pre-individual and individual and the collective' (p. 60).

From this perspective, both the experimental aesthetics of sonic memes and their infinite repeatability are necessary conditions for this individuation to occur irrespective of the kind of subjects that they individuate. Experimenting with rubber chicken and their sounds mediates the persona of a car mechanic, establishing him both as a creative and adventurous individual, but at the same time a member of a community of car mechanics. It acts as tool of empowerment against the classist stereotyping of handiwork as menial and unimaginative.

This kind of empowerment through individuation can also be found in participatory sound art. As part of my fieldwork, I have visited Kaffe Matthews' work *Ispod Fontana* (*Under the Fountain*) in Zagreb. Matthews makes sonic bikes—bicycles outfitted with loudspeakers and GPS units that play sounds from a sample bank depending on where the cyclist is in a previously mapped out space. In Zagreb, the bikes were taken for a ride by the city's Cyclists Sindicate—a volunteer association advocating for biking and cyclists. The participants told me that having sounds blast out from their bikes made them feel empowered in their collective identity as cyclists in a city that lacks much of biking culture and infrastructure. Some even viewed the ride as a political act, establishing their right to the city's public space.

In an earlier article, I have referred to this political modality of participatory sound art as 'postcritical', building on Rita Felski's (2015, 2008) proposition that art's political potentialities are not limited to 'didactic provocation' (Kester, 2011, p. 9)—nor to authorial intent in general (Keylin, 2020b). While sound artworks and sonic practices in participatory culture may present on their surface as apolitical, they create a space for the participants to articulate and share their political imaginaries, whether critical and antagonistic or remedial and reconciliatory. It is true that participatory art and culture are often critiqued—not necessarily unfairly—that in their aspiration to unity they gloss over difference and antagonisms existing in society (Bishop, 2006, 2004). I would argue, however, that sonic experimentation offers a way to circumvent this trap of complacency, re-introducing the difference and dissent in the discomfort of its listening experiences. Moreover, sound itself, according to Salome Voegelin (2019) 'Offers a portal into difference and the differently real and allows us to hear alternative slices on equal track' (p. 27).

Hadi Bastani's (2021) article *Experimental Electronic Sound as Playful Articulation of a Compromised Sociality in Iran* provides a succinct illustration of this political modality. Bastani traces the emergence and development throughout the 2010s of the Iranian experimental electronic scene—first on social media and then in physical performance spaces—naming three principal factors contributing to its proliferation. First, participating in international online communities of experimental soundmakers equipped Iranian artists—both trained musicians and amateurs—'with new imaginaries that enable alternative modes of sociality performed outside the reach and control of "traditional" authorities like the central government or nuclear family' (p. 388). Second, the abstract and technological character of electronic music and sound art helped circumvent state censorship, presenting their performances as ostensibly apolitical scientific workshops rather than art events. But most importantly, the aesthetics of experimental sound, not subjected to compositional rules and traditions, provided an outlet for the political subjectivity urging to escape the 'deep-rooted mechanisms of dogmatic ethical-moral judgement, crystallised for instance through the laws but also citizens' self-policing and policing of each other' (p. 391).

The joy of making noise

So far, I have concentrated on how the aesthetic value of participatory sonic experimentation is produced through emergent socialities that it facilitates and mediates. In this section, I want to return to Dewey to discuss the activity itself.

As I mentioned in the introduction, the principal unit of Dewey's (1934) aesthetic theory is aesthetic experience rather than aesthetic judgement. This experience is not limited to art perception but is a special kind of everyday experience, characterised by achieving a dynamic equilibrium between 'doing and undergoing' (p. 46)—affecting the environment and being affected by it.

Importantly, Dewey insists that the dialectic of doing and undergoing is fundamental both to art production and art reception. And although his writing precedes participatory art and culture per se, such practices present a very characteristic example of this dynamic, as they combine the processes of production and perception into one experience. In my previous research on participatory sound art, I have linked Dewey's aesthetics to affordance theory to show how the participants' actions—both elementary soundmaking acts and complex behaviours—and their experiences are conditioned by the artworks' materialities and at the same time reshape them (Keylin, 2020a). Similarly, both Goriunova (2014) and Born and Haworth (2017) discuss how the aesthetics of visual and musical memes is produced in convergence between the creators' subjectivities and the affordances of specific online platforms.

It naturally follows, that the dynamic equilibrium of doing and undergoing, of the actors' agencies and the affordances of material environments, is similarly constitutive of the aesthetic experience of sonic experimentation, irrespective of whether it occurs in the context of art or participatory culture. There are two aspects to it, however: the act of experimentation itself and the resulting sound.

Analysing the spoken comments of the man in the 'Rubber chicken exhaust mod' video, it is immediately evident that we are observing an experiment, as he constantly introduces his actions by using constructions like 'let's try' or 'let's see'. The language of uncertainty, curiosity, and experimentation is similarly present in the framing of many other sonic memes, whether in the videos themselves or in their titles and descriptions. This is not particularly surprising, as experimentation and play are, according to Henry Jenkins (2006), core values of participatory culture. The play ethos of online environment suspends critical judgement and encourages risk-taking—including aesthetic risks.

However, the ethos of play and experimentation has also always been a characteristic trait of participatory sound art. This can be evidenced, for example, by this quote by the Baschet brothers, French sculptors who in late 1950s originated sound sculpture by offering their experimental musical instruments to the public to play freely:

> Philosophically, we are happy when the public enjoys our exhibitions. And, as a general rule, it does.
>
> Musically, the result is often a sheer disaster. If one puts a small child in front of a piano, the child will pound on it as heavily as possible: 'I make noise, therefore I exist.' If one puts an adult in front of a sound sculpture and hands the adult a mallet, the same thing will happen: 'I make noise.' But I am sure that in both cases the hyperactivity stems from the pleasure of discovering. Next time both child and adult will behave better. It is a matter of education.
>
> *(Baschet and Baschet, 1987, p. 110)*[6]

Indeed, as I have noted in my previous publications, the very setup of participatory and interactive sound art encourages exploration and experimentation. Such artworks present new, unfamiliar, and sometimes deliberately obscured soundmaking mechanisms to audiences that typically have no previous musical training, while at the same time furnishing immediate sonic feedback to their actions (Keylin, 2019). Furthermore, the same can largely be said of the devices and technologies—be it rubber chickens, autotune and other vocal filters, or text-to-speech software to name a few examples—that are featured in sonic memes. Thus, the playful and experimental ethos of participatory culture is reinforced by the affordances of soundmaking tools.

How is then the aesthetic quality achieved in the experience of play and experimenting with sound? In his book *The Expressive Moment*, Marc Leman (2016) explores the cognitive mechanisms through which expression—particularly in interaction with music—can lead to empowerment. He names three principal components of this effect: the sense of agency derived from successfully predicting the direction that music is going to take, the feeling of arousal stemming from overcoming the effort necessary for interaction, and the energetic release found in expressive action. This in turn implies that to be rewarding and empowering the musical interaction must involve some degree of challenge and indeterminacy. Leman notes that music achieves this by relying on complex sonic patterns that are challenging to perform or respond to and unpredictable in the long term (though often predictable in the short term).

I would argue that sonic experimentation affords a similar effect. However, the challenge that it presents is more aesthetic than physical or mental (although figuring out the soundmaking capabilities of a sound sculpture or the settings of a vocal filter can of course be challenging in itself). In that regard the very quality of experimental sound that provokes hostility in other contexts—challenging entrenched ideas of the aesthetic—makes it rewarding in participatory situations.

At the same time, experimental sound offers a different kind of indeterminacy from traditional forms of music. By its very nature of being experimental it is unpredictable in the short term, arousing curiosity and reinforcing the frame of play. In other words, the lack of expectation of how something should sound encourages the participants to take risks. However, experimental sonic practices may also offer some predictability in the long term. People making rubber chicken or autotune videos online often have a rough idea of how the result would sound, having seen similar videos before. Likewise, audiences at participatory sound art exhibitions may repeat what they have seen others do with the artworks.

Finally, the balance of predictable and unpredictable is often enhanced through the audiovisual aspects of viral videos and sound installations. For example, in the case of *Speakers' Arena*, the extravagant look of the sculpture built of a multitude of motley recycled loudspeakers introduced a degree of novelty to the listening experiences of the participants who played their favourite music tracks through it. Conversely, in viral videos the sonic weirdness is grounded in the visual normality—mundanity even—of cars, rubber chicken, cats, and other objects and entities employed in soundmaking.

Conclusion

In *Art as Experience*, Dewey (1934, p. 134) defines the aesthetic form as 'completeness of relations' emerging only in lived experience of the artwork and necessarily including its situatedness. In this chapter, I have examined the networks of relations and mediations stemming from experimental sound practices in participatory forms of art and culture. Engaging in participatory soundmaking facilitates all kinds of emergent socialities, whether through heightening the sense of togetherness in the here-and-now or the processes of collective individuation in distributed online communities. These socialities—along with the liberating and empowering effects of sonic play itself—become integral components of the aesthetic experience, explaining the difference in aesthetic reception of experimental sound between participatory and non-participatory artforms.

But at the same time, what makes these emergent socialities possible is the materiality of experimental sound itself. Eschewing the established rules for musical production and thus the grounds for aesthetic judgement, it establishes a frame of free and playful aesthetic production. Being unconstrained and unpredictable, experimental soundmaking stipulates equality and horizontality in the relationships between the participants while not negating difference, revealing a potential for a postcritical treatment of the politics of sound.

Notes

1 For the original Russian-language video (https://www.youtube.com/watch?v=zwA7BbVuE10). The version dubbed into English (https://www.youtube.com/watch?v=FYRFQk2PjO8) at the moment of writing has 3.2 million views.
2 For a survey and analysis of critical reviews of *Soundings* see Lopez (2013).
3 While for the historical avantgarde this clash was intentional, a means of provoking the audience and upsetting the status quo, the same cannot necessarily be said of the post-war sonic practices.
4 Stirling notes however a difference in attitudes between the older Camden residents among whom 'bafflement and occasionally negative attitudes towards the work had prevailed' and 'a number of young arty individuals [who] excitedly stopped in their tracks' (p. 2, para 3).
5 Keylin, V. (forthcoming, 2023). Creative agencies in participatory sound art: Two case studies. *Organised Sound, 28*(1).
6 It is ironic how the Baschets flip the script here, accusing their audiences of the same lack of musicality that the audiences usually accuse sound artists of.

References

Baschet, F., & Baschet, B. (1987). Sound sculpture: Sounds, shapes, public participation, education. *Leonardo, 20*(2), 107–114. https://www.muse.jhu.edu/article/600464.

Bastani, H. (2021). Experimental electronic sound as playful articulation of a compromised sociality in Iran. *Ethnomusicology Forum, 29*(3), 379–400. 10.1080/17411912.2021.1896371

Bishop, C. (2004). Antagonism and relational aesthetics. *October, 110*, 51–79. 10.1162/0162287042379810

Bishop, C. (2006). The social turn: Collaboration and its discontents. *Artforum, 44*(6), 178–183.

Born, G. (2005). On musical mediation: Ontology, technology and creativity. *Twentieth-Century Music, 2*(1), 7–36. 10.1017/S147857220500023X

Born, G., & Haworth, C. (2017). From microsound to vaporwave: Internet-mediated musics, online methods, and genre. *Music and Letters, 98*(4), 601–647. 10.1093/ml/gcx095

Damrauer, C. (2008). *New Math.* [New Math] https://assortedbitsofwisdom.com/New-Math-1

Dewey, J. (1934). *Art as experience.* Minton, Balch & Co.

Fabbri, F. (1981). A theory of musical genres: Two applications. In D. Horn & P. Tagg (Eds.), *Popular music perspectives* (pp. 52–81). Gothenburg and Exeter: International Association for the Study of Popular Music.

Felski, R. (2008). *Uses of literature*. Blackwell.

Felski, R. (2015). *The limits of critique*. University of Chicago Press.

Goffman, E. (1986). *Frame analysis*. Penguin.

Goriunova, O. (2012). *Art platforms and cultural production on the internet*. Routledge.

Goriunova, O. (2014). The force of digital aesthetics: On memes, hacking and individuation. *The Nordic Journal of Aesthetics*, *24*(47), 54–75.

Jenkins, H., Purushotma, R., Weigel, M., Clinton, K., & Robison, A.J. (2006). *Confronting the challenges of participatory culture: Media eduction for the 21st century*. MIT Press.

Kester, G. (2011). *The one and the many: Contemporary collaborative art in a global context*. Duke University Press. 10.1515/9780822394037

Keylin, V. (2019). Medialities of participation in sound art. In B. Eriksson, C. Stage & B. Valtýsson (Eds.), *Cultures of participation: Arts, digital media and cultural institutions* (pp. 129–146). Routledge.

Keylin, V. (2020a). Crash, boom, bang: Affordances for participation in sound art. *SoundEffects-An Interdisciplinary Journal of Sound and Sound Experience*, *9*(1), 98–115. 10.7146/se.v9i1.118243

Keylin, V. (2020b). Postcritical listening: Political affordances in participatory sound art. *Organised Sound*, *25*(3), 353–361. 10.1017/S135577182000031X

Latour, B. (1994). On technical mediation. *Common Knowledge*, *3*(2), 29–64.

Leman, M. (2016). *The expressive moment: How interaction (with music) shapes human empowerment*. MIT Press.

Lopez, J.-D. (2013). *MoMA's "Soundings" exhibition: Critiquing the critics*. [Sounds Like Noise] https://soundslikenoise.org/2013/08/17/momas-soundings-exhibition-critiquing-the-critics/

Pariser, E. (2011). *The filter bubble: What the Internet is hiding from you*. Penguin.

Simondon, G. (1992). The genesis of the individual. In J. Crary & S. Kwinter (Eds.), *Incorporations* (pp. 297–319). Zone Books.

Stirling, C. (2016). Sound art / street life: Tracing the social and political effects of sound installations in London. *Journal of Sonic Studies*, 11. https://www.researchcatalogue.net/view/234018/234019

Tonelli, C. (2016). Ableism and the reception of improvised soundsinging. *Music & Politics*, *X*(2), 1–14. 10.3998/mp.9460447.0010.204

Turino, T. (2008). *Music as social life: The politics of participation*. University of Chicago Press.

Voegelin, S. (2019). *The political possibility of sound: Fragments of listening*. Bloomsbury.

PART IV

Sonic Disruptions: Creating Auditory Counter-Narratives

INTRODUCTION

Sarah Woodland and Wolfgang Vachon

A principal motivation for those working in community engaged arts and arts-led research has traditionally been to disrupt the hegemonic narratives of identity and culture that feed into dominant structures of power and oppression. While previously such struggles were focused on voice and visibility, the politics of sounding and listening works on an arguably more subtle and complex level, engaging with multisensory and vibrational knowledges as much as discursive ones (Bull & Cobussen, 2020). Our own field of applied theatre has long been concerned with the 'ethics of representation' (Prentki & Breed, 2020), prompting us to ask, how do community engaged sound arts or audio practices impact on the representation of marginalised subjects and support the creation of counter-narratives and counter-perspectives? What are the unique affordances of sound and listening in engaging participants and audiences aesthetically in such efforts? But also, following LaBelle (2020), how might we engage with sound, listening, and acoustics in navigating alternative social and bodily pathways and orientations?

All the contributions to this volume could be seen to be concerned with disruption in one form or another, but we see those collected in this section as retaining, despite their sonic inclinations, a strong focus on narrative. Specifically, the works here tend to highlight the operation of community engaged sound and soundscape alongside narrative forms such as drama, personal testimony, and oral history, inviting listeners and facilitators into the lifeworlds of young people in care, men in prison, Roma artists, and refugee youth.

Artist-researchers Maggie Inchley and Sylvan Baker in their chapter present the work of The Verbatim Formula (TVF), a performance collective who have developed verbatim theatre approaches to amplify the experiences and stories of care-experienced young people. The chapter describes TVF's involvement in an initiative aimed at examining the Scottish 'care system' and develop ways to improve it. The authors describe their participatory sonic-research approach

DOI: 10.4324/9781003164227-18

with a small group of 'care-leavers', which included developing auditory environments of sounds that were important to young people, sharing 'stigmatising stories' among the group, and using headphone theatre to explore topics and present information to parliamentarians and policy makers. Using ideas of sound and space they consider how young people in the care system experience their time and how the TVF process may contribute to challenging these minimising approaches.

Rosa Cisneros explores the role of podcasting in creating counter-narratives about Roma communities. She describes how her own podcast *Yellow Couch Convos*, invited Gypsy, Roma Traveller (GRT) artists to talk about their artistic practice and the unique issues facing their communities. The podcast aimed to take a contemporary approach to traditional storytelling and oral history to engage in cultural activism, challenging stereotypes of GRT communities, and transmitting and preserving important aspects their cultural heritage. Cisneros draws on Judith Butler (2004) to emphasise the 'humanising' potential of the podcast form, as it enables listeners to eschew the usual preconceptions and assumptions that accompany visuality in relation to interview subjects, inviting a space for inclusivity.

The contribution from composer and scholar Toby Young presents a narrative 'critical self-reflection', which highlights his own process of discovery creating a collaborative multimedia work incorporating sound art and orchestral composition with refugee youth. Readers are invited to consider the ethical and aesthetic tensions experienced by the author in working from a position of Western musical tradition and relative privilege, to engage with diverse cultural and social positionalities. Invoking Voegelin (2018), Young describes how 'the project draws heavily on the hyper-real geographies afforded by sound to help audiences experience the participants' stories of migration and displacement as tangibly as possible', without drawing directly from traumatic experiences. While some tensions remain within Young's project, this approach might go some way towards addressing the intractable problems with participatory performances created with refugees that so often solicit trauma stories, risk re-traumatisation, and trap refugees in narratives of struggle and survival for an avid 'settled' audience (see Wake, 2019).

The chapter by Gary Anderson and Niamh Malone focuses on *Odyssey on the Airwaves*, a project facilitated by the authors with men incarcerated in a UK prison during COVID-19 lockdown. In creating a performed reading of Homer's *The Odyssey* that would be broadcast on prison radio, the authors grapple with the 'classed' tensions of delivering 'high' culture inside a prison, and how to ethically navigate the power dynamics of 'incarcerated-free', 'uneducated-educated' and 'facilitator-participant'. Anderson and Malone offer a meta-narrative account of how the project unfolded in *Odyssean* terms, aiming to counter the traditional hierarchical 'praxis of judgement' held by both facilitators and participants by deploying Deleuzian 'affective encounters' (Braidotti, 1993) This is reflective of a more emancipatory integration of the ethical, aesthetic, and engaged.

References

Braidotti, R. (1993). Discontinuous becomings. Deleuze on the becoming-woman of philosophy. *Journal of the British Society for Phenomenology, 24*(1), 44–55. 10.1080/00071773.1993.11644270

Bull, M., & Cobussen, M. (2020). *The Bloomsbury handbook of sonic methodologies*. Bloomsbury.

Butler, J. (2004). *Precarious life: The powers of mourning and violence*. London: Verso.

LaBelle, B. (2020). *Acoustic justice: Listening, performativity, and the work of reorientation*. Bloomsbury.

Prentki, T., & Breed, A. (2020). Introduction to volume one: Considering the ethics of representation in applied theatre. In T. Prentki & A. Breed (Eds.), *The Routledge companion to applied performance: Volume – Mainland Europe, north and Latin America, southern Africa, and Australia and New Zealand* (pp. 1–6). Routledge.

Voegelin, S. (2018). *The political possibility of sound: Fragments of listening*. Bloomsbury Publishing.

Wake, C. (2019). Theatre of the real with resettled refugees: Old problems and new solutions in The Baulkham Hills African Ladies Troupe. *Performance Research, 24*(8), 20–30. 10.1080/13528165.2019.1718426

10

SOUND TRAVELS FASTER OVER WATER: SONICALLY RE-DESIGNING INSTITUTIONAL AURAL ARCHITECTURE WITH THE VERBATIM FORMULA

Maggie Inchley and Sylvan Baker

> *"How would you feel in my shoes?" Ryan.*[1]
> *"It's not the adults' future. It's ours. We can change it for the better," Kayleigh.*
> *(The Verbatim Formula Living Archive, 2015-)*

The Verbatim Formula (TVF) is a participatory practice research project based at Queen Mary University of London that explores ways to offer redistributions of power by disrupting institutionalised spaces and routine modes of service delivery. Developed in pre-pandemic times, TVF practises a version of 'connective aesthetics', a term coined by artist Suzi Gablick in the 1990s, to suggest the inter-relational qualities of listening-based art (1992, p. 2). According to Gablick, its participants can experience an empowering sense of being heard. Working with care-experienced young people and care leavers as co-researchers, as we have documented in other articles, TVF creatively centres the experiences of those who are by common statistical measures amongst the most disadvantaged in the UK (Baker & Inchley, 2020, pp. 171–172). Since 2015, the project has curated sonic, auditory and dialogic encounters based around young people's experiences of care in a series of events in various UK locations. These have brought TVF's young co-researchers together with social workers, civil servants and policymakers responsible for their care and education. Previously, we have written about TVF's activities as an 'aesthetics of care', practiced through the close aural attention to recorded audio clips that is required by our adaptation of headphone theatre techniques (Baker & Inchley, 2020). We have also described the project's interventions in university spaces in the UK education system, whose historical social inequalities are further reinforced by conventions and administrative structures that tend to suppress and stigmatise the experiences of care leavers (Inchley et al., 2019 and 2022).

DOI: 10.4324/9781003164227-19

Here, we want to explore sound and affect in relation to some of TVF's activities in Scotland, where we were invited to work with Our Hearings Our Voice (OHOV). The work of OHOV is to run an independent board for children and young people across Scotland who have experience of Children's Hearings, the integrated system that adjudicates matters of care and youth justice. In October 2019, the two of us (Sylvan and Maggie) travelled north to Glasgow along with a TVF facilitator, Alfie Kingsnorth, to meet and work with a group of young people who are members of the OHOV youth board, and with their adult youth worker project lead Jacqui Dunbar. Jacqui had invited us to work with OHOV in the facilitation of a two-day creative practical research workshop, which would lead to a one-week exhibition and event in the Scottish Parliament in Edinburgh for Members of the Scottish Parliament, civil servants, and key decision makers in the Scottish Children's Hearings System.

By sharing care-experienced young people's views and experiences of the Hearings system creatively, TVF would support OHOV in making a contribution to Scotland's Independent Care Review (ICR). Running from 2017–2020, the ICR listened to over five thousand young people and adults with experiences of the Scottish care system (Independent Care Review, 2021). In conclusion of this listening phase the ICR published *The Promise*, based according to its website, on what it had heard. It aims for Scotland 'to be the best place in the world to grow up' because children are 'loved, safe, and respected and realise their full potential' (Independent Care Review, 2020, p. 5). It states an 'unwavering commitment' to place Scotland's care experienced community 'at its very heart' (p. 4). Change will be predicated on 'a full and proper understanding' of how 'the "care system" feels'. *The Promise* insists that 'Children must be listened to', and a 'compassionate, caring, decision-making culture' be in place that is 'focussed on children and those they trust' (p. 13). The TVF-OHOV workshop sessions to prepare for the parliamentary activities also coincided with an event in Glasgow that was deeply relevant to the young people we would be working with. This was the Love Rally, where many hundreds of people took part in a joyous march through the city to show their support for people in the Scottish care system. Organised by the charity Who Cares? Scotland, the march was being held to bring attention to the need to build Scotland's care system 'on love', and to hold the government to its promises to do so (Browne-MacLeod, 2019). In their different ways, both the Care Review and the Love Rally paid attention to affective dimensions of the experiences of care. Our attendance of the Love Rally, as we describe below, entered and permeated the sonic work we developed with the young people.

In this chapter, we relate and reflect on the TVF-OHOV practice research work we undertook during our two-day workshop in Glasgow, tracing the activities that culminated in the event at the Parliament. In doing this, we mark the spatial and sonic as significant and political dimensions in young people's experiences, and explore some of the ways that creative sonic and audio practice can provide modes of listening for adults in spaces where young people are not always heard on equal terms. Broadly speaking, TVF's auditory and sonic creative approaches to research attempt to explore dynamics and disruptions of links between lived experience, sound, and practices of systemic and political listening. The project works with young people to articulate the ways that they are made to feel in care and legal spaces and systems, and the affective consequences of environments, regulations, and routines that are determined by adults. We ask whether such systems protect and nurture young people's lives appropriately and carefully enough. If they sometimes do not, can sonic and audio practices be used to create or suggest more caring and generative encounters? How can such practices create connections and continuities between young people whose lives have been disrupted and the policymakers and 'corporate parents'[2] who remotely determine the situated processes and systems of care and justice that they must undergo?

We argue here that in this phase of our TVF practice research a sonic soundscape used at the TVF-OHOV event at the Scottish Parliament opened a politically inflected dimension of creative emergence, potential and change. Embracing a sense of trespass at this event, as 'rhythmanalysts' and as creative practice researchers, the TVF-OHOV collective released affect in ways that transgressed boundaries of rational debate and infused policymaking space with lived experiences of care and in/justice. This chapter explores how creative sonic practice might disrupt the habits and norms of listening even when compassion is assumed to be a given, stimulate and adjust the quality of attention paid by policymakers in spaces where young people are not normally heard, and encourage them to recognise the uniqueness and potentials of young people beyond their care identities.

Attending to voice, sound, and agency

In the 2000s, the work of sociologist Nirmal Puwar explored the embodied and psychic relations of people with their spatial environments. Her critique of public political spaces and cultures in *Space Invaders* (2004) including the UK Parliament, examined the often unnoticed gendered and racialised 'somatic norms' that lead to the exclusion and marginalisation of those who do not belong to the dominant culture (Puwar, 2004, p. 7). Puwar's usage of the phrase 'space invaders', borrowed from sociogeographer Doreen Massey (1996, p. 185), acknowledges the embodied sense of those who 'trespass' in places where one's right to be is not undisputed (Puwar, 2004, p. 8). Puwar's study draws on the work of French philosopher Henri Lefebvre, which examined the social production of space and the embodied and psychic relations of humans with their environments. For Lefebvre (2012), relations between bodies and space are dynamic and mutually productive. While it is through the body that space is perceived, lived and produced, the 'laws of space' act to 'govern the living body and the deployment of its energies' (p. 170). It is in the affective dynamics of such relations, in the mutually productive links of space, psyche, and soma, that the care-experienced young people with whom TVF has been working are all too expert. A significant part of this expertise lies in the acute sensitivities to the ways their voices are heard inside spaces and systems of care and justice.

Uncomfortable feelings are frequently experienced by young people who enter such spaces and systems, even when they are apparently designed to protect them. Not only must young people undergo, often repeatedly, traumatic and dislocating experiences of being uprooted and entering homes that feel strange to them, but they have no choice but to become the subject of meetings that dictate their futures and files that monitor their progress, and to systems and procedures that make very significant decisions affecting their lives. One of TVF's care-experienced consultants, Jade Kilkenny (2020), has written that she found the experiences of being moved from place to place 'terrifying'. How could it not be, Jade argues, when children are taught not to go into strangers' houses. Even when a young person's new living arrangements and foster carers are accommodating and sensitive, becoming a space invader in such enforced circumstances is almost unavoidably an experience that is textured with very acute affective dimensions of rejection, loss, fear, and shame. Yet very often, the affective aspects of their experiences become discounted or dismissed, given what is adjudged to be in the best interest of their care, and even when such decisions exacerbate negative experiences. Speaking of these matters is far from easy. Jade likened the experience of going into care to being in a porta-loo at a festival, 'it's not just your business, everyone can look in at you whenever they want and you can't choose to close the door' (Kilkenny, 2020, Para. 14). Further, even though listening is commonly emphasised as central to good care, and even when a young person is heard with compassion, this can be undermined by them being made to comply to the judgement of those in authority rather than arrangements being made that accord with their own desires and perceptions. One of our co-researchers testified that she felt 'scared and quite uncertain' during the estranging experience of a justice hearing when a solicitor she had never met arrived to 'break down' everything that her carer had put in place to make her 'feel safe and comfortable' (OHOV, 2020). This felt like a breach of trust, undermining the commitment of her carers to her privacy, safety, and wellbeing.

In the realm of politics, within which young people's voices are infrequently heard, an emphasis on compassionate listening contrasts with historical conventions. As Susan Bickford (1996) elucidates, Western democracy is traditionally based on the 'dissonance' of rational debate. 'Political listening is not primarily a caring or amicable practice' (p. 2). For philosopher Hannah Arendt, the danger of compassion is that it erases the differences between humans as unique beings. For Arendt, there must be a quality of attention in the public realm that pays proper regard to *who* a person is rather than *what* they are (Bickford, 1996, p. 80). The assumption that Arendt makes of 'freedom' in the public realm (1958, p.73) does not acknowledge the ways that people's voices may be valued differently depending on perceptions of the speaker's identity, nor that different people will feel more or less comfortable in spaces that provide public and political fora. Nevertheless, her argument for regarding the uniqueness of a speaker is one that is important for the work of TVF in facilitating dialogue between young people and adult policymakers. As Bickford (1996) summarises, habits of seeing and listening can encourage a 'stereotypic perception' that 'imposes a mask whose effect is to blur and muffle individuality rather than let it sound through'. It is therefore important to find ways to disrupt the 'systematic distortion of some people's appearance and audibility' (pp. 96–101). In the case of adults listening to care-experienced young people, there is a risk that adults will respond only to the young person's generic care identity, failing to listen for the nuances of each young person's unique potential and individual being.

TVF's coalition with OHOV allowed us to explore the potential of creative auditory and sonic practices to resist the conflation of a young person's experience of entering care to a fixed 'care identity', and to embrace a sense of *trespassing* in the public realm and its political spaces. We wanted to cultivate a listening that would encourage respectful attention and regard, while avoiding a potentially patronising humanitarian perception of young people solely as objects of compassion. For Brandon LaBelle (2018), there are political possibilities in 'the shifting flows of vibrancy and reverberance' that 'shape our interactions with the world and with others' (loc. 78). A kind of 'sonic agency' is possible through the interventions of sound in the sphere of appearance. A sonic sensibility gives challenge to the primacy of visual gaze that operates to 'define limits, to pinpoint those who may cross lines or borders, to delimit the permissible within the social field according to what or who appears' (loc. 68). LaBelle follows Salomé Voegelin (2014), for whom a sensibility that is sonically attuned demands an interactive engagement between those who are present—a 'negotiation between your invisible world and mine' (p. 24). Immersed as 'part of the soundscape', a listener participates in the production of the space around them, and is part of a geography that is composed of multiple histories and 'timespaces' which include affective textures such as 'memory, anxiety, and sentiment' (p. 24). Applied to encounters in spaces between adults and young people, such observations suggest the potential of sound to afford each person present a way to connect with another's material and affective experience in the world. That is to hear, we are arguing here, beyond the habits and rules of listening that might confer power unequally or act to confirm objectifying, hurtful, or unjustly normalising or narrowing perceptions.

The approaches we adopted in the TVF-OHOV workshops resonate with phenomenological methodologies alert to dynamic relations between the body and the world. In the account in the following section of TVF's activities in Glasgow and Edinburgh in our journey to the Scottish Parliament, we trace how our use of sonic and acoustemological sensibilities aimed to engage and use the alertness of care-experienced young people to the affective textures of embodied experience. In a sense, during our collaborations with OHOV the TVF team and its young co-researchers made use of Lefebvre's methodology of 'rhythmanalysis' which advocated an attention and sensitivity to the body as it lives and breathes within the tissue of its surroundings. A rhythmanalyst, Lefebvre (2013) argues, must 'catch a rhythm and perceive it within the whole'. In common with non-analysts, a scholar rhythmanalyst must 'arrive at the **concrete** through experience' (emphasis original). Alive to the sounds of 'the wind, the rain, the storms' the rhythmanalyst understands that 'nothing is immobile', and that 'objects are not inert' (pp. 30-31). As part of this analytical practice of alertness a rhythmanalyst acknowledges bodily responses such as 'palpitation, breathlessness, pains' to rhythms that can surge up and impose themselves. Such an approach is useful, we suggest, not only in capturing and conveying young people's situated experiences, but also in the creative use of sonic practice to intervene in practices of listening. By cultivating an acoustic 'knowing', we aimed to encourage in our Parliamentary listeners 'alternative ways of engaging with the world, and aliveness to the possibility of hearing other realities' (Rice, 2018, p. 3).

Glaswegian encounters

Arriving in Glasgow late on a dark, squally Friday night after a long train journey from London, we made our way to the hotel where the OHOV group were meeting. In the foyer, we were greeted warmly by Jacqui, who was with some of the young people we would be working with as TVF co-researchers. To them, the TVF crew were strangers—academics Maggie, a middle-aged white woman, and Sylvan, a middle-aged Black man, and Alfie, a white actor and youth worker in his late 20s. The young people themselves, 15–19-year-olds, were arriving from all over Scotland. Some had come from remote highlands and islands, some from the urban centres. They had met each other on a few other occasions before our creative weekend. In getting to know them, it fortuitously helped that Alfie was recognisable to some of the young people from being in an episode of popular TV show *Hollyoaks* (1995-). Drawing on her own knowledge of and sensitivity to how young people feel when required to encounter adults for the first time, Jacqui welcomed and introduced us to each other. As we interacted any reticence on either side quickly vanished. We all relaxed, and individual personalities, moods, and affects, including excitement and anxiety, quickly surfaced.

After breakfast the next day, we walked to the venue where the workshop would take place. Jacqui had hired a set of rooms over two floors in Glasgow's Royal Concert Hall. The huge high-ceilinged venue made us feel expansive—the kind of space that makes you want to leap. Our main room had a huge window overlooking Buchanan Street, Glasgow's main shopping thoroughfare, and over beyond the city. This felt like an empowering position—a bird's eye view of the city, almost worthy of the 'totalising' gaze from the World Trade Centre, that according to Michel de Certeau (1984) lifted the viewer 'out of the city's grasp' (p. 93). This view from the window was one that the young people returned to frequently. We started our workshop with icebreakers and games, getting to know and trust each other. Next, we introduced TVF's sonic and auditory approach. Maggie began by asking the participants to talk about the sounds they loved in their everyday lives, giving the example of the grind of her bicycle gears as she sped through London traffic. This was an attempt to use the 'auditory imagination' to engage with the rhythms of the daily lives of the young people we were getting to know, and to share our own, through a 'sonic mapping of place as the hearing of sound' (Voegelin, 2014, p. 25). The young people described the sounds of their homes, their animals, the natural environment they loved to inhabit and interact with, and the music they loved to play and to hear. As they spoke, a sense of what Lefebvre (2013) calls '*bundles, bouquets, garlands* of rhythms' (p. 30) of their daily lives and surroundings emerged through their descriptions of sounds, and a sense of the affective investments, attachments, and passions of their lives. The sound of the sea was a recurring theme, several of the young people as well as Jacqui describing the constant crash and wash of the waves many of them heard every day in the outdoors environment. Together, using voices and materials around us in the room, we made a soundscape that built into a growing storm of rain, wind and sea, becoming present through the wild night of our collective imagination and passing on into silence.

On this first full day of our workshop, Glasgow's Love Rally intervened in our schedule. As a group largely composed of care-experienced young people and adults, we decided to join it, and include it in our sonic fieldwork by recording its sounds. Descending from our elevated position in the Concert Hall, we passed the statue at the base of its steps of Scotland's first First Minister, Donald Dewar (1999–2000). The statue seemed to point backwards to the history of Scottish democracy, and forward to our forthcoming trip to the Edinburgh Parliament. Weaving through the streets of Glasgow, we joined the procession on its way to George Square, the location of another large statue, of British Hanoverian King George III. There, a line-up of speakers with inspiring and supportive words for the care sector was to place care-experienced young people on a platform with Deputy First Minister of Scotland, John Swinney. On the way, amidst the throng of people, we noticed the beauty and grandeur of the buildings in The Merchant City—built, Sylvan commented, with profits made through the city's connection to the transatlantic slave trade. But we did not only experience Glasgow's history through its buildings and monuments; on that day the histories and presence of its peoples were *audible*, reverberating off antiquated buildings, carried in voices and music, moving and marching in the space around us. As a procession we moved to the beat of Sikh *dhol* drums, a reminder of Scotland as a destination for Sikh migrants, who have themselves welcomed all comers to the food kitchens of Glasgow's Gurdwara temples.

Writing about the excitement of crowds and musical protest, Marie Thompson and Ian Biddle (2014) explore how sound creates particular 'ambiance, or atmosphere, via the induction, modulation and circulation of moods, feelings and intensities' that can be felt and heard (p. 5). On this brightening October day in Glasgow, immersed in the unpredictable swirl of the crowd, the rhythm of the *dhols* lifted our hearts, making them pound with a sense of anticipation, pride, celebration and joy. Our excited voices became part of a sonorous flow, as the talking, shouting and chanting of young people, families and those in solidarity with care-experienced people surrounded and moved around us. All around us now, people waved flags and carried banners: 'We need love. Not a system' and 'Communities That Care'. Reaching the temporary stage erected for speeches, the band played the traditional anthem Scotland the Brave on bagpipes and *dhols*. The blend was significant, a world music mix of a multicultural community, producing anew a vibrant social space in Scotland with the sound of the crowd and its multiple but distinct communities.

After slipping away from the speeches to snatch a hasty lunch, our group returned to our work together. Back in our splendid suite of rooms at the Concert Hall, the TVF team shared with our Scottish co-researchers some of the testimonies that had been gathered during our previous workshops with care-experienced young people. We used the headphone theatre technique, where anonymised audio clips are made from recorded interviews (Baker & Inchley, 2020, pp. 181–182). Sharing these stigmatised experiences of the dehumanising aspects of care with the young people in Scotland triggered a sense of connection for the Scottish young people. They were deeply moved, feeling an intense affective connection through these testimonies to the 'invisible worlds' of their own lives. In private conversations with Maggie, the young people shared and recorded their own experiences. One 15-year-old, Paige, told us that her social worker had turned up at school one day in his car with her stuff in bin bags. She was being moved she learnt, without any warning whatsoever, to a new 'placement'. The social worker hadn't brought her make-up, which to Paige was her most important personal possession (OHOV, 2020). Another girl, Kayleigh, spoke of how as one of the 'compliant' and uncomplaining young people in need of care she was often ignored once she had been 'placed' (The Verbatim Formula Living Archive, 2015-). Her self-esteem had suffered an ongoing knock, and she still had difficulty going outdoors to unfamiliar places on her own. We also heard from the young people about their experiences of the Children's Hearings system, the difficulty of speaking with solicitors they did not know, the intimidation of court procedure and places, and having no say whatsoever in the rules decreed by adults that determine who can attend a Hearing, who can be heard, and how and when.

As well as raising pointed challenges to the workings of 'justice', many of these testimonies seemed to speak to young people's feelings of being 'space invaders', of the anxiety and palpitations that surge up in strange spaces whose rhythms, rituals, and languages are alien. Tellingly, Jacqui summed it up with her description of the 'big scary table' that dominates procedure in the Hearings system, and makes the experiences of justice for many young people intimidating and alienating. Jacqui's phrase resonated strongly with the young people. Soon, Sylvan and Alfie were facilitating activities with movement and space, working with suggestions from the young people to discover moments and positions of power between people, and explore how different arrangements of furniture consolidate or protect power. During this activity, the young people also called out words that they said had been used to label them, fixing their identities as 'aggressive', 'unstable', 'troublemaker' (OHOV, 2020). We pored over a ground plan of the Burns Room, the space Jacqui had booked for the event in a few weeks' time, to which Scottish Members of Parliament and other professionals involved in Scotland's Children's Hearings System and Independent Care Review were invited. How would we orchestrate our event in this room in the limited time we had?

When asked what they most wanted the invitees to take away with them from this forthcoming event, two of the young people seemed to lay down a challenge, 'How would you feel in my shoes?' asked Ryan, gently and boldly insisting on the need for more empathy and sensitivity to young people who enter care. 'It's not the adults' future. It's ours. We can change it for the better', stated Kayleigh, articulating a moral commitment to political change, and a position of agency in creating the conditions of the future. Her assertion raises important questions about *The Promise* and institutionalised policymaking. A promise suggests a fixed point in which commitments are supposed to be fulfilled, though its fulfilment remains subject to the flow the events. The actual or approximate conditions that people in the future must live with are created in *the present*, and are contingent on whose experiences are attended to and whose voices get heard. Young people are for the most part excluded from democratic processes, especially those who are under 18 who cannot, as elsewhere in the UK, vote in Scotland. Yet, as we are discussing, in the case of young people who enter care their lives are profoundly affected by political decision making. Again, Kayleigh's comment, and the commitments made to listen to young people in the ICR, seemed to point to the visit we would make in a few weeks' time to the seat of Scottish democracy.

The Scottish Parliament

On the face of it, the architecture of the Scottish Parliament in Edinburgh suggests a process that is open and outward looking. It is located between Salisbury Crags and Arthur's Seat, the extinct volcanic hill visible from many parts of the city, and the busy streets Canongate and Holyrood Road. Taking its shapes from the world around it, according to RIBA Stirling Prize judge Charles Jencks the Parliament 'nestles' into the environment as an 'icon of organic resolution', avoiding the 'monumental' architecture so often preferred by national buildings (Jencks in Langdon, 2021). Such monumentality, according to Lefebvre's (2012) discussion of cathedrals, immerses visitors both ideologically and sensually, as they decipher the symbols around them, transmuting fear of the passage of time. Rather than suggest the 'timeless' promise of monumental space by closing out natural and urban environments, the Scottish Parliament gives a sense of its purpose to embrace its business in the light of the present. Even the windows in the debating chamber are positioned to give glimpses of the city and landscape beyond. A poem by Edwin Morgan (2004) which begins 'Open the Doors!', written for the Parliament's opening, is prominently displayed to the public. The building, it says, is not expected by the people to be filled with 'a symposium of procrastinators', but with 'thinking persons as open and adventurous as its architecture'. The poem's statement, that there is 'a commerce between inner and outer, between brightness and shadow, between the world and those who think about the world', in some ways suggests a remit of parliamentarians that is methodologically rhythymanalytical, a duty to be in touch with the natural and social world beyond the building's bounds, and to cultivate a sensibility that connects with what is hidden as well as what is in the light.[3]

As we mentioned above, Jacqui had succeeded in booking the Burns room, named after Robbie Burns, the Scottish national poet. In his poem 'To a Mouse' (1785) Burns apologises for the 'panic' that 'Man's dominion' causes in a wee creature's 'breastie' for turning over her nest with a plough. In the planning stage that had taken place before the Glasgow workshop, Jacqui, Sylvan, and Maggie had visited this room, and had been shown its impressive modern facilities. When we first entered on the day of our reconnaissance visit, we had experienced a feeling of expansive connection with the natural environment. We craned our necks to see upwards to its high ceilings, the beams of oak reaching into the sky, letting natural light pour inside. There had been no sign of mice, the wee creatures no doubt not daring, we had speculated, to creep inside. Jacqui had been in close contact with Parliamentary officials, and pleasantly advised of the strict rules that would apply to our event and to the exhibition that would accompany it. Regulations, Jacqui discovered, were strict. There were rules to be observed on the space the exhibition could occupy, how many people could staff it, the exact time we would be allowed to enter the Burns Room, and what we were allowed to eat and where. Security regulations had to be followed, passes had to be issued, and groups accompanied by authorised officials.

The centrepiece of the room is a large round table used for meetings, with each station for delegates equipped with the microphones that add a charismatic technological buzz to the voices of speakers, drowning out much of the ambient, incidental, and background hum of the every day. We wondered how we might disrupt the sense of human sonic dominion that this technology tends to enable. Though it allows voices to be heard, what might this amplified vocality drown out or silence? To us, the round shape of the table suggested a space for listening; the wooden platform that formed the sunken middle of the table, designed like a pond. As practitioners whose key artistic practices are aural, we were mindful of 'the potential of listening as a way of redistributing power, reorganising power relations between who is speaking and who is listening' (Farinati & Firth, 2017, pp. 18). How could our event disrupt the Parliamentary norms that give dominion and rank to adult speakers? How would our use of sound, which travels faster over water than land, we wondered, encourage our invitees to listen? These were questions we explored with our young co-researchers in the activities we described above, as part of the workshop in Glasgow during which we planned for the event.

By the time the day of the event arrived, Jacqui had made further detailed logistical preparations. In the morning we met at the nearby Scottish Youth Parliament with the young people, some of whom had travelled again from far flung sea-washed and wind-swept corners of Scotland. Jacqui had briefed the group, and we headed off in good time to pass through parliamentary security, to make our hasty preparations in the Burns Room in the short time we were allocated. By the time the adult invitees arrived, the young people were sitting at the central table, taking the position where one might normally expect to see adults. Some of them, especially those with autism, were using pens, paper, and fiddle toys to mitigate any danger of anxiety in those who might feel themselves perceived as 'space invaders' by the incursion of adults who seem to occupy such spaces with such uncontested ease. Drawing on our exploration of space and power in our Glasgow workshop, we had switched up the norms of meeting etiquette, not inviting our guests to the table. Instead, they were ushered towards the spaces at the bottom of the room. Here, they were asked to stand, or to sit around the edges. The young people, for this hour, would be sat at the table, setting the sonic agenda.

To start proceedings, we blocked out the natural light, attempting to disrupt any assumption of political and symbolic transparency carried in its physical architecture. By drawing the blinds and dimming the lights, we aimed to challenge the logistical norms of conventional meetings, and to stir affects and responses that might feel out of the ordinary in the everyday Parliamentary routine. We hoped that the unexpected darkness would ready our invitees for a reorienting experience of non-visuality, reordering the sensible to allow the unexpected to materialise. We played a soundscape from our portable mini speaker, amplifying the track through the meeting room microphones. This soundscape provided an aural record of our TVF-OHOV work in Glasgow, and melded sounds of the young people's lives and their thoughts far beyond the immediate time-space. The sound of a crashing sea filled the room, and as the soundscape unfurled in time, scraps of voices layered into the sea noise became audible. The phrases 'What would you do in my shoes', and 'it's not the adults' future, it's ours' emerged from the crash of its foamy brine.

Gradually, the seascape faded and merged into the sounds of the outdoors event we had attended in Glasgow, captured on Maggie's recorder as we had marched along. The Love Rally drums leapt into the Burns Room to quicken our hearts, resounding with the streets of Glasgow where, as part of the rally, we had experienced an intense feeling of renewal, of joint endeavour, and a sense that we marched to a rhythmic beat that stirred a sense of something to come. Sudden distinct or recognisable voices and phrases rose from the sound of the social ferment in which we had been immersed that day, only to merge again into the sonic concourse. These sounds, we hoped, would stimulate the 'auditory imagination' of our guests, shifting masks, opening them to spaces and times beyond the immediate.

Our acousmetelogical approach aimed to ensure that not only were invitees listening to the young peoples' testimonies, but that they had the opportunity to encounter aspects of the young people's worlds in a way that seeped outside of norms and assumptions of Parliamentary process. Our soundscape and audio performances were intervening in the 'sonic architecture' of Scottish Parliament, inveigling themselves between and inside its affective and rhythmic structures, encouraging our interlocutors to be embodied and present to us and to each other. We had invited attendees to pay attention to disruptions rather than to observe routines, to tune into the 'sonic sensibility' that according to Voegelin (2014) demands that we be open to 'the truth of the world heard in its unseen mobility', and which 'imports the affective into the science of place' (p. 25). Through the sonic medium, a plurality of 'sonic timespace environments' (p. 24) are produced that include affective dimensions, and which can trigger in each and every listener different personal connections, memories, and responses. For Voegelin, listening 'illuminates the undulating pool of sound that moves and shapes the landscape, to hear at its depth an alternative view of all it is and all it could be, forcing new consequences onto our living in the world' (p. 12). We were aiming to rearrange the sonic norms of Parliamentary procedure, and in so doing inflect the future, as it was emerging from the bodies and minds of those who make it, with the lived experiences and creative presence of care-experienced young people.

After letting back in the light, Sylvan, Maggie and Alfie shared the young people's testimonies of the 'big scary table' and others that gave their experiences of entering care and undergoing Children's Hearings proceedings (OHOV, 2020). As in other TVF events, we used the headphone technique, where recorded testimonies of an anonymous interviewee are spoken by another, aiming to honour the experience of the original speaker (Baker & Inchley, 2020). A listener hears the exact words of the young person, but at a remove, through the body of another. This is intended as an act of solidarity that both protects privacy and evades the risk that the words are heard without a sense that some action is required of the listener. In addition to sharing their testimonies, the young people performed music and poetry, giving a sense of the affective qualities of their experiences and the unique and creative expression of them. Through these activities, the experiences of even the most withdrawn or shyer young people were shared. Significantly, this was a part of the young people's presentation of the work of OHOV in advising on improvements to the Hearings system. The young people asserted a sense of creative political agency, of themselves not solely as recipients of care but as political agents involved in a co-constructive policymaking process.

Promising care

Nearing the end of this chapter, we return to the dynamics and implications of working with care-experienced young people and the bold commitments that *The Promise* has made. Amongst them, voice is proposed as a foundation to be embedded into culture, policy and practice across Scotland. We hope this chapter has served to suggest some of the complexities of this aim, and to reinforce the awareness that big changes are needed for young people to feel that they are heard appropriately. Understood as a speech act, as defined in the theory of J. L. Austin, a promise carries an 'illocutionary' force (1962, p.101), creating a state of affairs in the world that did not exist before, a relation between you and I newly produced in social space. A promise is itself therefore productive of social space and structure in which the promiser is expected to make good on their commitments and be responsible for them. According to Austin however, a performative can become 'unhappy', failing because the conditions required for its action to succeed are not in place. Worse, it can become a 'non-performative' (Ahmed, 2012, p. 117), where a repeated failure to do what it says becomes its very action. It may be true however, as Ahmed writes, that when commitments do not work, the 'non-performativity' of promises becomes usable as a tool, making it possible to show that institutions *'are not doing what they are saying'* [original emphasis] (p. 121).

Arendt (1998) reminds us in *The Human Condition*, a promise is a safeguard against the 'basic unreliability' of people and 'the unpredictability of human affairs' (p. 244) and is reliant on forgiveness as an amelioration of the 'incalculability of the future' (p. 245). This, again, is an area in which care-experienced people are expert. TVF has worked with many young people who have testified to feelings of disappointment and ongoing experiences of uncertainty. The project has frequently encountered many who have intense and complex feelings towards family members, and feel let down by the systems of care that promise to help them. Under circumstances where no one seems to be listening, let alone *doing*, the 'unhappiness' of performatives remains stuck with the young person to whom promises were made. And yet it is the world's very uncertainty, and the possibility of going forward to a future un-foreclosed by an identity marked by the stigma of care, that young people embrace in their bravery and through their profound understanding.

Towards the end of our event in the Burns Room, a speech by Maree Todd MSP, Scotland's Minister for Children and Young People seemed to return us back to the normal routines of Parliament, where proceedings are dominated by those with the power of elected authority to speak on behalf of others. Todd acknowledged the responsibility this role brings with it to her and more broadly to all adults who have participated in Scotland's ICR. As we write this in 2021, the consultation phase of the Review is long over. A new Oversight Board responsible for overseeing the necessary changes has been established, committing 'To hold Scotland to account' (The Promise Scotland, 2020). A series of three plans up to 2030 have been proposed, in which 'Listening' is one of five fundamentals which also include, 'What matters to children and families', 'Poverty', 'Children's Right', and 'Language' (p. 7). Under listening, the plans promise that children and families will 'have their voices heard', and 'participate in decision making'. As well as this there is a commitment to a 'full and considered exploration' of how 'listening, shared sense-making and shared decision-making' can be 'embedded into practice', with 'mechanisms to actively use what is heard' (p. 15). Under 'Language', there is a promise that language will not be stigmatising, and words and terms will be used that are preferred by children and young people when their lives are being referred to (p. 19). Fulfilling these promises, which respond to the listening phase of the ICR, will require tenacity, labour and resources, as the Oversight Board openly acknowledges.

When we asked the young people about TVF's processes, Ryan told us that he had discovered a creative methodology that gave him a chance to share his experiences of care without being 'berated', and that he had appreciated the 'warmth' of our work together (Dar & Williamson, 2020, p. 18). In this comment Ryan responds in a way that shows insight into the affective and sensory conditions of communication and collaboration. How he is heard, how he is perceived, and how he feels are very significant aspects of Ryan's experience. As outlined above, TVF is interested in the linkages of environments with the affective lives of young people, especially where new spaces and places can feel intimidating or alien, or where norms and practices of institutions and systems tend to marginalise or suppress their voices and views. Working affectively and sonically, we agitate for a political system which hears and responds to young people, their agency, and their potentials. Trespassing at the Scottish Parliament, our space invasion introduced the sound of the outside world and the fluctuating depths in which young people live their lives, opening the policymaking routine to worlds hidden or beyond normalised perceptions and the immediate context in space and time. As researchers, we recognise that we are complicit in *The Promise,* as we play our part in constructing shared social space and an expectation that young people will be heard. We hope that our work underlines that corporate parents and policy-makers must pay embodied and affective attention to young people's experiences of care in the situated contexts in which they unfold, and be open to their boundless potentials far beyond their care identities. They must ensure that young people are heard not only with compassion, sensitivity, and respect, but also as creative co-builders of shared social spaces, with due regard for their capacity to act in a world for which care is a collective endeavour.

Notes

1 All names of young people have been changed.
2 The term 'corporate parenting' refers to the bodies and services who have responsibility for meeting the needs of children in care and care leavers.
3 Sadly, the events of 2021 surrounding an alleged cover-up of First Minister Nicola Sturgeon's knowledge of complaints against her predecessor Alex Salmond have been a salutary reminder that Scotland's democratic processes are not necessarily as transparent as its Parliament's architecture suggests.

References

Ahmed, S. (2012). *On being included: Racism and diversity in institutional life.* Duke University Press.

Arendt, H. (1998). *The human condition* (2nd ed.). University of Chicago Press. (Original work published 1958).

Arendt, H. (2004). *The origins of totalitarianism.* Schocken Books. (Original work published 1951).

Austin, J. L. (1962). *How to do things with words.* Clarendon Press.

Baker, S., & Inchley, M. (2020). Verbatim practice as research with care-experienced young people: An "aesthetics of care" through aural attention. In J. Thompson & S. A. Fisher (Eds.), *Performing care: New perspectives on socially engaged performance* (pp. 171–186). Manchester University Press.

Bickford, S. (1996). *The dissonance of democracy: Listening, conflict, and citizenship.* Cornell.

Browne-MacLeod, K. (2019, October 19). Glasgow "Love Rally" for those who have experience of care. *BBC News.* https://www.bbc.co.uk/news/uk-scotland-glasgow-west-50109545

Burns, R. (1785). *To a mouse.* [Poetry Foundation]. https://www.poetryfoundation.org/poems/43816/to-a-mouse-56d222ab36e33.

Dar, S., & Williamson, D. (2020). *TVF impact report August 2020.* People's Palace Projects and Queen Mary University of London. https://www.cssd.ac.uk/media/1093/download

de Certeau, M. (1984). *The practice of everyday life.* University of California Press.

Farinati, L., & Firth, C. (2017). *The force of listening.* Errant Bodies Press.

Gablick, S. (1992). Connective aesthetics. *American Art, 6*(2), 2–7. https://www.jstor.org/stable/3109088

Inchley, M., Dar, S., Pujara, M., Siqueira, P., & Baker, S. (2022). Performing aural and temporal architecture: Re-framing the university through The Verbatim Formula'. In A. Lindelof, S. Janssen, & O. Harslof (Eds.), *Performing institutions.* Intellect. Forthcoming.

Inchley, M., Dar, S., Pujuara, S., & Baker, S. (2019). The Verbatim Formula: Caring for care leavers in the neoliberal university. *Research in Drama Education: The Journal of Applied Theatre and Performance (RiDE), 24*(2), 413–419. 10.1080/13569783.2019.1604128

Independent Care Review. (2020). *The Promise.* https://www.carereview.scot/wp-content/uploads/2020/02/The-Promise.pdf

Independent Care Review. (2021). *The care review.* https://www.carereview.scot/about/

Kilkenny, J. (2020, April 13). *Living with strangers.* [Resilience Learning Partnership] https://resiliencelearningpartnership.co.uk/living-with-strangers

LaBelle, B. (2018). *Sonic agency: Sound and forms of resistance.* Goldsmiths Press.

Langdon, D. (2011). *AD classics: Scottish Parliament building/Enric Miralles.* [ArchDaily] https://www.archdaily.com/111869/ad-classics-the-scottish-parliament-enric-miralles

Lefebvre, H. (2012). *The Production of space* (D. Nicholson-Smith, Trans.). Blackwell. (Original work published 1974).

Lefebvre, H. (2013). *Rhythmanalysis: Space, time and everyday life* (S. Elden & G. Moore, Trans.). Bloomsbury Academic. (Original work published 1992).

Massey, D. (1996). *Space, place and gender.* Polity.

Morgan, E. (2004, October 9). *For the opening of the Scottish Parliament* [Scottish Poetry Library] https://www.scottishpoetrylibrary.org.uk/poem/opening-scottish-parliament-9-october-2004/ [Accessed 12 May 2021].

Our Hearings Our Voice (OHOV). (2020). *The zine.* https://www.ohov.co.uk/about-us/projects/the-zine/

The Promise Scotland. (2020). *Oversight board.* https://thepromise.scot/oversight-board

Puwar, N. (2004). *Space invaders: Race, gender and bodies out of place.* Oxford: Berg.

Rice, T. (2018). 'Acoustemology', *The International Encyclopedia of Anthropology.* Accepted version. [pdf]. Available at: https://www.academia.edu/37710771/Tom_Rice_Acoustemology?auto=download. [Accessed 12 May 2021].

Thompson, M., & Biddle, I. (Eds.) (2014). *Sound, music, affect: Theorizing sonic experience.* Bloomsbury.

The Verbatim Formula Living Archive. (2015-). *Hear it, see it.* [The Verbatim Formula]. http://theverbatimformula.org.uk/hear-it-see-it/

Voegelin, S. (2014). *Sonic possible worlds: Hearing the continuum of sound.* Bloomsbury.

11

YELLOW COUCH CONVOS PODCAST SERIES: NAVIGATING IDENTITY POLITICS THROUGH COLLECTIVE VOICES AND COUNTERNARRATIVES

Rosemary (Rosa) Cisneros

Introduction

This article is concerned with the niche area of podcasting in relation to the Roma community's cultural heritage. The *Yellow Couch Convos*[1] (YCC) Roma podcast is freely available online, and was a one-year research project to create, curate, and disseminate Roma cultural heritage episodes in conversation with influential grassroots and heritage sector individuals, and institutions. The participants included artists, dancers, curators, musicians, writers, poets, among others, that self-identify as members of the cultural heritage Roma ecology. The YCC podcast honours Roma storytelling techniques and oral history traditions but modernises it through the use of digital platforms. The body-to-body experience that is typical of the Roma storytelling tradition is reimagined through technology. Storytelling in the Roma community is important as it is used to teach, honour traditions, and is a form of oral history. The series covered a variety of topics that ranged from cooking, theatre-making, dancing, filmmaking, education, dance, writing, and painting. Through the 'baton exercise' method, which will be explained further into this writing, the artists and educators helped curate and disseminate the series to Roma and non-Roma. As the producer and director of the project, and also the host of the podcast, I drew on my Roma cultural heritage and professional experience as an academic, artist, and filmmaker to execute the delivery of YCC. I relied on fellow Roma artists and the European Roma Institute for Arts and Culture (ERIAC) to bring the podcast to life. ERIAC has a unique and single mandate as the first transnational, European-level organisation for the recognition of Roma arts and culture (2021). The organisation aims to promote Romani contributions to European culture and talent, success, and achievement, as well as document the historical experience of Roma people in Europe. It was through ERIAC funding and the Digital Innovation grant,[2] that the YCC project was

DOI: 10.4324/9781003164227-20

produced and the only podcast they supported. In this chapter, I expand on the co-creative methods employed and frame the Roma community's tense social realities. At this juncture, to make sense of storytelling during a pandemic, I briefly touch on the inclusion of lived experiences within research environments.

Podcasting harnesses the power of the human voice, communicates embodied knowledge and lived experiences, and amplifies an individual's voice and their personal memories. With the aim of gaining a deeper understanding of the nature or meaning of an everyday experience, stories are tools to help critically reflect and make meaning out of lived experience. As McIntosh and Wright (2019) state, 'Researching lived experience can signify a strategy of recognition that is attentive to feelings, bodily states, interactions and identities that tend to be devalued or ignored' (p. 10). UK-based educators Jennifer Leigh and Nicole Brown (2021) expand on embodied research and sensory practices through embodied inquiry (EI), an approach that places the body at the centre of any research question and foregrounds how feelings, emotions, and experiences affect the manner in which we interrogate problems and generate knowledge and meaning making. EI therefore encourages a critical reflection on positionality and reflexivity in research. The method considers how a project or study could be designed to take the body and the individual voice into account and stakes a claim that multimodal and ethical creative approaches to research captures embodied experiences and buried memories. The EI method lends itself to analyse the storytelling approach of the series, as the YCC captured individuals critically reflecting on their lived experiences and creative practices.

This chapter demonstrates that a shift emerges when memories and embodied realities surface to consciousness through reflecting on artistic practices. Via an ethical and co-creative mode of working, diverse community voices can shine and begin to challenge dominant narratives and 'emancipate subjugated knowledges' (Foucault, 1980, p. 84). This writing engages in an exploratory manner with Leigh and Brown's (2021) EI method and the idea that research questions that relate to the participant's lived experiences are a prime example of democratic embodied methods of inquiry. I also aim to expand on the baton exercise method I developed and employed in this research project that allowed for an ethical co-creative process to emerge. The baton exercise approach celebrates the diversity of the community and is a form of creative activism, which makes audible a heterogeneous Roma identity. Such diversity is needed because an erroneous image of the community exists in media and mainstream population. Such misrepresentations need to be counteracted by more accurate and deeper reflections on the experiences of subordinate groups. With this paper I claim that the YCC Roma podcast series that was commissioned and funded as a result of the COVID19 pandemic, encouraged the guests, who were also research participants, to reflect on their embodied experiences and Roma identity. Through the process of participating in a carefully considered conversation that probed how artistic practice can draw on Roma traditions, the project had multiple foci: to document and make visible the Roma's diverse artistic practices and their lived experiences; to uncover if the production of this sound material could instigate a counternarrative; and to trial the baton exercise curatorial method. This research project was a form of collective and creative activism that supported the long-term goals of educating, raising awareness, promoting, and celebrating the Roma community's heterogeneous art practices and varied culture.

Background: The Roma and the ERIAC *Yellow Couch Convos* Podcast

The Roma community is Europe's largest ethnic minority, with an estimate of ten to twelve million Roma (Council of Europe, 2020). As more people move from country to country within the European Union, the question of identity surfaces and European identity becomes challenged and forced to reinvent itself and adapt to a changing environment. Europe's identity has always been manifold; hence (additional), minority groups should not pose a threat to its social and cultural cohesion. This is however not necessarily the case in a contemporary cultural climate increasingly ruled by fear and blame. Social cohesion and inclusion of communities become more important as divisions permeate various parts of society. To find and implement solutions that lead to more democratic spaces is essential as traditional stereotypes and racialised misconceptions of Roma dominate popular discourse. Accurate narratives of the Roma community are not circulated widely, and the plurality of voices and personal experiences are often tucked away and masked by erroneous images and disinformation. Aidan McGarry (2017) uncovers causes of racism towards Roma and goes on to offer positive and constructive ways to combat the grim reality of what he has termed 'Romaphobia'[3]. Roma are often vilified and discrimination against them continues as massive inequalities, social exclusion, and segregation takes place throughout the UK, Europe, and the Northern hemisphere. This reality is a consequence of multiple burdens of social discrimination for their ethnicity, for their socio-economic status and, in some cases, for their migrant background. Members of the Roma community suffer negative consequences that translate into low expectations for their educational careers, which reduces the opportunities for social inclusion (Topolanszky, 2018). The European Fundamental Rights Agency (FRA) suggest that the Roma are still being deprived of their basic human rights in Europe (2021). The Roma situation in Europe regarding education, health, housing, and employment is alarming: high rates of school failure, unemployment and poverty, and a high percentage of Roma who face educational segregation, or who live in the most deprived areas (ghettos), among others (FRA, 2014). According to multiple European Union (EU) institutions such as the Council of Europe, the European Commission, the European Fundamental Rights Agency, and others, stereotypes, and prejudices against Roma, also known as anti-Gypsyism,[4] is a persistent concern because there are determinant barriers to Roma inclusion.

EU Member States' policies and the politicians and systems in place rarely challenge Romaphobia (McGarry, 2017). The stubborn persistence of this reality shapes narratives that are linked with negative ascriptions. McGarry's (2012) book *Who Speaks for the Roma?* looks at political representation of the Roma community and attempts to understand the vicious cycle of under-representation. Rather than prescribe mechanisms to rectify this perceived injustice McGarry attempts to advance research and understanding of this situation through highlighting the various ways that Roma organise and speak for themselves. His book examines political arenas and institutions where the Roma have organised and also outlines the alternative platforms that have been used to create hubs of solidarity and networks with local charities as well as non-governmental organisations. McGarry asserts that until negative stereotypes are addressed, the Roma community will continue to be the most marginalised group in Europe. McGarry's work informed my research for the YCC podcast and instigated thinking around the representation of Roma in mainstream media. There are few mechanisms encouraging equal, direct, and open communication with Roma and thus, I wanted YCC to be a space where negative stereotypes could be dealt with directly.

Starting from this position, preparing for the YCC project, I researched what other podcasts on Roma cultural heritage had been produced. The European Roma Rights Centre's *The Romani Tea Room*[5] podcast and *Romanistan*[6] were two European podcasts that were focused on either policy or artistic practices. Neither had a storytelling element or broad thematic scope, rather, they were keen to discuss a topic rather than the person's life story. At the time of writing both podcasts have four episodes. There are singular episodes that cover some aspects of Roma culture, but I found few podcasts that are Roma-led and inclusive of a variety of guests that are from a number of countries. The development of podcasts focusing on Roma cultural heritage, grounded in Roma voices and cultural traditions, is not common. Therefore, YCC developed its thematic content ensuring that dancers, visual artists, educators, musicians, theatre makers, as well as chefs, producers, and academics could be included in the podcast (Figure 11.1).

Positionality and ethical research methods

I have outlined in the introduction that my ethnic background and experience as a Roma community member navigating the cultural heritage sector and research environments informed this project. Counternarratives for the Roma community are a key component of my research and creative practice. My previous experience in contributing and developing the Dance section of the RomArchive[7], an online digital open-access archive that makes Roma art visible to the rest of the world, informed this work. There are ten sections which include the Visual Arts, Dance, Film, Flamenco, Theatre and Drama, Literature, Music, Romani Civil Rights Movement, Politics of Photography, and Voices of the Victims. The RomArchive focuses on self-representation and the objects have been collated from private collections, museums, archives and libraries around the world. The Council of Europe (2021) states that the archive has as a main focus to offer counternarratives and to illustrate the contribution of Roma to European cultural history.

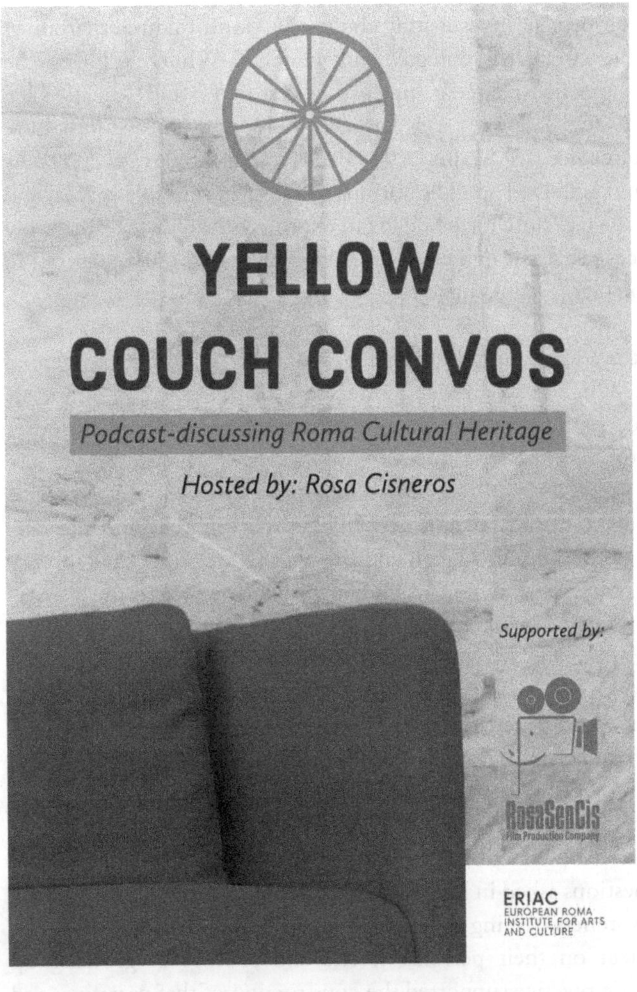

FIGURE 11.1 Logo for Yellow Couch Convos Podcast series

In addition to building a section of the RomArchive, I curated, produced, hosted, and disseminated a podcast series for a Coventry City of Culture project in Coventry, UK. Both projects allowed me to transfer my expertise in generating content with communities and know-how to the YCC project. Through my work and leadership of the Dancing Bodies in Coventry[8] I had the infrastructure set up to be able to deliver the YCC series. I also drew on my network with Roma in various European countries, extending from metropoles to smaller villages through the UK and Europe. The YCC project was produced professionally by myself and the materials were of high sound quality. The Zoom videoconferencing platform was used to generate and facilitate recordings. The audio platform SoundCloud[9] hosts the podcast series and it is freely available online.

YCC was a research project investigating the production of counternarratives and trialled co-creative curatorial strategies and practices. While exploring the differences between 'practice-related', 'practice-as-research', and 'practice-led research' falls out of the scope of this article, there is extensive literature pushing the discourse further (see Skains, 2018; Paillard, 2014; Sullivan, 2009). As an artist, researcher and activist, I straddle various roles, sit in multiple spaces, and critically reflect and navigate how ideas and creative projects intersect. I situate my research in the practice-based research camp and draw on Skains' (2018) definition to anchor my positionality. Skains suggests,

> the creative act is an experiment (whether or not the work itself is deemed 'experimental') designed to answer a directed research question about art and the practice of it, which could not otherwise be explored by other methods. (p. 86)

The YCC podcast falls under the practice-based research umbrella and the embodied collaborative method links to Leigh and Brown's (2021) embodied inquiry (EI). EI is centred in intersubjectivity, grounded in dialogue where the participation of all individuals is essential. EI lends itself to working with marginalised communities as it is participant-driven and is grounded in eliminating hierarchies. EI offers a co-creative approach that facilitates a co-construction of knowledge that is linked to the lived body and embodied experiences. The EI method draws on a premise that human communication is multimodal therefore knowledge also is multimodal, with the result that the different forms of communication, or modes, cannot be separated from one another. EI assumes that knowledge can be constituted through verbal interaction (Leigh & Brown, 2021). The method also acknowledges that questions asked in interviews throughout the life of the research can encourage self-awareness. Asking the podcast guests (also deemed the research participants[10]) to reflect on their personal stories and realities of growing up Roma, how their artistic practice supported the construction of that narrative, and how they chose to share that identity with others, is an example of the EI method in practice. The YCC podcast sound files are the creative artefacts and the basis of the contribution to the new knowledge.

The YCC question set was informed by EI principles and generated with the purpose of storytelling in mind. The questions were geared towards encouraging the participants to generate new knowledge drawing on embodiment and to link their life experiences with their artistic work. The YCC questions were given to each artist prior to the podcast recording and they were asked to prepare for the recording, to critically reflect on the questions, and to be self-aware. Each individual was also given the opportunity to communicate if they did not want to be asked a certain question or to steer clear of a certain topic. The eleven questions were:

1. Can you tell me a bit about your background? (country of origin, birth-place, family, etc.)

2. What is your art practice? What does that practice feel like in the body?
3. Was this passed on to you via a family member or elder from the community?
4. Is this practice exclusive to your culture? Country? Region?
5. Can you tell me how this art form has influenced your life? Profession?
6. How do you share this practice with others? Is there a difference in how you share this with Roma and non-Roma?
7. How would you describe the Roma community to non-Roma?
8. How would you describe Roma art? Does Roma art exist?
9. How do you see Roma cultural heritage in the future?
10. What advice would you give to young Roma? What have you learned from honouring an art practice?
11. What is your favourite colour?

To ensure that the questions were relevant and could instigate storytelling and critical embodied reflections, I asked a committee of five established Roma artists, respected or older people from the community to help me design the questions. They were known as 'critical friends' and were important throughout the life of the project as they helped me select the first four guests to invite. EI and its foundational principles, not only guided the development of the question set, but also framed my conversational style and approach to the actual interviews. Embodiment was central to the mode of engaging and interacting with the research participants at every step of the way.

Independent of EI, I developed the baton exercise, which involved a co-creative blueprint for the podcast—an egalitarian mode of working and of structuring the work. In practical terms, this meant that for the first four interviews, I approached a group of Roma artists, historians, and educators. This first set of invited guests were agreed by the critical friends committee and after their recording of the podcast, they were asked to suggest another Roma artist of their choice to invite to the series. That 'nominated' Roma artist would then be invited and a session would be scheduled, and once completed, then this individual was asked to nominate another Roma artist. This mode of working allowed for the series to be co-created and co-curated. This process and methodology allowed for a decentring of the individual and rather reimagined how a process could be more democratic and horizontal. This curatorial experiment produced a podcast that focussed on Roma cultural heritage materials and co-creatively generated and distributed the materials in an egalitarian manner. As architect, lead principal investigator and host of the project, this baton exercise method removed me from maintaining control and governing every aspect of the podcast. The approach, which was experimental and developed within this project, went on to inform another project entitled *Interactive Dialogues: Ethics and Equity in Practice Research—Dance and Theatre*[11] (2020–2021) run out of the Centre for Dance Research.

Each podcast guest was allowed to offer feedback soon after recording the episode. After each recording, the podcast episode was edited to varying lengths ranging from 25–45 minutes and the recording was then sent back to the individual to review and approve. A consent form was signed which clearly stated how I planned to share and store their data. It was only after this exhaustive process that the episode went public online and sat on the SoundCloud platform.

Another ethical dimension to the project is linked with remuneration for the participants. ERIAC supported the project with a small grant to cover costs of the production and to pay for the guest's time. Honouring the guest's embodied experiences and professional practices through remuneration is an ethical practice that values their lived stories. Each artist who contributed was offered a fee and, in some instances, the artists chose to donate the fee back into the project allowing another Roma artist to be invited.

For the purpose of the project, the artists interviewed, the critical friends committee and podcast listeners were asked for feedback on either participating and/or listening to the series, post publication of the podcast episodes. For those listeners who were not connected to the project, they were recruited through social media channels or emails. In some instances, listeners had previously sought me out to send me comments via direct messages through Twitter or the SoundCloud platform. Some of the podcast guests also followed the series and these listeners were invited to contribute to the research via a short survey or an interview. The testimonies included in the discussion sections below are anonymised as this ensured that participants could speak freely and openly about the series and their reflections on the work. The same is true for data from research participants and those offering feedback via interviews and surveys. For clarity, all of the responses embedded in this chapter have not been made public and all the data collected was gathered post publication of the podcast and consent was gained.

In summary, Leigh and Brown (2021) state that EI 'is very flexible and open and does not prescribe one particular way of carrying out research' (p. 64). While EI is not targeting bodies that are marginalised, it is using its 'critical-reflexive focus and offers an opportunity to reconsider positions of power, hierarchy and control and to see who is absent from a discourse' (p. 67). EI and the baton exercise gave authority to the Roma community to provide valuable content that reflects Roma cultural heritage and their own lived experiences.

Embodiment, representation, and the sonic

Meerzon (2021) claims that representation of a marginalised community is entrenched in tension and difficult to showcase without falling into blatant stereotyping, objectifying or further victimisation. To situate the reader, I briefly outline ways that Roma representation exists within the arts and cultural heritage sector. Literary scholar Mihaela Moscaliuc (2019) has examined how the fashion industry has recycled stereotypical tropes of Roma and has examined counter-appropriative practices that challenge and reclaim heritage and self-representation. She states:

> Exonymic representations of "Gypsies" have circulated for centuries across cultures and continents and have been reinscribed and disseminated through

> art, entertainment, literature, music, and now social media. Whether exoticizing or vilifying, such representations have operated together toward sustaining individual and national fantasies while also enabling gadje [non-Roma] to project their fears and justify discriminatory practices.
>
> *(Moscaliuc, 2019, p. 95)*

Roma woman, historian, and educator Sarah Carmona (2018) has explored decolonial methods in the field of museum education. Her research reflects on paintings in the Louvre in Paris, France and the Prado in Madrid, Spain. Carmona's analysis of Romani iconography in 13 selected works discusses the 'othering' of Roma both in society and in art. She claims that Roma are seen by thousands in these cultural heritage institutions but are not noticed because they remain in subjugated positions (see also Carmona (2020) for further description of the Louvre collection). The Roma historian expands on the notion of 'othering' in relation to the Roma and says that such viewpoints can be found in literature, media, art as well as other fields. Romani and gender studies academic Marina Csikós (2020) has studied intersectionality and feminism, and states that the majority of women of colour, including Roma women, have been excluded from social, cultural, and economic positions of power that would allow them to generate knowledge that shifts hegemonic discourses. She goes on to reflect that through art, several Roma women are creating their own spaces and also contributing to the development of Roma feminist knowledge and other forms of meaning-making.

Philosopher Judith Butler (2004) has challenged the aesthetics of representation of a vulnerable community and expands on how we can humanise and act ethically. Social scientists David Ben Shannon and Sarah Truman (2020) have drawn on their artistic practice as educators and composers to explore affect, privilege, and racialised and gendered spaces. In their article *Problematizing Sound Methods Through Music Research-Creation: Oblique Curiosities*, the pair offer a comprehensive literature review of scholars interested in critically enacting sound studies research. They reflect on qualitative research methods and turn to sound studies scholar Jonathan Sterne who claims that engagement occurs through the sonic. Sterne (2003) not only asserts that immersion into a world happens through hearing but goes a step further and says, 'vision removes us from it' (p. 15). Yet this immersion that Sterne is describing is not neutral and the social, cultural, and racial components must be factored into how we listen and engage with various environments. Shannon and Truman (2020) also reflect on the sonic backgrounding and foregrounding of environments and draw on the work of Poppy de Souza, an interdisciplinary researcher whose scholarship focuses on voice poverty and the politics of listening. De Souza has closely looked at inequalities and injustices in a number of spaces and explored the relationship between sound, race and the cultural politics of listening. Her research has described the acoustic violence of the 'white ear' (2018, p. 460) and considers the entangled relationship between sound, power, and violence.

In addition, Jennifer Stoever (2016), when exploring history of African American music and literature, states that listening is 'a form of agency, a technique of survival, ethics of community building, practice of selfcare, guide through racialized space, site of racialisation and mode of decolonising' (p. 17) and defines it as the 'sonic colour line'. For her, the 'listening ear' manifests itself as a disciplined and essentialised way of conforming to dominant power structures. Diametrically opposed to the hegemonic listening ear is Stoever's 'embodied ear', which represents the wholeness of the individual taking into account the historical context and their own physicality and totality of the person that is mindful of the sonic colour line. This embodied ear is a way to better understand the potentiality that exists with podcasts like the YCC that aim to engage listeners to experience Roma cultural heritage but in an egalitarian way. YCC attempted to challenge unitary listening practices and sought to challenge traditional hegemonic and stereotypical tropes that circulate in relation to the Roma. With this backdrop, through the absence of visual content a different kind of engagement, one that is unencumbered by the prejudices that come with the visual, the sonic allowed for the listener to experience the guest's story. The sound files, in this instance the human voice, allowed the listener to hear the story of the Roma community member drawing attention to the agency of the individual rather than focusing on the myths and stereotypes.

Theatre scholar Marie-Heleen Coetzee (2018) states that embodiment is surfacing in practice and research across contexts, paradigms, and disciplines. She also claims that our bodies have the potential to organise our 'felt politics' which hold transformative potential. Coetzee's foregrounding of embodiment frames the storytelling component of YCC, where the guests sensed themselves living in various tense situations and highly charged spaces. Such felt politics and embodied knowledge and experiences were shared in the YCC podcast. One guest said, 'Talking to you [the YCC podcast host] I felt I shared stories I had kept to myself or was afraid to talk about in public. It was perhaps too emotional but when I listened back, it really was my story' (YCC Roma guest, interview response, May 3, 2021). [12]

The YCC podcast was very careful with the sound content it created, curated, and disseminated. As an expression of embodiment, the podcast is far more than what might first be apparent to the untrained ear. The podcast portrayed a heterogenous Roma identity to tell stories and share that identity. One artist and guest said, 'It was a great pleasure for me to be a guest on the YCC last year. The conversations are widening the knowledge about the Roma people and provides an insight of who we really are as a person' (YCC Roma guest, interview response, 3 May 2021). Through this expressive storytelling, the guests all described how they belonged or didn't belong within mainstream society. Three of the artists referenced feeling disconnected from other Roma artists due to the pandemic and that the podcast allowed them to feel closer to their community. 'This [the YCC invite] came during a time when I was feeling very removed from my people [Roma], and I needed to feel closer' (YCC Roma guest, interview response, 23 May 2021). As the body and identity are interconnected (Leigh & Brown, 2021), the podcast reflected on the lived experience of the individual as analysing their social, political, educational and artistic environments. The podcast explored ways that the guests might share their lived experiences in an auditory manner. This contemporary storytelling offers the listener an opportunity to better understand the individual and the Roma identity being described. This dialogic approach allows for communicative daily life stories[13] to be captured; and the reflective and shared process of meaning-making that emerges through the use of the sonic facilitates an educational opportunity for the listener, and a shared sense of identity for the Roma. Relying on this sonic element provided the listeners a chance to engage with an embodied ear experience, as the visual component was eliminated. To return to Stoever's (2016) 'sonic colour line', the YCC podcast invited listeners to tune into sensory experiences as a way to better understand how racial politics play out for Roma in social, cultural, and educational environments. The guests shared their personal experiences and discussed racism and other inequalities, and did so describing their art practices via the sonic and without ever showing the listener any of their work.

Creative activism, counternarratives, and the YCC podcast

As discussed previously, the Roma community experience hate, social exclusion, and discrimination within society. Roma artists and activists are using their work to fight against and challenge the anti-Gypsyism and Romaphobia that is covertly and overtly directed at them. Digital platforms offer alternative avenues for counter-narratives and appeal to a younger generation. Examples of Roma artists challenging systems of repression are manifold. For example, ERIAC deputy director, curator, and art historian Timea Junghaus uses decolonial and race theories to challenge hegemonic language that has been used to describe Roma art and aesthetics, and the community. In 2014, Junghaus reflected on the various ways in which young Roma artists and academics are creating social and artistic networks, suggesting that digital platforms are allowing for activism to develop. Csikós (2020) listed various Roma artists 'such as Selma Selman, Delaine Le Bas, Mihaela Dragan, Alina Serban, Daniel Baker, Emilia Rigova, and many more that utilise their art to challenge discourses on whiteness, Romani identity, gendered roles, sexuality and politics' (p. 20). For some Roma artists there are clear political goals that reach beyond the field of arts and draw on their artistic practices to bring attention to and challenge injustices experienced by the broader Roma community.

Cultural heritage representation and identity are complex concepts, and their interrelationship deserves a more intersectional and intercultural analysis. Roma identity and intercultural dialogue can help demonstrate intersectionality, racism, and stigmatisation. The YCC podcast allowed the guests to express how anti-Gypsyism and Romaphobia can be tackled. Through artistic practices, the artist's replies also considered empowerment and how to raise aspirations of the younger Roma. The YCC podcast included the voices of research participants and its approach opened up space for the audience to encounter and interpret the cultural and artistic practices in different ways. In particular, I suggest that hearing the emotion and tone in a speaker's voice may 'evoke different affective engagements between the audience and the research content' (Kinkaid et al., 2019, p. 76) and instigate counternarratives. The YCC podcast set out to make the cultural heritage and values of diverse Roma communities visible to a mainstream audience, fostering empathy through highlighting similarities with other practices, as well as tackling stereotypes and anti-Gypsyism. YCC aimed to promote Roma aesthetics both as cultural heritage assets, and raise awareness, documenting and promoting authentic and original representations of cultures, through highlighting the diverse Roma communities as providers.

The podcast offered the Roma community a platform for political, social, and personal reflection and expression:

> Contemporary media coverage of the GRT[14] [Gypsy, Roma, Traveller] community, including recent TV series Here Come the Gypsies, still want to, by their narration and direction, seek to promote a stereotype of the (British) GRT community, as somehow obsolete, uneducated, archaic, criminal and antisocial. When media content is made about us, and narrated with objectification to promote a cultural stereotype, it adds to the collective (national/international) psyche for many years into the future; media has the ability to influence future generations into or out of prejudice. This is why YCC is so valuable, as it is GRT directed media with GRT content. Self-representation pushes against the media of settled culture's stereotyped prejudice propagation, emphasising that GRT community self-advocacy is the future of our cultural representation.
>
> *(YCC Roma guest, interview response, May 29, 2021)*

Creative activism allows and relies on different art practices to disrupt and articulate alternatives for socially charged topics. Harrebye (2016) suggests that creative activism is a form of meta-activism that 'facilitates the engagement of active citizens in temporary, strategically manufactured, transformative interventions in order to change society for the better by communicating conflicts and/or solutions where no one else can or will in order to provoke reflection' (p. 25). Rachel Lee Rubin's (2018) volume presents interviews with a variety of artists and critically explores how art and creative activism can be political. She uses the term 'cultural workers' and suggests that creative tactics and strategies can bring about social change. The YCC podcast demonstrates the potential for intersubjective and social transformation through art. The podcast can be seen as creative activism, as it relies on the combination of practice and the 'creative power of the arts to move us emotionally with the strategic planning of activism necessary to bring about social change' (The Centre for Artistic Activism (C4AA), 2018, para. 1). When a podcast guest was asked to reflect on the host's approach and how nuanced conversations were held, they replied, 'I really enjoyed your interview style and how empathetic and sensitive you are to those you speak with and interview'

(YCC Roma guest, interview response, May 3, 2021). Another Roma artist who was interviewed six months after her recording said:

> I really appreciated your conversational style of questioning and environment of trust. I loved listening to other contributors share with you as a colleague and friend, this style comes across as a warm and open discussion allowing a safe space to share and explore in creativity and community spirit.
> *(YCC Roma guest, interview response, May 14, 2021)*

The community spirit referenced here can be seen to reflect embodied inquiry (EI) principles since the technique focusses on bodily and embodied experiences. As the host, I tried to keep things casual, even though I was asking questions about heavily charged topics, because it was important to me that the guest feel comfortable, cared for, valued, and respected.

A podcast listener said, 'The artists and the host were articulate and connected dots and linked their work [artistic practice] with encouraging a change' (YCC listener, survey response, 7 May 2021). YCC aimed to show the public that there is a desperate need to share stories and use artistic practices to tell truths from the individuals who have experienced social injustices. The embodied knowledge articulated through a sonic environment facilitated a form of creative activism. Through YCC, the technology broadened the scope of advocacy, allowing more people to become involved in the reframing of Roma identity. When asked in a post-production questionnaire if YCC promotes Roma culture, one artist said:

> Yes, it significantly promotes GRT culture from within the community to the world, amplifying contemporary GRT voices and peer/colleague lead support of the YCC project; providing much needed bridging/healing experiences for the awareness gap of the dominant/settled culture misconceptions about the GRT community (as a whole). It promotes GRT culture as creative, open, approachable and valuable, and re-humanises our place in contemporary culture.
> *(YCC guest, interview response, May 10, 2021)*

As highlighted in the methods section I asked both Roma and non-Roma for feedback on the podcast. One non-Roma listener replied, 'Those interviewed really shared from the heart and I took a lot from the series' (non-Roma listener, survey response 12 May 2021). A Roma listener said, 'YCC is pushing the boundaries' (Roma listener, survey response May 12, 2021). Another Roma listener reflected on the podcast episode that spoke with artist and educator Dr Adrian Marsh: 'When I listen to him, Adrian has that 'something', has that specialness … he knows how to describe the connection with our people and traditional ways. I identify with, with him, and I want to learn from or follow him. Beautiful!!!!' (Roma listener, interview April 26, 2021). Referring to Lisa Smith, an English Traveller, another non-Roma listener commented, 'She was so emotional and honest … it was raw and bold. Lisa is courageous' (2021). Another non-Roma listener said that she learned so much from listening to the podcast and 'sitting with the guests. It felt like I was having coffee with them in my house and they were giving me advice' (non-Roma listener, survey response May 12, 2021).

What emerges from the project is the need for Roma community members to be given space to reflect on their life and artistic practice, through storytelling, critical reflections on anti-Gypsyism, Romaphobia, and cultural heritage as a tool to offer counternarratives. The YCC community aimed to point out and erode the anti-Gypsy narratives built about the Roma population over the centuries, while at the same time giving voice to the Roma themselves. The guests each shared their own personal stories and, in the process, generated their own narratives. Genuine cultural and artistic productions, their voices, and personal stories took centre stage of the podcast.

Conclusion

As a way to honour solidarity, channel positivity, and also build on the ERIAC network, the YCC podcast series aimed to bring together diverse Roma voices through encouraging established and lesser-known artists to discuss their practices in an ethically sound environment. The baton exercise was instrumental in my uncovering new community members and carrying out transparent and ethical research. Embodied inquiry allowed me to collate multimodal knowledge and stories and curate a podcast series that facilitated Roma members to share their craft in a sonic digital space. This podcast series not only addresses a gap in Roma and sonic studies but honours storytelling and oral history traditions using contemporary technologies. EI informed the types of questions that were co-created with the critical friends of the project. My technique as a host was informed by my embodied practices and thus facilitated conversations that were critical, and safe and respectful. The experimental baton exercise curatorial method decentred me in a number of ways and also placed other community members in the metaphorical driver's seat. In summary, EI and the baton exercise methods were central to developing a podcast that was more representative of the Roma cultural heritage scene, and also ethically sound and transparent. The ethical methods and reflexive focus on embodied experiences via an audio-focussed platform educated, promoted and celebrated the Roma communities' arts and culture. This research project is also a form of collective and creative activism that supported the broader ERIAC long-term goals of educating and raising awareness. Having a platform where vulnerable groups are coordinated and allowed to connect with other members from their own communities, and diverse other communities, is essential for shifting narratives and developing a sense of agency that feeds creative activism. Traditionally, Roma communities are seen as marginal groups or victims, however, YCC worked with Roma communities as protagonists in the construction of the project, fostering a space that challenged anti-Gypsyism. This project can serve as the basis for future interventions against anti-Gypsyism in Europe, based on the active and constructive participation of Roma people.

Notes

1 This project was supported by the European Roma Institute for Arts and Culture and was part of their Digital Innovation Programme. More can be found here: https://rosasencis.org/
2 Digital Innovation Grant: https://eriac.org/eriac-arts-and-culture-innovation-program/ In response to the global outbreak of Covid-19 Pandemic, which has also impacted severely the Roma arts and culture field, ERIAC aims to strengthen resilience of Roma cultural producers by supporting creativity, the spirit of innovation and self-sustainability. Through cultural initiatives, the Program will support the development of new arts and culture content, fostering knowledge-production and knowledge-transfer and promote creativity and self-sustainability of Roma artists, entrepreneurs and creators.
3 McGarry (2017) offers the term Romaphobia which is hatred or fear of those individuals perceived as being Roma and involves the negative ascription of group identity that can result in marginalization, persecution and violence.
4 Anti-Gypsyism is commonly used as a generic term for a broad set of negative feelings, stereotypes and discriminatory practices against the Roma.
5 The Romani Tea Room: http://www.errc.org/media/the-romani-tea-room
6 Romanistan: https://instagram.com/romanistanpodcast
7 RomArchive: https://www.romarchive.eu/en/
8 Dancing Bodies in Coventry Podcast: https://dancingbodiesincoventry.com/ Dancing Bodies in Coventry (DBiC) is a multimedia project telling the story of dance in the city. We are researchers from Coventry University's Centre for Dance Research (C-DaRE), and we produce films, podcasts and more with the aim of recording and celebrating the amazing variety of performance taking place in this city. The team provides dance artists, dance schools, festival organisers and local groups with a platform to share their stories with the wider community.
9 SoundCloud platform allows music and audio creators to both share and monetise their content with a global audience, as well as receive detailed stats and feedback from the SoundCloud community. www.soundcloud.com
10 The project was approved through Coventry University's Ethics System. Prior to the recording of podcast, the author prepared a project information sheet, a consent form, research instruments documentation to ensure that the project and all of its partners understood, adhered to and complied with all ethical regulations that are set out by Coventry University.
11 https://www.coventry.ac.uk/research/research-directories/current-projects/2020/interactive-dialogues/
12 All of the quotes that appear in the article were from interviews that happened independent of the podcast recording. The research participants who were interviewed to discuss and reflect on the podcast agreed to critically discuss the series but we agreed to anonymise their responses. At the time consent was gained we agreed to anonymise all the data collected. This decision was taken to ensure that participants could speak freely and openly about the series and their reflections on the work. The artists and academics who were guests completed a separate consent form that made it clear that their name and the episode would clearly identify them in all YCC SoundCloud materials.
13 Communicative Daily Life Stories is a communicative data collection technique where the goal is not to construct a biography, but rather to encourage participants to narrate and reflect by interpreting their lives: past, present, and future. The aim is to establish a collaborative dialogue on their reality by means of thoughts, reflections, and analysis. A communicative daily life story is a cooperative process of understanding that, unlike life stories that seek to obtain information, is oriented towards transformation (see Aubert et al., 2011).

14 GRT is an acronym used for Gypsy Roma Traveller communities. In the UK, the term is used and there is a distinction made between Roma, Gypsy and Traveller. Throughout the world, Roma is the preferred umbrella term.

References

Aubert, A., Melgar, P., & Valls, R. (2011). Communicative daily life stories and focus groups: Proposals for overcoming gender violence among teenagers. *Qualitative Inquiry*, *17*(3), 295–303. 10.1177/1077800410397808

Butler, J. (2004). *Precarious life: The powers of mourning and violence*. Verso.

Carmona, S. (2018). Decolonizing the arts: A genealogy of Romani stereotypes in the Louvre and Prado collections. *Critical Romani Studies*, *1*(2), 144–167. 10.29098/crs.v1i2.22

Carmona, S. (2020). *The representation of Roma in major European museum collections – Volume 1: The Louvre*. Council of Europe Publishing. Available from: https://book.coe.int/en/minorities/9157-epub-the-representation-of-roma-in-major-european-museum-collections-volume-1-the-louvre.html [Accessed 5th August 2021]

The Center for Artistic Activism (C4AA). (2018, April 9). *Why artistic activism?* https://c4aa.org/2018/04/why-artistic-activism

Coetzee, M.-L. (2018). Embodied knowledge(s), embodied pedagogies and performance. *South African Theatre Journal*, *31*(1), 1–4. 10.1080/10304312.2018.1488524

Council of Europe. (2020). *Committee of Ministers adopts new action plan for Roma and Traveller Inclusion*. https://www.coe.int/en/web/roma-and-travellers/-/committee-of-ministers-adopts-new-action-plan-for-roma-and-traveller-inclusion [Accessed 5th August 2021]

Council of Europe. (2021). *RomArchive*. https://www.coe.int/en/web/roma-genocide/romarchive-digital-archive-of-the-roma

Csikós, M. (2020). *Doing Roma feminist knowledge production in order to challenge hegemonic knowledge production: A case study of the exhibition "Roma Women Weaving Europe"* [Master's thesis]. Central European University.

de Souza, P. (2018). What does racial (in)justice sound like? On listening, acoustic violence and the booing of Adam Goodes. *Continuum*, *32*(4), 459–473. DOI:10.1080/10304312.2018.1488524

European Fundamental Rights Agency (FRA). (2021). *Monitoring framework for an EU Roma strategic framework for equality, inclusion, and participation: Objectives and indicators*. https://fra.europa.eu/sites/default/files/fra_uploads/2020-portfolio_of_indicators_working-paper_en.pdf [Accessed 27th August 2021]

European Fundamental Rights Agency (FRA). (2014). *Education: The situation of Roma in 11 EU member states*. Available from: https://fra.europa.eu/sites/default/files/fra_uploads/fra-2014-roma-survey-dif-education-1_en.pdf [Accessed 27th August 2021]

European Roma Institute for Arts and Culture (ERIAC). (2021). *About ERIAC*. https://eriac.org/about-eriac

Foucault, M. (1980). Two lectures. In C. Gordon (Ed.), *Michel Foucault power/knowledge: Selected interviews and other writings 1972–1977 by Michel Foucault*. Pantheon Books.

Harrebye, S. (2016). *Social change and creative activism in the 21st century*. Palgrave Macmillan.

Junghaus, T. (2014). Roma art: Theory and practice. *Ethnographica Hungarica: An International Journal of Ethnography*, *59*(1), 25–42. 10.1556/AEthn.59.2014.1.3

Kinkaid, E., Emard, K., & Senanayake, N. (2019). The podcast-as-method?: Critical reflections on using podcasts to produce geographic knowledge. *Geographical Review*, *110*(1-2), 78–91. 10.1111/gere.12354

Leigh, J. & Brown, N. (2021). *Embodied inquiry*. Bloomsbury Academic.

McGarry, A. (2012). *Who speaks for the Roma? Political representation of a transnational minority community*. Bloomsbury.

McGarry, A. (2017). *Romaphobia*. Bloomsbury and Zed Books.

McIntosh, I., & Wright, S. (2019). Exploring what the notion of 'lived experience' offers for social policy analysis. *Journal of Social Policy*, *48*(3), 449–467. doi:10.1017/S0047279418000570

Meerzon, Y. (2021). Precarious bodies in performance activism and theatres of migration. In Y. Meerzon, D. Dean, & D. McNeil (Eds.), *Migration and stereotypes in performance and culture* (pp. 21–38). Palgrave Macmillan.

Moscaliuc, M. (2019). Accessorizing (with) "Gypsyness" in the twenty-first century. *Critical Romani Studies*, *2*(1), 92–114.

Paillard, C. (2014). Artist, researcher, artist-researcher, researcher-artist, researcher, artist. *p-e-r-f-o-r-m-a-n-c-e.org*, *1*. Available from: http://www.p-e-r-f-o-r-m-a-n-c-e.org/?p=355

Rubin, R. (2018). *Creative activism*. Bloomsbury Academic.

Shannon, D. B., & Truman, S. E. (2020). Problematizing sound methods through music research-creation: Oblique curiosities. *International Journal of Qualitative Methods*, *19*, 1–12. 10.1177/1609406920903224

Skains, L. (2018). Creative practice as research: Discourse on methodology. *The Journal of Media Practice and Education*, *19*(1), 82–97. 10.1080/14682753.2017.1362175

Sterne, J. (2003). *The audible past: Cultural origins of sound reproduction*. Duke University Press.

Stoever, J. L. (2016). *The sonic color line: Race and the cultural politics of listening*. New York University Press.

Sullivan, G. (2009). Making space: The purpose and place of practice-led research. In H. Smith & R. T. Dean (Eds.), *Practice-led research, research-led practice in the creative arts* (pp. 41–65). Edinburgh University Press.

Topolanszky, A. (2018). *The situation of Roma women*. [European Economic and Social Committee] https://www.eesc.europa.eu/en/our-work/opinions-information-reports/opinions/situation-roma-women-exploratory-opinion-european-parliament

12

MANY WORLDS IN ONE PLACE: COMPOSITION AS A SITE OF ENCOUNTER

Toby Young

Introduction

> A city is a particular kind of place, perhaps best described as many worlds in one place. It compounds many versions without quite reconciling them … a compendium of perspectives, a snatching out of the infinite ether of potential versions a few that will be made concrete and visible.
>
> *(Solnit, 2010, p. vii)*

In musicology, there is a growing interest in the social impact of intercultural music-making on marginalised communities, largely in the context of an increasing concern with cultural diversity within classical music (Cottrell & Impey, 2018). These musical projects have the potential to form rich sites of exposure and relationality for participants, facilitating engagement between diverse cultural traditions, social groups, and aesthetic experiences. One area of particular richness in this field is work involving refugee communities, for whom 'sound as a contested social terrain of citizenship' (Western, 2020, p. 300) is often deeply imbricated with their lived experiences of migration. Music-based activities have been shown to be of particular benefit to these individuals, through their role in supporting the ongoing development of emotional and social resilience, cultural identification, and a positive sense of self (Marsh, 2012).

As Joseph Browning (2020) acutely observes, the prominence of scholarly work 'extolling the critical, political, and philosophical significance' of art practice within musicology 'has often eclipsed consideration of other, more localized, tangible, and mundane activities' involved in the making of complex artistic interventions (p. 498); these activities being an important but under-shared form of knowledge in the arena of socially-engaged artistic practice. Specifically, where research to-date has focused largely on the therapeutic benefits of exposure to

DOI: 10.4324/9781003164227-21

music-making activities for refugee and migrant communities (Storsve et al., 2010), the body of work examining the creative outcomes of artistic co-production with refugee participants is minimal. This chapter offers a discussion of one such co-production: 'Journey', a project undertaken with refugee students aged between 16 and 18 at the Oxford Spires Academy (OSA) using the co-composition of a musical work[1] for string orchestra and soundtrack (i.e. a pre-recorded electronic component) as a space for the exploration and expression of migrant identity.

Rather than offering yet another attempt to codify the social mechanisms by which these sorts of encounters promote lived change, my approach in this chapter is one of critical 'reflection-on-action' (see Moon, 2005), analysing my role in this project by focusing on the interaction between my own artistic practice as a composer and audio producer—through the acquired frame of reference this work has given me—and my attempts to negotiate the complexity and messiness of co-produced intercultural composition. I begin by outlining the project and its aims, before offering a brief account of the workshop process and the issues they raised. I will then consider the implications of my role as composer and mediator, recounting my working process and reflecting on the techniques and challenges this afforded to explore the emergent ways that my agency and function as a creative practitioner was challenged and negotiated during the composition process. It is my hope that in considering the dialogue between my own artistic identity, assumptions, processes, and feelings and the practical experience of working within the complex networks and negotiations of power imbalances (at both systemic and local levels) that framed this work, this chapter might offer practitioners ways to reflect on the entanglements between artistic positionality, emotion, technique, and agency in their own work.

Project overview

The City of Oxford is home to thousands of refugees and asylum seekers, for whom integration and social cohesion is an ongoing challenge. Building on a successful previous project with young refugees (Displaced Voices, 2018), 'Journey'[2] aimed to break down the socio-cultural barriers preventing intercultural interaction between the young asylum seekers and the wider Oxford community through the co-creation of a new composition for orchestra and acousmatic soundtrack to be performed by professional musicians from the Orchestra of St John's (OSJ), and shared with the community at three live performances, including a concert in the prestigious St John Evangelist Arts concert series in June 2021. Led by OSJ's Associate Conductor, Cayenna Ponchione-Bailey and me in my role as Composer-in-Residence with the orchestra, together with the support of undergraduate students from the University of Oxford acting as project assistants. Participants were a group of young asylum seekers students on a specialist year 12 programme for legal migrants known as Steps Towards English Proficiency, and were refugees to the UK for many reasons (including territorial, political, juridical, and economic).

In collaboration with programme-leaders from the school, we co-devised two central goals for this project that we collectively believed would benefit both participants and the community around them:

1. Co-compose a piece of music[3] that would communicate knowledge of the participant's refugee experiences to other students in the school and wider community.
2. Develop a creative-practice community to connect participants with other students from OSA creating a sense of belonging at the school, and empowerment to seek support from their colleagues in the process of resettlement.

These goals firmly situated the project within Georgina Born's (2017) notion of socially engaged arts as real-world social (rather than purely aesthetic) interventions: 'Politically informed interdisciplinary practices in which "external" social and political realms become the arena within which art stages … [activities] intended to influence the "real" world' (Born, 2017, p. 37). By creating environments for participants to explore and express themselves emotionally, the project sought to support the articulation and legitimisation of their cultural identities: a 'shared space in which … [participants] could safely construct, negotiate and enact hybrid identities … [outside of] personal, cultural, lingual and institutional borders' (Dieckmann & Davidson, 2018, p. 172).[4] But by including an orchestral element—something alien to most of the participants—this project performed a transgressive site of encounter, shaped by 'instances in which something is "unexpectedly broken open" or destabilised' (Wilson, 2016, p. 609), supporting the participants to challenge themselves and inviting them to connect actively with their own creative agency. Nevertheless this planning process was not without challenge, and unfortunately due the long time-scale required to seek funding for this project and rapidly fluctuating nature of the group, we were not able to consult the participants before the planning phase as we had initially intended.

First encounters

Refugees and newly arrived immigrants 'face a range of social, emotional, and cultural challenges related to geographical and cultural displacement and trauma experienced … in the process of relocation and resettlement' (Marsh, 2012, p. 173), including educational deprivation, language disparity, loss of identity, and difficulties of cultural acclimatisation. It was hardly surprising to Cayenna and me that at our first meeting many of the students presented tense and closed body language—sitting back into their chairs with crossed arms and legs and hoods pulled down over their faces—and little overt desire to communicate with us. In the early sessions Cayenna and I worked hard to create a relaxed and informal 'vibe'—using first names and speaking individually with students where possible—that situated our workshops as a relational space, outside the formal interactions with teachers or other adults at the school. In doing so we hoped to engender a space that felt socially and psychologically safe and would hopefully become associated with enjoyment and emotional release over the course of the workshops. Our activities were generally held in non-typical spaces within the school (such as a theatre auditorium) and were uncoupled from the normal timetables that govern the school day. Sitting outside the daily rhythms of school life helped us in creating an 'other' space for the students—a heterotopia both within and outside quotidian expectations and challenges for them to feel able to open up, explore, play, and be challenged (see Foucault, 1971)—but also led to practical challenges when students had to leave at awkward times to go to other lessons or activities.

We tried to adopt an attitude of sharing from the start by playing numerous warmup games, trying to put the participants at ease and develop the sense of trust necessary for creative work (Watson & Ward, 2013). Working collaboratively in small groups, the first first two or three sessions focussed on enabling participants to explore and understand each other's musical heritages, share the journeys which brought them together, and discuss music's role in their individual and collective identities. In one activity for example, participants were tasked with researching their musical heritages and sharing examples of music from their i) traditional heritage, ii) parent's generation, and iii) country today. Each group then gave short presentations to everyone about what they learned, sharing audio and video recordings found on YouTube to illustrate their discussions. Nonetheless there was an awkwardness during the first two sessions in our engagements with the participants, as well as between each other (since many of them had little in common with even each other beyond the broad narratives of their migratory experiences). Participants were initially reticent to talk about themselves or their cultures in any explicit terms. As we later found from individual sharing, this was, for some participants, due to a deeply felt trauma and insecurity over their ability to communicate effectively that was only relieved after weeks of repeated assurance that we valued their ideas and voices far more than their grammatical aptitude. We were also conscious of the fact that our presence as white outsiders may have been associated with colonial forces or even experiences of trauma for some participants, and so were not surprised by the slowness of self-revelation. To combat this reticence, we worked hard to welcome participants to share with each other, whilst emphasising explicitly that they were not obligated to contribute at any point.

There were also several practical challenges to navigate, such as lack of attendance, focus and direction in the second session, as well as more participants than we had originally planned for, negating the ability to spend meaningful individual time with each participant in each session and facilitate the rapport-building we would have liked. We persevered with activities involving play, problem-solving and curiosity: techniques used commonly in trauma therapy[5] to empower individuals to feel psychologically safe enough to move from the low-level stress reaction (also known as 'fight-or-flight' state) into their more protected 'social engagement' zone (Porges, 2014). Focusing more on facilitating the inter-student integration than the integration between them and us felt a lot more productive, as did using techniques like active listening to draw out and reflect participant's contributions in front of the whole group. These moments of verbal validation prevent participants from becoming isolated or silenced and draw out commonalities between participants, motivating desires to understand more about each other through 'micro-moments of connection' (Fredrickson, 2001, p. 11) that might help them to feel seen and understood by others. We started to notice participants became more engaged in this work, and by the third or fourth session many seemed comfortable talking about themselves in the space.

Navigating uncertainty

As we continued the process of collectively gathering musical and cultural artefacts, participants began to seek opportunities to share elements of their heritages more actively, and through workshop activities and games designed to support creativity and world-making (e.g. the creation of sonic 'mood boards'). Through this, participants began to be more skilled at using basic aural analysis and technical language to talk about the sounds of these artefacts, demonstrating enhanced knowledge of different musical instruments and the techniques players were using to produce sound, whilst also using increasingly adjectives to describe the timbres of a variety of sonic objects. It was important in the project design that the artistic form and genre of the final piece was not pre-existing or fixed, in order to enable everyone the feeling of traversing a creative unknown together.

Like many participatory arts projects during the COVID-19 pandemic, 'Journey' was forced to adapt to changing circumstance when our work was drastically ruptured in March 2020. It had been our intention that the subsequent workshops would have focused on more skill-based activities, such as teaching the participants recording techniques to capture audio material (such as their own voices) to add to the 'basket of ideas', but when in-person sessions resumed after lockdown ended, there was no longer the time to facilitate this. At this point in the process, with limited time and access to the group, Cayenna and I had to make some difficult decisions about what was possible within the limitations, ultimately deciding to prioritise the first of our objectives (sharing the participants' experiences with the wider community) over the second without being able to consult the participants. To ensure the maximum aesthetic impact on an audience it was important to have the participants' voices included in the final composition, so I was forced to record these voices myself to avoid the 'challenge to professionalism' of 'publishing amateur or substandard sound … [and] undermining the seriousness of … the project' (Feld & Brenneis, 2004, p. 470), but leaving me with a sense of sadness and frustration that I had not been able to achieve the active 'skilling' and deep sense of collaboration I had hoped for.

A further effect of COVID-19 was a shift in membership of the group, which drastically ruptured the sense of belonging and trust we had fostered. In our first online session, trying to pick up the rich complex discussions of music, sound, and emotion from before the break was almost impossible. The dramatic new limitations imposed by geographical distance and technological mediation also meant that the in-depth collaborative discussions I had envisaged about how I should structure the piece became unmanageable, and we refigured the final sessions to involve less creative decision-making and instead focus the participants' agency and energy on smaller, less 'high stakes' and more manageable tasks such as collecting more material for the 'basket'. Focusing on clearer tasks and goals, rather than creating more open-spaced dynamics, resulted in a significantly higher engagement from the students, but led to ethical questions for me around whose voice and agency was really being represented in this piece. Whilst it was always my role to curate and structure the participants' ideas into a final composition, this change in the amount of creative agency and input the students had in the final work forced me to refocus my activities, and led me to think more critically about the role of a composer in collaborative projects like this.

Wayfinding, curating and translating

As I began the compositional process there were four distinct activities that emerged. The first involved going through the artefacts collected during workshops—including drawings, audio recordings, YouTube links, and notes taken by project leaders—using my role as listener and curator to locate the most powerful and evocative sounds from the participants' creative data. These artefacts are not simply 'raw material' needing to be compiled, but a reproduction of complex social relations that are intimately and inescapably imbricated within the participants' own cultural and emotional identities. Rather than 'mining' the data, I attempted to feel my way through the material with careful listening and observation, drawing on the distinction Tim Ingold (2011) makes between navigation and wayfinding as two alternative techniques of orientation: the former revolving around the collation of accurate knowledge about the route towards a destination, while the latter presenting a more circuitous process that 'depends on the attunement of the traveller's movements in response to the movements in his or her surroundings' (p. 242) to help the traveller form a reciprocal relationship with the environment in which they travel. This process of wayfinding provided a helpful framework for me to make sense of the complex tasks of deciding what to include and not include, and the implications of these decisions for me, the group, and the project.

The next stage of composition involved the collation and assembly of material into a musical narrative (*composueram* in Latin translating as 'putting together'), following a broad structural narrative designed in conjunction with the participants: intended to initially 'share our cultures', then 'help the audience feel our sadness' before finally 'leave them feeling hope'. Hagar Kotef (2018) rightly observes that structures 'can provide a political and theoretical foundation … if they are built into a dialectic of change, of detailing, of differentiations, they also tend to blur the details or side-line those which do not fit well' (p. 218). When choosing which material from workshops to use and where, I felt it was important to avoid my initial instinct to reduce the idiosyncratic ways in which participants expressed their experiences of migration (solicited through broad questions in the earlier workshop sessions) into broad themes; instead, I attempted to think of this process as one of affective 'world-building' and rather than architectural structuring in order to maintain a sense of plurality in my process and avoid assimilating outlier voices within a more obviously programmatic narrative. To support this development of commonalities I adopted a theme and variation structure, where various threads of more abstract music weave around fixed points of concrete world-painting (e.g. the sounds of an Indian tabla, Syrian qanun or Ethopian masenqo, described by participants as being particularly evocative of their countries of origin) in order that these specific materialities and emotions evoked might prove points of aural convergence for the listener without forcing the non-Western sounds to operationally 'fit' into my classical composition system and structures (Robinson, 2020, p. 6).

In the orchestral component of the score, this is achieved through strategies of developing variation (the repetition of material with subtle ongoing variations that draw elements). In the soundtrack, ongoing processes of sonic mediatisation and reappropriation spin out only two or three elements of sonic material: participant's voices, non-human sounds from the workshops (both processed and left intact) and sounds of musical culture. These ranged from traditional instruments mentioned above—all sourced and licensed for use from pre-existing recordings apart from a sarod sample which I recorded for the project—through to a heavy East-African inspired trap beat constructed largely from the recycling of 'unwanted' noises from participant recordings (such as microphone clipping, table banging, and so on). In this way, my approach sits somewhere between the 'objective representation' of soundscape composition that 'use[s] footage in a reconstructive manner to convey what it feels like (or at least sounds like) in a given place' (Demers, 2010, p. 120) and the practices of musique concrete which abstracts found sounds in order to de-signify and re-contextualise them in the final work. Already acutely aware of the potential imbalance of power and agency in this project (see 'The Challenges of Agency' below), I felt a strong ethical obligation to only add or change material if it created a rhetorical force within the composition that might offer an audience—of predominantly Western classical music—a 'way in' to the participants' experiences; whilst also being acutely aware of the tension inherent in feeling the need to 'smooth the path' for Western audiences into non-Western material.

Intercultural art, observes Kelly Oliver (2014), potentially allows us to:

> act as if we inhabit the same world, as if cohabitation is possible. And yet … [it] is also where this phantasm of cohabitation comes up against its limit insofar as it shows the multivocity of language and languages, the untranslatability from one island to another, which at the same time demands translation and interpretation. (p. 129)

As a vehicle to convey meaning, music and sound exemplifies this inbetweeness; acting both within a quasi-semantic system of direct and determinate utterances

(Middleton, 1990) and a 'kind of template against which feeling, perception, representation and social situation are created and sustained' (DeNora, 2000, p. 43) by a subject individually. Where much of the soundtrack material remained close to the original artefacts, recontextualising elements of the global music traditions the participants had directed me towards within the orchestral score presented a creative opportunity. Whilst not directly attempting to translate the participants' artefacts into 'equivalences' in the Western art tradition, the possibility of creating musical affordances—'the uses, functions, or values of an object … to a perceiver' (Clarke, 2003, p. 117)—to help frame and direct an audience's interpretive processes was appealing.

With a focus on musical composition and aesthetics than ethics, I considered how each the string orchestra's sonic palette could work to support the soundtrack with its more tangible and direct connection to the participants' lived experiences. Pitched and melodic material (with questions about tuning systems and temperament), harmonic language, instrumental playing techniques and performance traditions, concepts of time/tempo, rhythmic topoi, and systems of notation all needed to be considered alongside the piece's social function, and the context of its performances. Within the interpretive framing I sought to create, direct emotional conveyance provided an obvious central role for the string orchestra in providing what DeNora (2003) calls a 'grammar of emotion' that outlines both general moods for the listener and more specific 'peaks and troughs of emotions' that suggest 'how, when, and with what intensity feeling may climax or crystallise' (p. 140).[6]

Dialogic editing

As outlined above, aspiring to an 'authentic' representation of the participants' ideas was central to the project. To evaluate this, I adopted Steven Feld's (1987) process of dialogic editing to structure a final workshop[7] before the public performances. Dialogic editing is a process in mediated composition where the power is returned to participants for whom other voices are speaking on their behalf, and ensures that 'the impact of … [participant] voices on what I tell you about them in my voice' is included in the final work, acknowledging too 'how their take on my take on them requires reframing and refocusing my account' (p. 191). This process attempts to redress the political imbalances of an external agent 'producing selections and passing them off as authentic and genuine' (Feld, 1987, p. 192) by recentering the work within the original frame of focus. Throughout the workshops we tried to nurture reflective listening practices to skill participants with the vocabulary and aural analysis skills needed to express their auditory experiences and nurture 'modes of engaged attention' within the group (LaBelle, 2018, p. 7). Several of the participants verbally engaged with this process and expressed pleasure at 'finding' themselves in the music located in specific instrumental sonorities or musically genres, but not all reacted so actively.

It is true the aim of dialogic editing is 'not only to gather data, but also … initiate a conversation' (Talianni, 2020, p. 67); however, with limited English language acquisition many of the students found verbally communicating their response to complex opaqueness of the string orchestra material to be challenging. Furthermore, as discussed above, each level of world-building within the single musical assemblage was intended to target different audiences and different listening approaches, so that the piece could 'work' on a listener regardless of performance environment, coded patterns of response, current mood or prior knowledge of the project. Participants seemed to initially find it difficult to connect with orchestra since they lacked the elements of the coding required to situate much of the specificities of Western musical meaning-making within their own listening subjectivities; for example, the way that certain chord sequences or sonorities might typically evoke certain emotional states for audiences who grew up withing the listening practices of Western art music. Rather than focus participants' feedback towards the soundtrack alone and further occlude the orchestral component—which was, in itself, sonically overwhelming for many of the participants who had not heard live orchestral music before—we tried to help them access the space of abstraction and multiplicity through carefully scaffolded discussions and creative methods to best solicit feedback about the participants feelings towards how my musical and sonic choices had reflected their lived experiences.

Supported by local artist Merlin Porter, participants were given art materials and encouraged to paint their responses to the music alongside verbal expression; both facilitating a more obvious creative and phenomenological listening approach and offering an alternative way to present their reactions to the sounds. These artworks ranged from figurative drawings of the musicians to more abstract pieces, but regardless of subject typically employed intense and dramatic choices of colour or brushstrokes. During informal discussions with the participants, it was clear that they had largely experienced intense emotion experiences as they heard elements of themselves in the work and were directed to reflect on their experiences. However, they largely agreed that this was not an unpleasant experience; just an unusual one. In the first performance of the piece in front of a live, public audience, the participants—many of whom had chosen to attend the concert of their own volition—expressed great pride at hearing a piece they had helped create about themselves, and when they saw and heard the audience's clear positive responses to these affordances (including extended applause and verbal affirmations after the concert) many of them reported feeling deeply happy that the process of conveyance had seemingly been successful.

Many worlds in one place

Inspired by the work of cartographer Emma Willard—whose ground-breaking maps made the hidden (hi)stories of marginalised communities concrete and visible (Laskow, 2018)—the final composition unfolds as a sort of 'imaginary mapping' that superimposes different narratives over each other, creating imagined spaces with multiple layers of rich memories and experiences that are overlayed to draw out commonalities between them. Entitled 'Many Worlds in One Place', the composition explores what it means to be a displaced citizen on one hand and a resident of Oxford on the other. The first movement ('Hi, hello!'[8]) presents introductions from the group, both formal and informal, building into a celebratory samba movement combining celebratory street sounds with the expressive sounds of a South American rainforest. A revving motorbike shifts us into the chaotic soundscape of Bengal, as honking horns and sounds of a chaotic fish market herald the start of the defiant second movement ('We don't just eat curry. We don't smell'). The chaos dies down into an imagined wedding scene, full of rich colour and expressive heritage, which melts away into a calm pastoral landscape. Movement three is a brief interlude ('In the temple there exists a bell') which reflects on several participants' experiences of the *Adhan*—the Islamic call to prayer—and its ability to 'make everyone go silent in a really satisfying way'. The sound of a babbling brook emerges from the still tranquillity in the fourth movement ('In my country there are garden of fruits'), evoking the ancient tradition of poetry recitations in Afghanistan which build from emotional singing into the upbeat rhythms of the *Dabke*: a circle dance to promote the growing of bountiful crops.

So far in the piece, the role of the string orchestra has supported the soundtrack; sometimes joining in with the action as they play along with a passing samba band in the first movement, other times providing a static accompaniment to underpin the sonic landscape of dusk in movement three. This time however, as we start to hear the destructive sounds of war and anger overhead, the dancing builds up to a triumphant climax in the orchestra in strong moral resistance to the devastation around it. Movement five ('I left Syria because of the government') begins starkly with the cacophony of modern trap music—a genre of hip-hop associated with urban violence—where the sound of guns has quite a different meaning. As the anger begins to subside, the scene shifts into one of hope centred around celebration of Eid. In the sixth movement the warmth of this community gathering is starkly ruptured by austere thunder and rain ('I came to London by plane'), as we hear participants talk about the disruption to their lives as they were forced to leave their home countries. The rain continues to fall, and the solitude and emptiness experienced by some participants is felt strongly in the soundtrack; but as we move into the final movement ('I want to stay') there is a change of tone in the commentary as the participants' voices acknowledge that despite everything, 'in England there is no war'. Gradually musical themes from the rest of the piece come into focus and begin to intertwine, creating peaceful and hopeful conclusion representing the participants' recurrent description of their new home affording them relief, happiness, and peace.

As this description highlights, 'Many Worlds in One Place' draws heavily on the hyper-real geographies afforded by sound to help audiences experience the participants' stories of migration and displacement as tangibly as possible. These journeys are brought to life through sound's 'almost effortless' ability to invoke space, place, and location (Demers, 2010, pp. 120–122) rather than reduced through literal mapping. 'Recordings of [soundscapes]', observes Thomas Brett (2016), 'are kinds of virtualities that realistically reproduce the sonic patterns of physical worlds, audibly similar enough to them that they create for the listener the sensation of being physically present in them' (p. 469). In other words, the representational modality of field recordings as the primary site of sonic encounter in this work creates a direct and productive sense in which we might imagine ourselves as nomadic citizens. This illusion of haptic presence has the effect of invoking the common genre of a soundwalk, but here transforming it from a literal sonic voyage taken on foot to a virtual journey around a global landscape folded into one place, whilst also making space for the participants' emotional geographies—such as nostalgia and pride for their home countries, experiences of anxiety on the journey, and a complex mixture of relief, sadness and uncertainty arriving in the UK—to be mapped onto their real, lived movements.

But as well allowing us to 'travel' across the globe in seconds, this virtuality also works the other way: by enabling the soundtrack to be uncoupled from the specific continuities and temporalities of lived experience. Salomé Voegelin (2018) suggests that 'a geography of sound has no maps, instead tracing invisible trajectories and configurations between people and things, unfolding in the dimension of … an invisible terrain that leaves no trace' (p.76). This quality proved particularly important during the dialogic editing process, especially when trying to avoid the triggering of traumatic memories. By disassociating the soundscape from literal journeys ridden by planes, lorries and boats, the music was able to invoke the participant's topographical ruptures and relocations in less explicit ways whilst still maintaining the sense of travelling through real terrain. This decoupling also works to bring the audience in closer, with those 'shared spaces that belong to no single public and yet which impart a feeling for intimacy' (LaBelle, 2010, p. xvii); an intimacy which has the potential to invoke empathy from an audience more powerfully than literal sonic representations of migration ever could.

The challenges of representation

In *Teaching to Transgress*, bell hooks suggests that marginalised actors must themselves locate the reality they seek to transform, rather than adhering to outsider narratives of change. 'Only through such practice', she notes, 'can the act of helping become free from the distortion in which the helper dominates the helped' (Hooks, 2014 [1994], p. 54). In my discussions with participants, a strong mandate emerged to celebrate and share knowledge about their home countries and cultures with the audience, emphasising for me that my role in this project was an agent for the participants rather than an 'authoritative representative of [their] subaltern consciousness' (Spivak, 1994, p. 53). As discussed earlier, within the soundtrack the representation of identity through the inclusion of localised sounds, languages and instruments does this work with relative ease, but in 'translating' (or at least 'interpreting') these experiences into the context of western art music for the string orchestra's accompanying component, the role of representation was far more complex.

Even aside from the complex intercultural issues present in this project, translation holds with it a heavily politicised 'tug between faithfulness to, and emancipation, from the original author' (Hutchens et al., 2021). I was acutely aware throughout both the workshop and composition processes how easy it would have been to unintentionally misrepresent the participants, and involuntarily perform the colonial violence often synonymous with intercultural translation (see Ross, 2016). Inevitably in selecting the ideas or material to incorporate in the piece through the lens of my artistic subjectivity I prioritised some ideas over others, choosing not to include certain pieces of music or statements for practical reasons of sonic or communicative clarity. Whilst I attempted to engage in this process as cautiously as possible, the weight of the task's potential for miscommunication, loss of context or the obfuscation of information was challenging. Furthermore, as Subramanian Shankar (2012) reinforces, there is a danger in attempting aid to intercultural contact that:

> Translation can be used, not to communicate or make contact, but to alienate the text from its source culture and to relocate it within a new one, to pluck it from one aesthetic system … and thrust it into another … In such cases translation becomes a deliberate technology of othering—of excising the foreign, the alien, the other from the text so that it can be accommodated to new social, political, and aesthetic contexts. (p. 123)

The challenge to navigate here is that there is also power in aesthetic relocation, as the participant's work becomes canonised in the artwork, acquiring cultural capital and prestige within the field of Western art music, and validating their voices and opinions. The line between sharing on one hand and appropriating on the other is painfully thin, especially within the complex network of power imbalances already pre-existing in the school environment (see Samuels et al., 2010). At no point did I consciously 'take any musical content—referring to compositions, themes, styles, motifs, and other musical structures—for … [my] own use' (Young, 2010, p. 176); being clear both in my mind and the way I communicated the work to others that I was writing this piece collaboratively with the students, and to achieve their aims rather than my own. I was conscious and active in communicating these complicated cross-understandings to the concert audiences, employing paratextual material such as the performance score and instrumental parts, programme notes, and marketing materials (such as event posters) to perform this framing-function; but neither was I able to fully extricate any creative decisions from a background awareness in the benefit of this project and its public exposure on my career, as well the fact that I was receiving monetary exchange for my creative labour. Furthermore, due to issues of safeguarding given the participants' ages (and in some cases real threats of physical danger), we were unable to credit them by name as co-composers in any of the publicity of this project. Referring to students anonymously represents a highly problematic element of a project that was designed to empower them and serves as a stark reminder of the multitude of ethical challenges raised by collaborative work with vulnerable participants.

There are complex proprietary questions in play with community co-composition, including legal and moral issues of intellectual property of musical content. In this case, given the participants were undertaking this project as part of their education it was deemed inappropriate by the school to offer renumeration, but given the nature of this activity I felt that the participants' time and ideas were given in reciprocity of my own emotional investment in my artistic advocacy of their creative voices. Equally, in trying to affect an emotional connection between the audience and the participants it was difficult to choose what material to emphasise, without letting the easy emotional signification of certain more 'marketable' stories overshadow those without the overt 'aura[s] of tragedy and victimhood displacement' (Western, 2020, p. 304) that are often taken as synonymous with refugee voices (see Wake, 2015). 'I want to hide the sad stories and just tell the happy stories' said one participant, clear that they did not want to repeat their narratives of trauma for public consumption. Even by ensuring that I used at least one piece of material from every student, regardless of how emotive or communicative I found it to be, I still had to regularly be actively mindful of this issue, and use care when working out how mediated or clear each element was in the final mix.

Other challenges of representation emerged during the process, for example how to structure and delineate the participants' narratives without essentialising their experiences and contributing to the distortion or assimilation of their minority cultures. Observing the rich overlaps in musical traditions between many of the heritages represented by our participant group, I decided early in the compositional process to structure the piece around geographical groupings in order to draw out the overlaps and interrelationships. It became apparent however when describing the piece to friends and colleagues that the simplicity of this framework for an audience would be reductive, and that an 'uninformed or naïve listener [might] mistake … it for the real thing, coming away with a distorted view [that] perpetuates harmful stereotypes' (Shelby, 2009, p. 191). Instead, I grouped material by emotional and experiential narratives such as shared experiences of the festive dinners at Eid or the destruction of war, which happily offered an even richer creative seam than geographical delineation and avoided any trace of those 'binary oppositions … [that] construct a hegemonic relationship between Western cultures (as allegedly stable and homogeneous entities) and "other" cultures (as marginalized and exoticized)' (Maier, 2016, p. 173). This recalibration served as a powerful reminder that as artists we have a duty of care to be critical of how the tacit implications and biases within project frameworks or activities have the potential to engender both positive and harmful modes of working together that can lead to reductive framings of identity-positions as easily as they can to creative autonomy.

Conclusions

When the project finished, my initial reflections were largely focused on the specific mechanisms and process by which music and sound design channelled the participants' and my sonic agencies—LaBelle's term for the creative negotiation of 'systems of domination through listening and being heard' (LaBelle, 2018, p. 4)—to create multiple meanings. Where soundscapes 'can be used to imitate reality and impart truth in an unambiguous manner' (Demers, 2010, p. 122), music with its abundant system of codes and expectations offers a mode of affective conveyance that makes room for a plurality of interpretations and subject positions. Between the faithful recreations of the participant's expressions and my 'translated' interpretation is a rich space for an audience to explore, creating new meanings and connections out of the interplay, regardless of prior experience or knowledge within a concert hall setting.

Writing this chapter some months after the project's conclusion, I find myself far more struck by ways that the deep ethical challenges at the heart of this work have left a lasting influence on my artistic practice. As modes of engagement charged with emotion and imagination, composition and sound design are valuable tools in navigating these real-life borders and facilitate the delicate and complex task of building trust with vulnerable participants. However, in doing so we have the potential for harm through misuse of power or representation. Whilst practitioners may use some of the techniques outlined earlier to try and let a community speak for itself, there will always be mediations in both the social and aesthetic realms that have the potential to impose interventions on a group's behaviour or values. If activities like these are to meaningfully foster a meeting of cultures, we need to constantly ask ourselves what types of social and artistic relations are being produced, for whom, and why. How are 'diversity and cultural "in-between-ness" to be celebrated without eroding core identities?' asks Martin Stokes (2008). 'Who are to be the gatekeepers, the explainers, the interpreters, the gobetweens, the intellectuals? Who are to be the guardians of propriety and fairness … [in] vulnerable local communities?' (Stokes, 2008, p. 4). For instance, it does not escape my attention of the contradiction here between our aim to break down barriers between us and the participants, whilst undertaking this operation through powerful white institutions within the cultural establishment.

The manifestation of tensions and intimacies within the collective composition process serves as a stark reminder that social collaboration is not a 'clean' process, but rather a messy way of 'being entangled with communities' (Hickey-Moody & Willcox, 2019, p. 44). Projects like these are real, complex spaces filled not only with actors expressing a range of objectives, agendas and agencies, and require us to interrogate our own presence as both an artistic mediator and social agent. Intercultural collaboration in particular requires a substantial amount of active artistic, emotional and communicative labour to maintain and engage with participant's contributions and agencies in a meaningful and ethical way, and break the habitual mould of this work to 'fit' non-Western ideas into the compositional systems of Western frameworks 'as a contribution to—or an enrichment of—art music performance' (Robinson, 2020, p. 6).

This work prompts us to reflect on our interactions at every level: Where does my agency as 'artist' begin and end? Is this work valuable for the community? Are my voice and skills really needed in this space, or is my inclusion part of a solipsistic desire to help? Does my presence as (in this case) a white, cis-male body with its associated power negate the real human connections made through this project? Asking questions like these as part of a critically reflective praxis helps us begin to navigate the complexities of power in these projects and perform our moral responsibility to channel any privilege of materials into meaningful forces for actual social change; but equally they should not put us off from engaging with, and embracing, challenges of this work. Art, writes Simon O'Sullivan (2006):

> does not so much offer up a set of knowledges as set up the conditions, we might say the contours, for future knowledges still to come. It is in this sense also that art involves the posing of new questions and as such will always make demands on any already existing audience. It is precisely the imagination, or simply the ability to 'go beyond' what appears to be reality (that is, our habitual perceptions and practices), that constitutes this fundamental power of innovation and invention. (p. 146)

We need to avoid thinking about projects like these as magic, emancipatory 'moments of political subjectivisation in which the excluded … put forward their claim to speak for themselves' (Ranciere, 2006, p. 69). As artistic agents and mediators, we are just as likely to celebrate the constitution of entangled agencies as they are to disturb the 'powerful … expression of defiance', that defiant silence that some victims use as 'a statement of agency' in its own right (Nooshin, 2009, pp. 30–31). We need to instead embrace the good work they do as sites of social and cultural encounter; facilitating dialogues of knowledge production and sharing between both participants and leaders, and participants and the wider community through the actual and intimate making of art. Provided we create work out of real and meaningful relationships with care and attentiveness, the bonds we make by listening together, composing together, and performing together could be the key to ensuring encounters such as this one lead to a reimagined and equitable future.

Notes

1 It is recommended that this musical composition is viewed alongside reading the chapter. It can be accessed at https://tinyurl.com/mujuz3m7.

2 This project was funded by Arts Council England and The Oxford Research Centre in the Humanities.

3 The new composition was designed to be performed by the OSJ, rather than participants.

4 As part of the funding for this project, an external team of researchers was employed to formally evaluate the project and measure the success of these goals, and their work will be published shortly.

5 The author of this chapter is an accredited counsellor and coach, with a speciality in trauma counselling.

6 For example, in the harmonic language of the piece I was able to integrate different minor-altered intervals from the modal music of participant's home cultures (including the *Raga Thaat Bhairavi* with its flattened second degree of the scale and the *Dastgāh-e Homāyoun*—a scale common in Arabic and Iranian music—with its flattened second and sixth). This also enabled me to explore musical connections beyond the literal, such as bringing these sonorities into dialogue with the phrygian mode with its strong resonances of deep sadness—used famously in Samuel Barber's 'Adagio for Strings'—but also the altered major phrygian mode used by Oxford-based band Radiohead (e.g. in their 'Pyramid Song') and even the phrygian dominant mode found in much of the Arabic hip-hop music that many of the participants listened to.
7 Initially we were due to have two workshops for the 'editing task' to ensure amendments could be made and evaluated by the group, but due to COVID-19 difficulties this stage was reduced to a single session.
8 Movement titles were taken from the participants' spoken material featured in the soundtrack.

References

Born, G. (2017). *After relational aesthetics: Improvised music, the social, and (re)theorizing the aesthetic*. Duke University Press.

Brett, T. (2016). *From environmental sound to virtual environment enhancing*. Oxford University Press.

Browning, J. (2020). Meeting the garden halfway: Ethnographic encounters with a sound installation microculture. *Ethnomusicology, 64*(3), 498–526. https://www.jstor.org/stable/10.5406/ethnomusicology.64.3.0498

Clarke, E. (2003). Music and psychology. In M. Clayton, T. Herbert, & R. Middleton (Eds.), *The cultural study of music: A critical introduction* (pp. 113–123). Routledge.

Cottrell, S., & Impey, A. (2018). *Community music and ethnomusicology*. Oxford University Press.

Demers, J. (2010). *Listening through the noise: The aesthetics of experimental electronic music*. Oxford University Press.

DeNora, T. (2000). *Music in everyday life*. Cambridge University Press.

DeNora, T. (2003). *After Adorno: Rethinking music sociology*. Cambridge University Press.

Dieckmann, S., & Davidson, J. W. (2018). Organised cultural encounters: Collaboration and intercultural contact in a lullaby choir. *The World of Music*, 7(1/2), 155–178. https://www.jstor.org/stable/26562979

Feld, S. (1987). Dialogic editing: Interpreting how Kaluli read sound and sentiment. *Cultural Anthropology, 2*(2), 190–210. https://www.jstor.org/stable/656355

Feld, S., & Brenneis, D. (2004). Doing anthropology in sound. *American Ethnologist, 31*(4), 461–474. 10.1525/ae.2004.31.4.461

Foucault, M. (1971). *The order of things*. Vintage Books.

Fredrickson, B. L. (2001). The role of positive emotions in positive psychology. *The American Psychologist, 56*(3), 218–226. 10.1037/0003-066X.56.3.218

Hickey-Moody, A., & Willcox, M. (2019). Entanglements of difference as community togetherness: Faith, art and feminism. *Social Sciences, 8*(9), 264. 10.3390/socsci8090264

Hooks, B. (2014 [1994]). *Teaching to transgress*. Routledge.

Hutchens, J., Paz, A., Vogt, N., & Wakeford, N. (2021). Working and not working with you. *OAR: The Oxford Artistic and Practice Based Research Platform*, (4). https://www.oarplatform.com/working-working/

Ingold, T. (2011). *The perception of the environment: Essays on livelihood, dwelling and skill*. Routledge.

Kotef, H. (2018). Migration, settlement, movement. *Cultural Dynamics*, *30*(3), 214–219. 10.1177/0921374018795232

LaBelle, B. (2010). *Acoustic territories: Sound culture and everyday life*. A&C Black.

LaBelle, B. (2018). *Sonic agency: Sound and emergent forms of resistance*. MIT Press.

Laskow, S. (2018, October 15). *Maps have the power to shape history*. [Atlas Obscura] https://www.atlasobscura.com/articles/first-atlas-of-the-united-states

Maier, C. J. (2016). Sonic modernities: Listening to diasporic urban music. In J. G. Papenburg & H. Schulze (Eds.), *Sound as popular culture: A research companion* (pp. 173–181). MIT Press.

Marsh, K. (2012). *Music in the lives of refugee and newly arrived immigrant children in Sydney, Australia*. Oxford University Press.

Middleton, R. (1990). *Studying popular music*. McGraw-Hill Education.

Moon, A. (2005). *A handbook of reflective and experiential learning: Theory and practice*. Routledge.

Nooshin, L. (2009). *Music and the play of power in the Middle East, North Africa and Central Asia*. Ashgate Publishing.

Oliver, K. (2014). The poetic axis of ethics. *Derrida Today*, 7(2), 121–136. 10.3366/drt.2014.0085

O'Sullivan, S. (2006). *Art encounters Deleuze and Guattari: Thought beyond representation*. Renewing philosophy. Palgrave Macmillan.

Porges, S. W. (2014). *Clinical insights from the polyvagal theory: The transformative power of feeling safe*. W. W. Norton.

Ranciere, J. (2006). *The politics of aesthetics*. A&C Black.

Robinson, D. (2020). *Hungry listening: Resonant theory for Indigenous sound studies*. University of Minnesota Press.

Ross, V. (2016). *Framing intercultural music composition research*. Routledge.

Samuels, D. W., Meintjes, L., Ochoa, A. M., & Porcello, T. (2010). Soundscapes: Toward a sounded anthropology. *Annual Review of Anthropology*, *39*(1), 329–345. 10.1146/annurev-anthro-022510-132230

Shankar, S. (2012). *Flesh and fish blood: Postcolonialism, translation, and the vernacular*. University of California Press.

Shelby, T. (2009). *We who are dark: The philosophical foundations of Black solidarity*. Harvard University Press.

Solnit, R. (2010). *Infinite City: A San Francisco Atlas*. University of California Press.

Spivak, G. C. (1994). Responsibility. *Boundary 2*, *21*(3), 19–64. https://www.jstor.org/stable/303600

Stokes, M. (2008). On musical cosmopolitanism. *Macalester International*, *21*(8), 3–26. http://digitalcommons.macalester.edu/macintl/vol21/iss1/8

Storsve, V., Westby, I. A., & Ruud, E. (2010). Hope and recognition: A music project among youth in a Palestinian refugee camp. *Voices: A World Forum for Music Therapy*, *10*(1). https://voices.no/index.php/voices/article/view/1853

Talianni, K. (2020). The soundscape of Anthropocene. *Airea: Arts and Interdisciplinary Research*, *2*(2), 63–76. 10.2218/airea.5037

Voegelin, S. (2018). *The political possibility of sound: Fragments of listening*. Bloomsbury Publishing.

Wake, C. (2015). Performing authenticity in the refugee determination process. In E. Cox (Ed.), *Staging asylum: Contemporary Australian plays about refugees*. Currency Press.

Watson, A., & Ward, J. (2013). Creating the right 'vibe': Emotional labour and musical performance in the recording studio. *Environment and Planning A: Economy and Space*, *45*(12), 2904–2918. 10.1068/a45619

Western, T. (2020). Listening with displacement: Sound, citizenship, and disruptive re-presentations of migration. *Migration and Society*, *3*(1), 294–309. 10.3167/arms.2020.030128

Wilson, H. (2016). On geography and encounter: Bodies, borders, and difference. *Progress in Human Geography*, *41*(4), 451–471. 10.1177/0309132516645958

Woodruff, J. (2020). Composing sociality: Toward an aesthetics of transition design. In S. K. Groth & H. Schulze (Eds.), *Bloomsbury handbook of sound art* (pp. 41–70). Bloomsbury Publishing.

Young, J. O. (2010). *Cultural appropriation and the arts*. John Wiley & Sons.

13

ODYSSEY ON THE AIRWAVES: A JOURNEY FROM HMP TO HOPE

Gary Anderson and Niamh Malone

> *The story of a man whose grand adventure is simply to go back to his home … For this hero, mere survival is the most amazing feat of all.*
>
> *(Wilson, 2018, p. 2)*

Odyssey on the Airwaves is a radio project facilitated for adult learners in a local prison[1] in 2020–2021 during the COVID-19 pandemic lockdowns. This audio project quickly revealed that incarcerated men find Odysseus's return journey home to Ithaca, in Homer's epic poem *The Odyssey* (circa 800 BCE), something that resonates powerfully with their lived and living experiences. The project was planned to be completed within six weeks, however, ironically, like the constant stream of setbacks experienced by Odysseus on his journey home, COVID-19 blew us off course and forced us to rethink and re-do. The implementation of strict lockdown restrictions (23 hours a day in cells every day) in the local HMP however, did not seem to lessen the determination from prisoners and university student volunteers, and finally we managed to make it home.[2]

Malone had already established a strong working relationship with the local prison and had run two previous applied theatre projects as part of the *Forgotten Futures and the City (2018–2019)* programme. These projects had taken place pre-COVID-19 and we had been able to use the recording studio within the prison to work with the men on audio production. In 2018–2019, the men recorded their own creative writing, including poetry, prose, spoken word and music. Due to COVID-19 *Odyssey on the Airwaves* required us to rethink our entire delivery strategy. Such constraints were responsible for initiating a whole new way of delivery—albeit primitive—to enable the men to record their contributions. The only approach available was for us to visit the prison, sit in an isolated office space, and phone directly into the cells of the participating men via the in-house telephone facility. The men would then read their self-selected passages of the *Odyssey* into the phone, while we

DOI: 10.4324/9781003164227-22

held a Dictaphone up to the phone receiver to catch their voices. Later, the files were then downloaded from the Dictaphone and transferred onto editing software within the prison. Files from the university students were then emailed into the prison to the local educational facilitator who undertook the complex editing task of bringing all the recordings together. All these creative projects between the university and the prison, have been enabled by Novus[3] (a social enterprise) responsible for educational provision in 52 of the 117 prisons in the UK.

Odyssey on the Airwaves is now complete and 'live' on the radio inside the local prison. However, the project was fraught with setbacks practically and theoretically, and this essay is an account of that journey. It concludes with some theoretical suggestions from Deleuze and others for applied theatre in prisons, specifically, how affective encounters represents a useful theoretical tool for approaching cultural work with incarcerated men.

Telemachy, Apologoi, Mnesterophonia

The decision-making process of arriving at the Odyssey was surprisingly quick, but not without its problems. One of the participating men, apparently in cahoots with some of the other participants on his wing, when asked what he wanted to focus on via the in-cell telephone call, said he wanted to learn about 'something proper!' We were then asked to list a series of 'proper' pieces of work. Caught unawares, not quite knowing what was at stake, we started with some Shakespeare plays, a few Ancient Greek plays and some relatively well-known poetry from the Romantic period. Unimpressed he continued:

> "What's the first ever great piece of work?"
> our response "What do you mean?"
> "You know, what's the first best thing ever written?"
> Off guard, we replied "In the West, I suppose it would have to be Homer?"
> "Who?"
> "Homer. H-O-M-E-R. Homer."
> "Alright, fuck it. Let's do that. Homer? Like Homer Simpson?"
> "Yes. Same spelling."

After some resistance from the other men, which we will discuss below, and together with prisoners and student volunteers, we got to talking about the structure of *The Odyssey*. Collectively we realised that it was structured like a good argument with three key sections: the problem, the actual struggle and the conclusion. Classical philology scholars Erwin Cook and Ruth Scodel call it the 'Telemachy', the 'Apologoi', and the 'Mnesterophonia' (Cairns & Scodel, 2014; Cook, 2014). The Telemachy, named after Odysseus' son Telemachus, offers the problem of *The Odyssey* from a particular point of view. Telemachus, in conversation with the goddess Athena, gives us a handle on the central issue: Odysseus has been gone for 20 years and all his rivals are about to take over his kingdom (including his wife Penelope—something which chimed strongly with the prisoners). In other words, we get a critical perspective on events from which to begin our journey as readers/listeners. Our Telemachy here is the first section of the essay, which sketches out our theoretical approach to the project. This belongs to the growing field of affect theory in applied theatre and will draw on theories of 'affective encounters' and 'becomings' after twentieth century French philosopher Gilles Deleuze.

The second section focuses on the events themselves: 'the Apologoi'. This is the first-person narrative which talks about what has happened so far. In *The Odyssey*, Odysseus retells his story to the Phoenicians, mapping it as he goes from the sack of Troy to his escape from Calypso. For us this translates as the literature review and the mapping of the field of applied theatre in prisons. Closely related to this is a description of the ways Odysseus tells the Phoenicians how he chose to handle the problems he faced (e.g. his fight against the Cyclops, how he travelled to Hades and back, how he escaped Calypso); without pushing the metaphor too far, this is Odysseus' methodologies section. In this part, we focus on the methods deployed with the prisoners at the local HMP, the role our student volunteers played and our part as two university teachers facilitating the project.

The final section is the Mnesterophonia. In *The Odyssey* this is the 'slaughter of the suitors' or the 'revenge'. This is where Odysseus, using the wisdom from his journeying defeats his rivals to his kingdom, thereby concluding his story. We use this section to make claims about applied theatre in prison and the necessity for the movement away from a regime of judgement towards an openness to the 'affective encounter' in prison work.

Telemachy: Re-unlocking the theoretical gates

Theoretically the essay draws on 'affect theory' of 'encounters and becomings' which belongs to the field of academic inquiry associated with Deleuze studies, after French philosopher Gilles Deleuze (1925–1995). We draw on two seminal papers from the 1990s from two of the leading scholars in Deleuze studies: Feminist philosopher Rosi Braidotti's (1993) *Discontinuous Becomings. Deleuze on the Becoming-Woman of Philosophy* and philosopher and social theorist Brian Massumi's (1996) essay *Becoming Deleuzian*. These texts, academically speaking, are not recent, but serve for us several purposes: first, as an accessible antidote to the more or less ubiquitous privileging of conservative politics in literary criticism (most especially in USA's formerly leading literary critic Harold Bloom and his influential 1994 magnum opus 'The Western Canon'). Bloom identified a group of scholars who were interested in placing the work of art or literature in its immediate socio-historical contexts, thereby politicising it, as the 'school of resentment'.[4] Deleuze and his followers would be thought of as 'resenters' in Bloom's estimation. See, for example, the much-quoted mantra of the Bloomists from the 1990s—the following straight from the horse's mouth:

> We are destroying all intellectual and aesthetic standards … in the name of social justice. What has been devaluated is learning as such, as though erudition were irrelevant in the *realms of judgment and misjudgment.*
>
> *(Bloom, 1994, p. 79 [our emphasis])*

More importantly, because of our experiences in prison with men, who for understandable reasons might have allergic reactions to 'high' art/culture, we are sometimes forced to reproduce a position of exclusion from 'high' art for people suffering multiple deprivation. The paradox is fascinating to us and something we felt needed to be tackled. Certainly not from a position that 'high' art is good for you, or the suspect academic debates of the 1990s which argued for a democratisation of culture (meaning everyone should know Shakespeare, not just the privileged). Yves Evrard (1997) outlines the two positions well in his seminal essay. Maybe, for many of us, including Evrard, the Bloom arguments have already been fought and lost, but in prison it feels like the battles need to be waged and won again. There is much work to be done. Indeed, many of the men, including the mixed gender student volunteers, defaulted to a sort of Bloomist position when asked about *The Odyssey*. Debates from the 1990s are not necessarily up to date for contemporary academics, but they are still very much alive in applied settings. Reminding ourselves of these debates can help situate the applied practice appropriately. Debates in prisons, in the UK at least, tend to develop at a different pace to academia and 'old' debates are arguably more relevant than current ones. Or, perhaps more accurately, elements of old debates can prove useful when brought to light together with more recent insights—which is what we are attempting here. In other words, has Kantian Bloomism been debunked enough in applied settings? This essay contributes to a growing field of affect theory applied to lived-in settings by claiming it has not, and that there is vital work required.

Prisoners' responses, after the decision was taken by them to focus on *The Odyssey*—no doubt influenced by the original suggesters on their wing—tended to reproduce positions Bloom might recognise with some sympathy: either the work was too precious to be messed about with, or the adult learners themselves were not educated enough (i.e. correctly schooled in the 'realms of judgement and misjudgement') to appreciate properly. Here is a sample of their immediate responses to exploring the text:

> "How long ago? Nearly 3000 years? That's for snobs mate" (Prisoner#1)
>
> "3000 years ago. You taking the piss?! That's for knobs in uni and that. Let them do it." (Prisoner#2)
>
> "What? Homer. Yeah, I've heard of him. It was an episode on *The Simpsons*. That one about the wooden horse and then he ends up throwing the spear though Smithers and Mr Burns coz they fancy Marge? Is that the same story? I'll watch the cartoon, mate. Don't reckon I'll be up for the proper book though … Sounds a bit … dunno … too much" (Prisoner#3).[5]

Typical responses were Bloomist: high art is of such aesthetic value that to apply it to the everyday, either in terms of its production or reception, would be to misrecognise it and do it a disservice. When asked about whether we could read the poem in the vernacular accent (most men involved self-identified as 'scouse'[6]), and to make it about prison life, typical responses were:

> "You can't do that with poems, can yer? You'd have to write a new play wouldn't yer?" (Prisoner #5)
>
> "Scouse! That would just make it sound funny mate." (Prisoner #2)
>
> "Will we have to put on accents?" (Prisoner #1)

After running sessions on politicising Homer's poetry, and how it was being used to establish Athens as the dominant city in the region during the Peloponnesian wars 2500 years ago,[7] the men were immediately more interested and responded accordingly:

> "So, it's like propaganda? Yeah, I'll have some of that!" (Prisoner #3)
>
> "Cheeky bastards coming across like they're all posh, when they're as dirty as them in charge of the country today!" (Prisoner #4)
>
> "Yeah, I'll have a look at the original if you want. Is it hard though? Is there loads of funny words?" (Prisoner #1)

The student volunteers, from the university, seemed more difficult to convince, flagging up reasons of 'respecting the cultures of the ancient world', 'great names' such as Homer, the Greeks et cetera, and 'We have been taught to respect them, not really question them'. Pure Bloomism! *Odyssey on the Airwaves* is an applied theatre project which seeks to challenge the conservative cultures of the 'Western canon'. In this way it keeps company with many similar initiatives which may be more reliant on more conservative understandings of cultural work (or the 'praxis of judgement', see below) most notably the laudable but problematic *Shakespeare Behind Bars*.[8] In other words, an effective counter to Bloom and the politically conservative approach to the arts is affect theory. Furthermore, to reference theoretical debates from the 1990s is to refresh our memories of the radical intentions of affect theory, before 'the affective turn' in social theory in the noughties.[9] It is also to strengthen James Thompson's insights in applied theatre around the 'end of effect' and the foregrounding of affect (Thompson, 2011, 2014, 2017). From the 1990s affect theory in anglophone contexts has developed, and with it an exponential increase in discourse. Patricia Clough's *The Affective Turn* (2007) and Linda Zerilli *The Turn to Affect and the Problem of Judgment* (2015) are two excellent examples. But as the field develops and proliferates and produces new insights, arguably some of the radical intentions have been left behind. Neuroscience has developed apace with affect theory,[10] but the

Deleuzian thinkers Braidotti and Massumi, have maintained affect theory's intellectual resistance to hierarchy and judgement—as set out in their work from the 1990s. The sites of applied theatre (e.g. hospitals, care homes and prisons), not unsurprisingly, have had less chance to engage in debates around affect theory, and in our recent experience prisoners have unwittingly tended to default into Bloomism when asked explicit questions about high art/culture. Bloom's shadow on cultural work, including on our undergraduate student cohort, is long, dark and difficult to escape from. We did not go into the prison thinking this, we learnt it from the men on the wings.

We found Deleuze, Braidotti, and Massumi's relevance refreshed and directly applicable to the prison work we undertook. In many ways it saved the project from folding. It became essential for us to keep 'the realms of judgement and misjudgement' at bay in our own work and to approach the difficult world of incarcerated spaces 'ethologically',[11] rather than phallogocentrically; non-judgementally rather than judgementally.[12] Theory really mattered to us, more than we had anticipated.

Crucial to the 'affective turn' is the concept of 'the encounter', originally a Spinozist term referenced by Deleuze in his lecture series of 1980–1981, and picked up by Braidotti and Massumi later on. Deleuze states:

> Spinoza, in the *Ethics*, uses the Latin term: *occursus*; *occursus* is exactly … the encounter. I encounter bodies; my body never stops encountering bodies. Well then, the bodies that it encounters, sometimes have relations which compose, sometimes have relations which don't compose with it.
>
> *(Deleuze, 1980–81)*[13]

For us this is a good definition of the radical basis of the 'affective turn', and one which we, as practitioners took up in this project. Encounters do not prioritise judgement, in fact a strict Deleuzean-Spinozism would conclude that encounters model experience differently from the Kantian ethos of 'disinterested judgements' (of which Bloom is an obvious disciple) and instead make judgement suspect. The 'praxis of judgement'[14] is based on Platonic and Kantian systems of understanding—where there is a perfect example of something—the recognition of which constitutes a perfect 'judgement'. This judgement structures experience hierarchically thereby necessitating the structural privileging of one thing over another in order to make sense of the world and reaffirm its moral order. Ontologically, Spinoza, then Deleuze, turn that structure on its head and proffer a single, infinite substance responsible for all modes and manners of being. Spinoza calls this God or Nature, Deleuze calls it radical immanence.[15] The impact of this theory is shocking: nothing is by nature better or worse than anything else. Things in the world are not there to be judged, but to be encountered. Your subjectivity (your unfinishable identity) either agrees or disagrees with the encounter. In other words, if it suits you, continue. If it does not suit you, move onto something else. There's a radical liberation in that. Massumi's (1987) seminal preface to Deleuze and Guattari's (1987) *A Thousand Plateaus* called 'The Pleasures of Philosophy' ends:

> The question is not: is it true? But: does it work? What new thoughts does it make possible to think? What new emotions does it make possible to feel? What new sensations and perceptions does it open in the body?
>
> *(Massumi, 1987, xv)*

Deleuzian thought is useful in prison settings particularly in arts' processes. James Thompson writes, 'Deleuze, and in particular his disciple Massumi, allow us to consider the encounter in the arts process as the key terrain from which to consider its power—and its radical potential' (Thompson, 2011, p. 125). The radical potential of becomings and in Thompson's argument 'affect' is the possibility of not just undertaking, but actively living a non-judgemental approach.

If we take Massumi and Thompson seriously then the moral orders are in a state of disintegration, but what about the people incarcerated? How can we approach prisoners non-judgementally? Braidotti puts the definition of becomings succinctly:

> The body is not an essence, let alone a biological substance: it is a play of forces, a surface of intensities—pure simulacra without originals.
>
> *(Braidotti, 1993, p. 44)*

Here, Braidotti disavows that identity can be understood in terms of production, or of the dynamics of opposition, which is, as she goes on to note, a non-Hegelian way to theorise the self. Instead, identities or more precisely subjectivities, can be more affirmatively understood as unfinished and in-transit. In other words, identities are the various becomings of our subjectivities. This is relevant to the project *Odyssey on the Airwaves* because without it we as prison educators, risk falling into the trap of 'the praxis of judgement'. This helped us overcome our negative assumptions about incarcerated spaces and those who inhabit them, as well as providing a base upon which we could build a post-Bloomist educational space.

Yet another reason for relying on Deleuzian theory is that it is so effective in undermining unhelpful binaries, some of which feel very active in prison settings. For example, 'criminal/law-abiding citizen', 'malevolent/benevolent intentions', 'bad person/good person'. These binaries felt heavy for us upon entering prison and meeting people. Add to that the affective power of the site, in this case a purpose-built, Victorian prison housing more than one thousand prisoners and prison employees. These heavy binaries have an effect on what it is possible to think and feel whilst engaged in prison work. In our experience, unhelpful binaries in education such as 'educated/uneducated', 'expert/novice', 'canonical/vernacular knowledge' tend to be intensified in prison settings, but 'freedom/non-freedom' is perhaps the most misleading binary we have encountered. Massumi writes:

> Deleuzian freedom is immanence to a collective creative process necessarily connecting individuals' belongings and unbearabilities to the singularity of

> their becomings (ethics, ontology, politics, and aesthetics rolled up into one). Deleuzianism is a monist philosophy of the collective expression of singularity.
>
> *(Massumi, 1996, p. 405)*

In other words, what we ended up responding to, were the incredibly intense encounters with site and prisoners, removed from the moral order which tends to reproduce oppression and stymie freedoms. Massumi's 'collective expression of singularity' will be different for all participants, but for us, perhaps most surprisingly, we encountered (and hopefully overcame) our own prejudices and our uncritical reproduction of false binaries. We ended up not entering the prison to teach undereducated prisoners about antiquity, but rather to encounter the dynamic loss of our own unbearabilities—or the struggle we had with abandoning our own apparatus of judgement-making. There's no room in this process for Bloom's authority. Nor is there room in Braidotti's notion of subjectivity, 'The embodied subject is a term in a process of intersecting forces (affects), that is, spatio temporal variables that are characterised by their mobility, changeability and transitory nature' (Braidotti, 1993, p. 45). For us Massumi and Braidotti's work from the 1990s with its radical edge, became increasingly central to the way we saw the project evolving.

Aylwyn Walsh (2014), using Foucault's *Discipline and Punish* (1977), identifies the false binary of 'containment/freedom' in her essay which maps the prison as a site *and* a field of performance,

> I suggest that it is also possible, through examining everyday performances and the habitus of incarcerated populations, to witness the subjective agency of prisoners as manifesting prisons as sites of openness and possibility, and not merely as sites of containment. (p. 42)

Arguably not much has changed in the prison system since the publication of *Discipline and Punish* in English in 1977 except the exponential increase in number of inmates, 'The prison population of England & Wales quadrupled in size between 1900 and 2018, with around half of this increase taking place since 1990' (Sturge, 2020, p. 3). Most sources cite an increase of around 400 percent across UK prisons from the end of the nineteenth century. With this, according to Walsh and Foucault, is an intensification of population control both inside and outside prison settings, across public institutions, and into the private sphere. Governmentality is a form of control via the uneven power relations present in any, and every, given situation. This is keenly felt in prison settings according to Walsh, thereby necessitating the site of prison as a performance category in and of itself.

In conclusion to the Telemachy, we wish to make a case for a vernacular, localised empowerment which foregrounds the prison as a site of encounters between bodies according to the subjectivities who use the space on a daily basis. In other words, deploying Deleuze (via Braidotti and Massumi) is preferable to deploying a Kantian Bloomism which prioritises 'the realms of judgement and misjudgement'. This builds on the work of Thompson (2011), McAvinchey (2011, 2020) and Walsh (2012, 2014) in particular, so that this project foregrounds the vernacular, in order to potentially enable incarcerated men to make decisions about their own learning processes and what is of value to them, not us. Although, in the end, we would have to admit that the empowerment of prisoners is more of a hope than an actuality with projects such as our own, but we can speak for ourselves as facilitators and the usefulness of the theoretical approaches we took to making this project possible in the first place.

Apologoi: *The Odyssey* as High Art, Rehabilitation, Prison Radio

The prisoners chose to study and perform *The Odyssey* after a short series of conversations, most of which rejected any 'high' art out of hand, as mentioned earlier. The fact that many of them had seen it as an episode of *The Simpsons* was a key factor in its selection, but it all happened fairly quickly and without the sort of build-in time we would have hoped for (discussions, for and against arguments, 'why this and not that?') and before we knew it, about two project days in, the men had come to a decision—they wanted something proper and they decided that Homer's *Odyssey* was 'proper enough!' We had raised money to buy books and equipment by walking Hadrian's Wall in the summer of 2019,[16] so used what was left to buy the most up to date—and arguably easiest to read—version of *The Odyssey*. This was by classics scholar Emily Wilson (2018) who is the first female translator of Homer's *The Odyssey*.

The Odyssey is recognised as belonging to the canon of literature in the cultures of the global north (see Bloom, 1994). Its sister work, *The Iliad*, by the same author, recounts the trials and tribulations of the ten-year Trojan war—believed to have taken place in the late Bronze Age 1500BCE in what is now Canakkale on the Gelibolu (Gallipoli) Peninsula, Western Turkey. The *Odyssey* narrates the return journey of Odysseus after the sack of the town and belongs to the oral tradition—sung or recited out loud for others to hear. These canonised texts, while their origins lie in the every-day vernacular, 'and from Greek dialects associated with different regions' (Wilson, 2018, p. 11), have been institutionalised by educational establishments over the centuries. This has resulted in their re-positioning as 'high' art and the property of those more formally educated who are, perhaps predictably, privileged enough to see it as apolitical. As Bloom (1994) memorably claimed: 'Reading the very best writers—let us say Homer, Dante, Shakespeare, Tolstoy—is not going to make us better citizens. Art is perfectly useless' (p. 36).

For the contemporary reader, the archaic language and elevated register, reaffirms the effect of estrangement for readers unexposed to Bloom-like privilege, complicated by the additional element of translation. According to Wilson (2018), 'The style is, from a modern perspective, strange: it is full of repetitions, redundancies, and formulaic expressions' (p. 2). However, Wilson has an eye for the contemporary vernacular, which was preferable to other more archaic-sounding translations such as Shewring's 1980 Oxford University Press or Riue's 1946 Penguin Classics editions, both still in print and popular: 'My Homer does not speak in your grandparents' English, since that language is no closer to the wine-dark sea than your own. I have tried to keep to a register that is recognizably speakable and readable, while skirting between the Charybdis of artifice and the Scylla of slang' (p. 191).

A significant proportion of the prison population in the UK have a lower-than-average level of educational attainment. According to Coates (2016) "42% of adult prisoners report having been permanently excluded from school" (p. 3). Therefore, acknowledging the immediate discrepancy in education levels between adult learners in the prison and students registered on a BA (hons) at university, the decision to collaborate on such a text poses particular questions about our intentions and our approach. This section will now consider how a project of this nature, not only contributes to the growing number of artistic practices that are being implemented in prisons across the UK, as part of the field of arts intervention in rehabilitation programmes, but also how the mechanisms of pedagogical practices can be used to provide an opportunity for a modest sense of liberation for both the facilitator (pedagogue) and the participant.

The first state-built penitentiary was in Millbank, London in 1816. A brutal journey through two centuries of severe physical and psychological punishment, the recognition of prisoners benefitting from rehabilitation efforts was slow to be trusted, with some eventual acknowledgment in the mid to late twentieth century (Foucault, 1977). The 1980 and 1990s saw the official establishment of theatre companies such as Clean Break Theatre Company (est. 1979) by Jackie Holborough and Jenny Hill; Geese Theatre Company (est. 1987) by Clark Baim and John Bergman; and Theatre in Prisons and Probation (est. 1991) by James Thompson and Paul Heritage. Developments have also flourished with the establishment of more theatre companies who dedicate themselves to the use of arts within prisons and other rehabilitation centres, such as Synergy Theatre Company and Fallen Angels Dance Company, who continue to develop the work pioneered at the end of the twentieth century. The *Odyssey on the Airwaves* project is an attempt to build on the work of those who have gone before us, and to expand the portfolio of work that is currently being implemented in the UK. Withdrawing from the "praxis of judgement" to a more affirmative methodology of "the affective encounter" has proved crucial for us as facilitators.

Applied theatre history in general (Prentki & Preston, 2009) and arts programmes in prisons in particular (McAvinchey, 2020; Walsh, 2014) testify to the long and trying journey arts intervention projects have had to travel, often against a raging sea of challenges and limitations. The neoliberal frame from which prisons are currently operating, characterised by contracting out essential services, requires the arts to produce work which is "value for money". Leaving this to one side, there continues to be an appetite for participating in the arts when incarcerated. In other words, what justifies a continuation of the arts in prison settings is a clear understanding of their soft benefits, no matter how difficult they may be to quantify. According to Shailor (2011):

> Artistic expression introduces introspection and self-awareness that impacts the performer, writer, creative artists directly, and audience indirectly. Without realizing it, the inmate begins to examine their life, learn about themselves, recognize themes, and begin to heal deep personal wounds. (p. 14)

We would argue that this applies, perhaps even more readily, to the facilitators. This observation from Shailor recognises a sense of freedom that is granted to prisoners that participate in artistic programmes, which in turn often releases, or indeed uncovers creativity, leading to self-discovery and self-expression. Such an unearthing can then help nurture a sense of personal agency and an ability to look to the future more positively. Shailor (2011) expands this notion by seeing such work as 'revolutionary' which has 'the potential not only to change inmate lives, but also to humanize the culture of corrections' (p. 8). The applied theatre approach in prisons, according to Walsh (2019) 'positions the work as "doing" something, claiming transformation by examining behaviours "before" and "after", as well as describing the processes of creative participation "during" workshops and rehearsals' (p. 10). Thompson (2011) however is quick to remind us that applied theatre limits itself 'if it concentrates solely on effects—identifiable social outcomes, messages or impacts—and forgets the radical potential of the freedom to enjoy *beautiful radiant things'* (p. 6, author's own emphasis). Perhaps it is for this very reason that the adult learners in the prison decided to work on the epic poem *The Odyssey*, resonating all the more with Thompson's observation, that the value for them was not about transforming their behaviour or attitude, but it was about the joy of working on a collective project. The issues we encountered initially were the Bloomist views most prisoners had internalised about what beauty or joy was and who had access to it. The noun 'beauty' extends in a practical sense here too, where the project could continue, albeit in an altered state, despite the restrictions of COVID-19, not forgetting that the men have to spend 23 hours a day in their prison cells. The prisoners desired to explore this classic text, with all nine participants having heard of it, but none having engaged with the actual text at any level prior to this project. Their enthusiasm about exploring this story was due, in part, to their familiarity with the numerous reference points circulating in popular culture, and they now valued the opportunity to read and perform a passage from the poem, as part of a collective, for the in-house prison radio.

Likewise, for the student volunteers, all studying drama and theatre studies at BA level, when approached to participate in this project, had heard of *The Odyssey*, but had never read the text, or performed it. As mentioned earlier, prisoners were only contactable via a phone call into their cells with no visual communication. Student volunteers were communicating from their own homes via Zoom. Both sets of learners decided to read the whole book through, then select one of 24 chapters to read extracts either in one sitting, or, as in the case of the prison, on a weekly basis. Luckily all the student volunteers had access to voice recording technology, and delighted in recording it themselves, and emailing it to us. The setting in the prison was more complex, with both facilitators on site, inside the prison every Friday to call the cell telephone number, discuss the material selected and then record the voice through a dictaphone held up to the earpiece. There was no communication between prisoner and student, except that both parties knew of the others' engagement in the project. The task then was to get the full 24 books of *The Odyssey* recorded, with varying degrees of quality, and bring the recordings together to create a radio version of *The Odyssey* to be played as part of the prison's in-house radio programming.[17]

It has been important for us to remember Thompson's (1998) questions in his seminal work *Prison Theatre: Perspectives and Practices* around the intentions of arts programmes in prison settings. One relevant question for us was 'Does [theatre] provide the means to transform a person's life or does it transform the whole community?' (p. 10). *The Odyssey on the Airwaves* project is a modest attempt to answer this question practically: an arts project which seems to function well at the level of the individual in the prison and the student in the university, while deliberately blurring the boundaries between students who are incarcerated, and those who are not, thus interrogating what constitutes an academic community. Presenting such a binary between the adult learners in the prison and those students in the university, prompts caution on behalf of us as researchers, not to, albeit unintentionally, reaffirm already problematic established power binaries between those who are incarcerated, and those who are 'free'. Such a false binary is also in danger of endorsing an unjust distinction between those who wield cultural capital (in this instance, university students for the most part) and those who do not (adult learners in the prison). If we apply Bourdieu's (1979/2010) lasting theory about the social advantages to those of us who have a certain level of cultural capital—that, inevitably often reflects social class—and those who have a much lower level, then the employment of a classical text, such as *The Odyssey*, would suggest favouring the university students over the prisoners. Such a position is strengthened further by Rachel Dobbs' (2019) statement where she reports that out of the '82,000 inmates, just 100 were studying at A5-Level or above in 2018' (n.p.). Despite such a glaring discrepancy, we found the opposite: that the adult learners in the prison, with none of the participants having third level education, embraced the opportunity, doing extra work by engaging in background research to support their reading/performance. Most of the prisoners had spent a considerable amount of time rehearsing their sections and were ready and relatively confident for their recording. The university students, while producing some good work, saw it more as part of their course of study (even though their participation was voluntary), and the discussion generated exploring the text and its historical relevance and contemporary status, was second to the enthusiasm that the prisoners demonstrated. Such a welcome response from the majority of the adult learners in the prison, does not testify however to some unflawed model of applied practice. It does keep us as researchers focused on what the actual intention of this project is and whether or not we have met it?

Thompson (2011) again is right to caution that working with a particular minority group 'has a politics that needs to be questioned, but it is, not by some default process, one that is inevitably critical, or one that can claim an automatic contribution to social change' (p. 5). As observed earlier, the majority of applied theatre projects are understandably fixated on demonstrating an element of 'social change' however modest that may be, for this project our intention became clear towards the end: encounter bodies affirmatively and co-produce moments of mutually realised joy (in the Spinozist sense) despite the haunting of the praxis of judgement from all sides, including (perhaps even especially) the site. And further, is it possible, because of the site, to produce moments of joy? For the facilitators, yes. For the prisoners, we think so and we have evidence for that in terms of what they told us. Methodologically speaking though, we will never really know what the participants underwent or continue to as they listen to themselves on the in-house radio. Testimony is probably always unstable, produced under conditions which shape it, and then reproduced in contexts which determine it. We can only speak for ourselves as authors of this essay: for us it was revolutionary. We managed, on occasion, to *Have Done With the Judgement of God*, to cite Antonin Artaud's final piece of work, which happens to have been a radio play.[18]

While offenders' access to radio depends on a number of factors, and varies from prison to prison in the UK, on the whole, it is one of the more accessible means of communication and is available to much of the prison population. 1994 proved to be a pivotal year in elevating the role of radio in the prison setting, with the establishment of Prison Radio Association (PRA).[19] It was the brainchild of two local residents of HMP YOI Feltham, who recognised the need for a form of intervention for offenders when they are at their most vulnerable, especially at night, where suicidal thoughts and self-harming often occur. In 2006, Prison Radio Association gained charitable status, and was now in a position to offer their expertise and guidance to prisons across the UK on setting up their own radio projects. Their radio channel is now available in 100 prisons across England and Wales, with 76 percent of the prison population having listened to/or listen to National Prison Radio.[20]

Radio as a medium, offers an invitation to explore an assortment of content, with the form accommodating a variety of styles from interviews to scripted plays. It can also protect identities, which can be an issue for some of the prisoners, while educating the learners on both production and presentation skills. As the radio is only aired within the prison walls, this means it is advert-free, unlike mainstream radio in the UK, but crucially has control and understanding of the demographic of the listening audience. The role of radio seems to be two-fold. If we reflect on the origins of PRA, with its initial therapeutic and interventionist impetus, we can also recognise radio as an effective learning resource. Charlotte Bedford (2016) is an avid believer in the power of radio as an educational tool, especially with a prison population where levels of educational attainment are well below average. She states, 'Radio has been shown to be an effective and innovative method for both disseminating information and engaging people back into education, particularly those with previously negative experiences of schooling' (p. 129). Bedford's observation is relevant for us, where part of our overall aim for the prisoners was to demystify third level education and present it as a real opportunity for the adult learners on their release from prison.

In our local HMP, an internal radio system is available to the vast majority of prisoners, exceptions notwithstanding. Novus, (the education providers that enabled access for us to the adult learners on site), at the time of writing, had just installed all-new software to make the radio system more efficient. The Novus employee in charge of the recording studio produces 24 hours of content per week. He works with, and trains up, many of the prisoners, where they work collaboratively to decide on form and content. *Odyssey on the Airwaves* has been a welcomed contribution to their programming and has extended their remit to include 'the classics'—something we see as double-edged sword, but could, in a post-Bloomist way be very useful. As researchers, we have found Radio, as a creative medium, to be particularly well placed when working within the constraints of the prison system. This was particularly true during the COVID-19 lockdown, and the retaining of men in their cells for 23 hours a day. They could work in a solo capacity on their research, practice and perform *The Odyssey,* with their labour contributing to a collective group project. We think, and hope, that our project *Odyssey on the Airwaves,* combined both the educational (as noted by Bedford) and the therapeutic (as emphasised by PRA) for all our participants—but again, we cannot claim this with authority. The best we can say is that we think it was better than nothing. For us as facilitators, it was much more. Thompson (2011) suggests that all applied theatre projects 'reveal certain aesthetic and political choices that need to be considered' (p. 5). Our approach, no doubt dripping with educated privilege, is Deleuzian in that the aesthetics and the politics are part and parcel of the encounter. No claim is made for the project except on behalf of the facilitators who have benefitted enormously. It is our view that the praxis of judgement haunts prison work, perhaps more immanently than other areas of applied theatre work, and like Thompson warns, great care must be taken in assumptions, conscious or otherwise, brought into the prison.

Mnesterophonia: 'slaughter of the suitors'

At the end of *The Odyssey*, Athena tells Odysseus just before his final victory over his enemies, as he doubts himself, 'Odysseus, you are adaptable; you always find solutions'. (Wilson, 2018, p. 946). We found adaptability the single most important requirement for working in prison—adapting Bloomism to Deleuzeanism, classrooms full of adult learners to one-on-one telephone conversations, and a relatively well-functioning society to COVID-19 lockdowns. Although negativised by many, it is Odysseus' cunning (or theoretical methodology) that wins him his kingdom back in the end, and he does it by being adaptable. Wilson observes that '*The Odyssey* puts us into a world that is a peculiar mixture of the strange and the familiar. The tension between strangeness and familiarity is in fact the poem's central subject' (p. 5). For us, the project, and the unexpected adaptations to the changing environments, was both strange and familiar. Strange as we were forever reminding ourselves to think of it as 'affective encounters' and 'becomings'; familiar as we involuntarily deployed the praxis of judgement on both 'high' art and the prisoners/student volunteers. Despite being conscious of the pitfalls, we inevitably fell into them again and again. That too, was familiar. The familiarity of driving up to the local prison every Friday morning, parking outside the long high brick wall rounded at the top with concrete and black grease, walking in through the reception doors, providing proof of identity, restating the purpose of the visit, going through the double-check on who we were meeting, sitting on the waiting couch, then being escorted through 13 doors (each of them locked after us) and reaching the no-windowed room with the internal phone system and the polystyrene low ceiling, being handed a box with the excel spreadsheets of the long phone numbers and codes to the different wings in tiny print, and the old prison-owned dictaphone that looked like it had seen spy action in the Cold War. It became almost impossible to feel affirmative, in the Deleuzean sense. Our instincts were forever convincing us to leave and stay away. A real mixture for us of the strange and the familiar.

In this context, our powers of action were diminished to such a point, we thought we would end up being a negative influence on the prisoners. Then the phone conversations started, and with them *joy*. Pure joy at what the men said, how they read, what they'd been thinking about that week, how they had ordered a philosophy book from the prison library that we had put there at the beginning of the project 'because of that conversation we had the other week', (prisoner#3) how they were thinking about applying to university when they got out (some imminently, others in years to come). The way the men sang or stumbled through the text, polluted with phone crackle, their raised voices, the different ways in which they held their in-cell telephones, the distances they kept between mouth and phone receiver—no standard form, no basic level, no uniformity. The delivery of the Odyssey down the phone line to an immediate audience of two (Malone holding the Dictaphone to the receiver, Anderson listening in as best he could) was a revelatory sonic experience. The interruptions from prison staff warning the reader to get back to the laundry or the regular jangling of keys or sliding of barred doors, the coughing of the prison officers who arrived in the 'phone room' to get the days' newspapers from the staff on the computers, to squeaks on the floor of the wing, laughter from somewhere else on the corridor all form the background soundtrack to the work. Is this in keeping with the traditions of the oral forms of delivery—the messiness, the unpredictable, the haptic? On reflection we found the experience richer than if we had the 'perfect' sound recording studio conditions from which to record and broadcast. If the prison is a performative site in and of itself which is better suited to the affective encounter than the moral judgement, then it makes sense to have it play a role in the project. The telling of this tale, ancient in origin, might now send a gentle echo to its communal beginnings via the Dictaphone recordings of unplanned for sounds as opposed to our modernised and arguably sanitised insistence on 'clean sound'. For us, this gave the project a particular kind of added energy, something which helped convince us to be more receptive to the affective encounters available. The soundtracks remain the private property of the men and the prison radio system—to hear it, you would have to be locked up in the local prison.

As Neal (2011) puts it, 'You must not enter with an attitude of showering enlightenment but must be hyper-vigilant to your desire to share … your work as an artist with others. Most importantly listen and learn' (Neal cited in Shailor, 2011, p. 10). Or, to repeat Thompson (2011), 'Working in a particular site or community does not inevitably lead to an attention to social justice'. (p. 5). In our case, the opposite (the Bloomist approach) runs the risk of being true. The theoretical move away from the praxis of judgement—everywhere a temptation—was what kept the project alive. Or, understanding the prison visits and their consequences as affective encounters, the result of which was the experience of 'joy' (an increase in our powers of action) or 'sadness' (a decrease in our powers of action), was what we ended up being convinced about. For the men on the inside, we can't say for certain whether they felt an increase or decrease in their powers of action. That is for them to say in a context of their choosing, and not one to which we apply a box ticking agenda.

The radio (being the site of choice for Artaud in 1947), is also the site (or the excuse) for our momentary, affective liberation from unhelpful binaries and unhelpful theory. It is the site where our own praxis of judgement collapses and we go back 'outside' a little bemused, perplexed, but elated if regretful that we have to work at university, in comparison, a site dripping with privilege and without the opportunities for joy prison affords us.[21] Theoretically speaking, prisons, at the moment, suit us better than universities because they help us understand that the praxis of judgement is a very limited form of encountering life. Like Foucault (1977) observed about governmentality being everywhere—just most apparently in the prison system. In a sort of inverse trajectory, the prison for us, becomes a site of opportunity to move beyond judgement and, if only for fleeting moments, be free of it. That leaves us with the awkward and troubling fact that probably the only people who would rather be there than anywhere else on Friday mornings, is us.

Notes

1 For security and clearance reasons, this essay will refer to 'the local prison'.

2 Our thanks to Novus for endless support in facilitating our project; our student volunteers for getting their teeth stuck into the project, but most of all for the incarcerated men who participated in the process with wisdom and patience.

3 Working with Novus since 2018, our courses have been co-designed by the prisoners based on conversations around what particular people would like to learn. The usual hook for us is 'Who wants to do a course on Philosophy and Performance?' that way we tend to get interest from those interested in philosophy and drama. From there it evolves into whatever the prisoners determine in terms of the shape of the course, how long it runs, and who will and will not be covered in terms of thinkers, artists and theorists. This is only possible with support and understanding from the Novus team who facilitate everything.

4 See an enlightening essay on the 'School of Resentment' by Schneidau (1995).
5 *Homer's Odyssey* (Kogen et al., 1990). See also, Wolodarsky (2001).
6 'Scouse' is the identity label for somebody who is Liverpool-based. Many of the men in the local prison fall into this category and would self-identify as such.
7 See Josho Bouwers' online lecture 'Greek warfare and Homer' (Wolfson College, 2020, July 28)
8 See a very useful film on this project also entitled *Shakespeare Behind Bars* (Rogerson, 2005).
9 See Linda Zerilli's (2015) excellent positioning essay *The Turn to Affect and the Problem of Judgment.*
10 See Barrett and Russell (2014).
11 Deleuze in his Spinoza lectures at Vincenne is the early 1980s talks about how Spinoza would prefer a 'list of abilities' to describe a human being, rather than a 'moral compass'. Deleuze was often looking for ways in which an 'ethics' approach would be preferable to a 'moral' approach. Deleuze calls this an 'Ethology' or an 'Ethics'. See Deleuze (1980–81)
12 'Ethologically' means we look at what is possible in any given situation, what each person/place/things' power of action might be. Phallogocentrically would be to impose a master discourse on the events, from a position of authority—judgement being a central tool in this operation
13 A very welcome resource for Deleuze scholars and enthusiasts is now live online at Purdue University, Indiana, USA under the title of The Deleuze Seminars, which includes an archive, translated into English, of many of Deleuze's lectures/seminars from the 1970s and 1980s. https://deleuze.cla.purdue.edu/. All websites last accessed 30 March 2021 unless otherwise stated. This includes a seminar series devoted to Spinoza called 'The Velocities of Thought'. See Deleuze (1980–81).
14 Driving to the prison and back every Friday morning we ended up calling all the Bloom, Kant and Plato moral orders, which chimed well with the prison set up, the 'praxis of judgement'. In this sense the war was fought against the praxis of judgement.
15 See Deleuze (1998) essay *To Have Done with Judgement* but also much of his work has focused on 'the plane of immanence'—a sort of non-hierarchical conceptual space which equips a thinker with ways to undo moral orders, specifically those of German Philosopher Immanuel Kant.
16 See Anderson and Malone's walk https://www.justgiving.com/crowdfunding/booksforprisoners
17 The recording is not available to the public. This is due to internal identity protection procedures, but also because every participant agreed that was how they wanted it. 'Inside for the insiders' was a phrase commonly repeated during conversations.
18 See Deleuze (1998) on Artaud's radio play *To Have Done with the Judgement of God.*
19 See National Prison Radio (n.d.).
20 See National Prison Radio (2020, August 24).
21 To that effect and in conversation with the men remaining in custody we are planning a scouse reading of Foucault's *Discipline and Punish* and a performance of Antonin Artaud's 1947 radio play *To Have Done with the Judgement of God.* Equipped with Deleuze's essay *To Have Done with Judgement,* it is something we as facilitators are eager to develop further. This will come under the support of the Live Art Development Agency and the Study Room in Exile project. A dedicated study space will be provided by the Study Room in Exile through Novus and the prison library in collaboration with Liverpool Hope University, under the general title 'HMP to Hope'—a project run by us to encourage prisoners to take up university study upon release. To this end and under the general title 'HMP to Hope', we have run a level 3 (A level equivalent) course in Performance and Philosophy since 2018.

References

Barrett, L. F., & Russell, J. A. (Eds.) (2014). *The psychological construction of emotion*. Guilford Publications.

Bedford, C. (2016). Penal reform: 'Making waves behind bars'. *Alternative Law Journal, 41*(2), 129–130. https://heinonline.org/HOL/P?h=hein.journals/alterlj41&i=134.

Bloom, H. (1994). *The Western canon*. Harcourt Brace and Co.

Bourdieu, P. (1979/2010). *Distinction: A social critique of the judgement of taste*. Routledge.

Braidotti, R. (1993). Discontinuous becomings. Deleuze on the becoming-woman of philosophy. *Journal of the British Society for Phenomenology, 24*(1), 44–55. 10.1080/00071773.1993.11644270

Cairns, D., & Scodel, R. (Eds.) (2014). *Defining Greek narrative*. Edinburgh University Press.

Clough, P. T. (2010). The affective turn: Political economy, biomedia and bodies. In M. Gregg & G. J. Seigworth (Eds.), *The affect theory reader* (pp. 206–228). Duke University Press.

Coates, S. (2016). *Unlocking potential – A review of education in prison*. Ministry for Justice. https://assets.publishing.service.gov.uk/government/uploads/system/uploads/attachment_data/file/524013/education-review-report.pdf

Cook, E. (2014). Structure as interpretation in the Homeric Odyssey. In D. Cairns & R. Scodel (Eds.), *Defining Greek narrative* (pp. 75–102). Edinburgh University Press.

Cull, L. (2012). *Theatres of immanence: Deleuze and the ethics of performance*. Palgrave Macmillan.

Deleuze, G. (1980–81). *Spinoza: The velocities of thought*. [The Deleuze Seminars] https://deleuze.cla.purdue.edu/seminars/spinoza-velocities-thought

Deleuze, G. (1998). To have done with judgement. In G. Deleuze (Ed.), *Essays critical and clinical* (pp. 126–135). Verso.

Deleuze, G., & Guattari, F. (1987). *A thousand Plateaus: Capitalism and schizophrenia* (B. Massumi, Trans.). University of Minnesota Press.

Dobbs, R. (2019, March 26). Key to prison education handed to Governors. *The Guardian*. https://www.theguardian.com/education/2019/mar/26/key-to-prison-education-handed-to-governors

Evrard, Y. (1997). Democratizing culture or cultural democracy? *The Journal of Arts Management, Law, and Society, 27*(3), 167–175. 10.1080/10632929709596961

Foucault, M. (1977). *Discipline and punish: The birth of the prison* (A. Sheridan, Trans.). Allen Lane Penguin.

Kogen, J., & Wolodarsky, W. (Writers), & Archer, W. (Director). (1990). "*Homer's Odyssey*" (Season1, Episode 3) [TV series episode]. In J. L. Brooks, S. Simon & M. Groening (Executive Producers), *The Simpsons*. Gracie Films; 20th Television, Inc.

Massumi, B. (1987). Translators foreword: Pleasures of philosophy. In G. Deleuze (B. Massumi, Trans.), *A thousand plateaus: Capitalism and schizophrenia* (pp. ix–xv). University of Minnesota Press.

Massumi, B. (1996). Becoming-Deleuzian. *Environment and Planning D: Society and Space, 14*(4), 395–406. 10.1068/d140395

McAvinchey, C. (2011). *Theatre & prison*. Palgrave Macmillan.

McAvinchey, C. (2020). *Applied theatre: Women and the criminal justice system*. Bloomsbury.

Prentki, T., & Preston, S. (2009). *The applied theatre reader*. Routledge.

Prison Radio Association (2020, August 24). *Who listens to National Prison Radio, and why? The numbers are in* ... https://prison.radio/who-listens-to-national-prison-radio-and-why-the-numbers-are-in/

Prison Radio Association (n.d.). *About*. https://prison.radio/about/

Rogerson, H. (Director). (2005). *Shakespeare behind bars* [Film]. Philomath Film.

Schneidau, H. N. (1995). Harold Bloom and the school of resentment; or, Canon to the right of them. *Arizona Quarterly: A Journal of American Literature, Culture, and Theory, 51*(2), 127–141. 10.1353/arq.1995.0024

Shailor, J. (Ed.) (2011). *Performing new lives: Prison theatre.* London: Jessica Kingsley.

Sturge, G. (2020). *UK prison population statistics.* UK Parliament. https://researchbriefings.files.parliament.uk/documents/SN04334/SN04334.pdf

Thompson, J. (1998). *Prison theatre: Perspectives and practices.* Jessica Kingsley Publishers.

Thompson, J. (2011). *Performance affects applied theatre and the end of effect.* Palgrave Macmillan.

Thompson, J. (2014). *Humanitarian performance: From disaster tragedies to spectacles of war.* Seagull Books.

Thompson, J. (2017). No more bystanders: Grandchildren of Hiroshima and the 70th Anniversary of the Atomic Bomb. *TDR/The Drama Review, 61*(2), 87–104. 10.1162/DRAM_a_00649

Walsh, A. (2012). Performing prisons, performing punishment: The banality of the cell in contemporary theatre. *Total Theatre Magazine,* Summer, 13–14.

Walsh, A. (2014). (En)gendering aabitus: Women, prison, resistance. *Contemporary Theatre Review, 24*(1), 40–52. 10.1080/10486801.2013.858327

Walsh, A. (2019). *Prison cultures: Performance, resistance, desire.* Intellect.

Wilson, E. (Trans. & Ed.) (2018). *The Odyssey by Homer.* W.W. Norton & Company, Inc.

Wolfson College. (2020, July 28). *Dr Josoho Brouwers: Greek warfare and Homer* [Video]. Vimeo. https://vimeo.com/442296261

Wolodarsky, W. (Writer). (2001). *The Simpsons season 1 DVD commentary for the episode "Homer's Odyssey"* (DVD). 20th Century Fox.

Zerilli, L. M. (2015). The turn to affect and the problem of judgment. *New Literary History, 46*(2), 261–286. 10.1353/nlh.2015.0019.

PART V

Sonic Resistance: Soundscapes of Protest and Activism

INTRODUCTION

Sarah Woodland and Wolfgang Vachon

We believe the progression of works through this volume builds inevitably towards considering the role of sonic engagement in supporting the various struggles facing communities and publics in the face of crisis, tyranny, and oppression. As LaBelle (2018) suggests, 'Sound and listening form a supportive base from which to nurture a broader intelligence in approaching the pervasive realities of crisis' (p. 3). But it can also be intentionally deployed and manipulated to engage directly in acts of resistance. In their chapter, Dont Rhine discusses with Mitchell Akiyama and Marcus Ware (drawing on Paulo Freire, 1970), how there is an urgency to move artistic practice from collective analysis to collective action; and it is only through ethical and meaningful engagement with community members in the construction of meaning and, we suggest, aesthetic framing itself, that this can be achieved. Following the thread of 'acoustic performativity' (LaBelle, 2020), we can conceive 'protest aesthetics' as having potential beyond visual and performance cultures as discussed for example in McGarry and colleagues' (2019) volume. In embracing the sonic, protest aesthetics becomes concerned with how noise, music, vibrations, voices … and silences, might facilitate embodied, transformative movements towards action. Sound is increasingly being deployed as a central element in activism and protest: from the subversive broadcasts of voices and sounds out of the refugee detention centre on Manus Island (Manus Recording Project Collective, 2019), to the 'forensic listening' undertaken with survivors of Syria's Saydnaya prison in order to reconstruct 'acoustic memories' and testimony of their confinement in darkness and silence (Hamdan, 2019), to the deployment of noise and sound in protest described by Yoganthan in this volume.

We begin this section with a contribution from sound artist and acoustic ecologist Leah Barclay, which draws us into the operation of participatory acoustic ecology in engaging local and global communities in ecosystem monitoring and

DOI: 10.4324/9781003164227-24

conservation. The protest or activist aesthetics here may seem gentle by comparison to some of the other examples cited but is no less urgent given the unfolding global ecological crisis. Through case studies of three interdisciplinary projects in Australia and Mexico, Barclay explores acoustic ecology as 'an interdisciplinary field embracing the artistic and scientific possibilities of listening to the environment and encouraging and facilitating participatory collaborations with local communities', with a view to mobilisation towards climate action.

The second chapter in this section, sound artist and scholar Nimalan Yoganathan provides a theoretical analysis of noise and silence as forms of resistance and protest, building on Nancy Fraser's (1990) idea of 'subaltern counterpublics' to develop his theory of 'aural counterpublics'. Through social movements such as Idle No More and Black Lives Matter, as well as the 'riots' in the French banlieues, and the 'performative silence' and refusal to speak of NFL football player Marshawn Lynch, Yogananthan explores the implications of noise and silence as tools of resistance, 'acoustical agency' (Rice, 2016), efficacy, and impact. Yoganathan draws largely on protest aesthetics to argue that social action and resistance though demonstrations are forms of performance, with understandings by the actors regarding how they will be received and read, framing 'aural counterpublics' as 'a response to the systematic silencing imposed on them by the state and corporate media'.

The theme of sonic resistance concludes with a contribution from Mitchell Akiyama in conversation with veteran sound artist and activist Dont Rhine (co-founder of the international art collective Ultra-red) and scholar, artist, and activist Syrus Marcus Ware (a core team member of Black Lives Matter Toronto). The chapter presents an edited conversation between three uniquely situated artist-activists involved in a sound project called *Street Hassle*, exploring the 'opioid crises' in Toronto, Canada. It works as a collective reflection on many of the themes in this book: community, trust, aesthetics, ethics, participation, the role of arts, politics, and responsibility among others. With an ear to sound theory, the authors consider engagement through notions of noise, dissonance/resonance, consensus/dissensus, and how these function within political artmaking in (and for) community.

References

Fraser, N. (1990). Rethinking the public sphere: A contribution to the critique of actually existing democracy. *Social Text*, (25–26), 56–80. 10.2307/466240

Freire, P. (1970). *Pedagogy of the oppressed* (M. B. Ramos, Trans). Continuum.

Hamdan, L. A. (2019). Saydnaya (the missing 19db). In J. Parker & J. Stern (Eds.), *Eavesdropping: A reader* (pp. 42–59). City Gallery Wellington, Liquid Architecture, Melbourne Law School.

LaBelle, B. (2018). *Sonic agency: Sound and emergent forms of resistance* (Vol. 1). MIT Press.

LaBelle, B. (2020). *Acoustic justice: Listening, performativity, and the work of reorientation.* Bloomsbury Publishing USA.

Manus Recording Project Collective (2019). How are you today. In J. Parker & J. Stern (Eds.), *Eavesdropping: A reader* (pp. 172–215). City Gallery Wellington, Liquid Architecture, Melbourne Law School.

McGarry, A., Erhart, I., Eslen-Ziya, H., Jenzen, O., & Korkut, U. (2019). *The aesthetics of global protest: Visual culture and communication*. Amsterdam University Press.

Rice, T. (2016). Sounds inside: Prison, prisoners and acoustical agency. *Sound Studies*, *2*(1), 6–20. 10.1080/20551940.2016.1214455

14

ENGAGING COMMUNITIES IN LISTENING TO ECOSYSTEMS: CASE STUDIES FROM ACOUSTIC ECOLOGY RESEARCH IN AUSTRALIA AND MEXICO

Leah Barclay

Introduction

Engaging communities in practices of listening to changing ecosystems is one of the most powerful ways to inspire a presence and connection to place. Deep listening processes can reveal the interconnected nature of our ecosystems and inspire communities to engage with conservation. Accessible field recording technologies that amplify our aural perception have rapidly become a valuable tool for inspiring environmental stewardship and ecological engagement. This research explores these opportunities through the development of a series of large-scale interdisciplinary projects that position sound artists and acoustic ecologists within multi-platform collaborations working directly with communities. The projects are introduced through the lens of contemporary acoustic ecology, a socially embedded interdisciplinary field using participatory methods to interrogate our relationship with place through sound. Understanding how sound influences our environment and wellbeing is central to the field of acoustic ecology, but these digital audio recordings are also addressing major gaps in our knowledge about ecosystems by using non-invasive technology to monitor ecosystems and document the temporal and spatial complexities of changing environments. The interdisciplinary intentions of audio field recording through these research endeavours raise many ethical considerations, particularly in relation to the role of aesthetics and human perception, with various epistemological tensions across the arts and science divide. This chapter introduces a series of case studies working directly with communities in Australia and Mexico through the international research projects *Biosphere Soundscapes, Biosphere Open Microphones (BIOM)*, and *River Listening*.

DOI: 10.4324/9781003164227-25

Listening to ecosystems

The impacts of climate change are becoming increasingly visible and audible in ecosystems across the planet. Scientists continue to sound the alarm and we urgently need more effective ways to engage communities in conservation and climate action. There is still much to learn about the interconnected nature of the planet's ecosystems and many remote locations still lack basic species inventories. Conventional environmental monitoring that requires scientists to travel for field work is highly invasive and has not been practical or possible during the COVID-19 global pandemic. This style of environmental monitoring is also constrained to restricted areas and manual processing of observations which can often generate biased results (Linke et al., 2018). Listening to changing ecosystems with remote audio recorders offers a transformational opportunity to monitor remote environments and engage communities in the process.

Rapid advancements in acoustic technology mean we can now monitor changing environments through accessible, affordable, and non-invasive acoustic sensors. Innovations in hardware and software mean that acoustic monitoring can be conducted by local communities and offers a viable method for measuring ecological changes and monitoring biodiversity (Deichmann et al., 2018). While the technical advancements and opportunities for community engagement are recent developments, the scientific field of acoustic monitoring has historically been defined as 'bioacoustics' and has been active for several decades with a specific focus sounds produced by wildlife (Krause, 1987). The term 'ecoacoustics' has recently been adopted to define a new interdisciplinary field that studies sound along a broad range of spatial and temporal scales to understand environmental changes (Sueur & Farina, 2015).

Ecoacoustics has seen rapid engagement across the planet and calls for collaboration with other disciplines including electronics, remote sensing, big data, humanities, and social sciences (Sueur & Farina, 2015). Despite being a relatively new scientific technique, there is clear evidence that measuring the acoustic diversity at a site provides information on the health of an ecosystem (Deichmann et al., 2018; Fuller et al., 2015; Staaterman et al., 2017; Sueur et al., 2008; Tucker et al., 2014). In addition to advancements with recording technologies, software advancements mean we can now process large qualities of acoustic data and create algorithms for automated analysis and species recognition (Duan et al., 2011). These advancements in audio technologies and acoustic sensors are 'revolutionising environmental monitoring and leading to new thrust in environmental research and education, including ecological forecasting' (Farina & Gage, 2017, p. 9).

High-quality audio recorders and acoustic monitoring equipment have historically been expensive, which has reduced the accessibility and opportunities for community engagement. The recent advancements in affordable technology mean that communities can now actively participate with various accessible devices now available. For example, the AudioMoth is a low-cost full spectrum acoustic sensor that is the size of a credit card and can operate for long-durations in the field. The AudioMoth can effectively capture acoustic data from remote locations and has made adopting ecoacoustics methods accessible for communities who found previous acoustic sensors cost prohibitive (Hill et al., 2019). The ability for audio recordings to be collected from many places simultaneously but analysed remotely later makes it a valuable tool that can actively engage many communities across a diversity of ecosystems (Ribeiro et al., 2017; Sueur et al., 2008). This process of acoustic data collection does not require specific scientific knowledge or skills—meaning the deployment and management of acoustic monitoring can easily be conducted by community members, including youth, and returned to a central repository for analysis. This provides pathways to empower communities to take active leadership in environment management and accessible ways to engage local communities in conservation.

While collaborations are emerging, ecoacoustics is yet to recognise the value of community engagement. Local communities have an existing relationship and connection with the soundscape and can bring specialist knowledge about the environment that can inform decisions for field work and the locations of field recorders. In the past, acoustic sensors have often been deployed in the field by scientists at predetermined locations without drawing on local knowledge that could offer significant improvements in the project and outcomes. Active listening in the environment and drawing on local knowledge can improve ecoacoustics research and inform every stage of the process—from selecting locations in the field, deploying acoustic equipment and analysing the resulting audio recordings. As scientific developments in ecoacoustics become interdisciplinary and align with sound studies more broadly, the value of seeking permissions from traditional custodians and drawing on Indigenous knowledge systems will become increasingly critical in ecoacoustics research projects.

The emergence of ecoacoustics is considered a new science, yet the concept of listening to ecosystems to understand environmental changes is not a new proposition. Practices of listening to understand the environment and seasonal changes are central to Indigenous knowledge systems across the planet. Ecoacoustics also draws on acoustic ecology, a well-established interdisciplinary field that emerged in Canada in the late 1960s. Canadian composer and scholar R. Murray Schafer founded the discipline with the premise we should attempt to hear the acoustic environment as music and take responsibility for its composition (Schafer, 1977). Acoustic ecology has been active across the world with regional affiliates of the World Forum for Acoustic Ecology drawing together artists, scientists, and researchers for regular gatherings, projects, and events. Ecological sound artists emerging from the acoustic ecology movement have propelled many of the most significant discoveries that have informed the development of ecoacoustics. Their nuanced understanding of our sonic environment and knowledge that sound can engage communities and listeners at a deeper, more attuned level strongly aligns with the interdisciplinary intentions of ecoacoustics and the foundations of acoustic ecology. Immersive sound installations and performances drawing on recordings from ecosystems have been successful in evoking empathic and philosophical responses to climate change that can inspire ecological action in communities (Burtner, 2011; Monacchi, 2013).

Ecological sound art presented in a collective listening experience can create metaphors to help communities connect with environmental issues on a deeper and more personal level. These creative works can immerse communities in a remote ecosystem and inspire awareness and respect for invisible or inaccessible areas of the environment that they would not traditionally be able to access (Gilmurray, 2017). Salomé Voegelin (2014) believes that when communities engage in these collective listening experiences it is possible to become submerged in a 'sonic possible world' where personal listening experiences can inspire communities to think differently about the environment. Ecological sound art and participatory listening experiences can also inspire environmental empathy in a way that is not possible through other media. The temporal presence and connection can make listeners feel part of the ecosystems generating an environmental empathy that evolutionary ecologist Monica Gagliano (2018) believes is the missing link in many responses to climate action.

Community engagement and participatory research has been integral to ecological sound art since its inception. Composers, including Alakasan-born Matthew Burtner, have actively engaged with communities through education and field recording, and the establishment of organisations that have advanced the field. In 2008 Burtner created Eco Sono, an activist network for ecological sound art that hosts field trips in remote locations. Burtner (2011) believes that

> by creating responsible interactions between the natural world and audiences, those involved may come to think in new ways about the environment. They may convince others of the need for sustainable practices, or they may join preservationist societies and activism networks. (p. 244)

In a similar mission, Italian composer David Monachi launched Fragments of Extinction in 2001 with the intention of recording the world's undisturbed primary equatorial forests to highlight disappearing soundscapes (Monacchi, 2013). Fragments of Extinction has evolved into a non-profit organisation that works in collaboration with artists and scientists to produce immersive installations to increase public engagement and scientific knowledge of the acoustic biodiversity of equatorial rainforests (Monacchi, 2016).

The emerging opportunities with accessible technology and intersections of ecoacoustics, acoustic ecology, and ecological sound art have been the catalyst for a series of large-scale interdisciplinary research projects that have launched in Australia over the past decade. These projects position sound artists and researchers within multi-platform collaborations working directly with communities. This research adopts acoustic ecology as the umbrella term for a interdisciplinary field embracing the artistic and scientific possibilities of listening to the environment and encouraging and facilitating participatory collaborations with local communities. Contemporary acoustic ecology explores how sound influences our environment and wellbeing, and by intersecting with the emerging science of ecoacoustics, the field is now addressing major gaps in our knowledge about ecosystems by using non-invasive technology to monitor ecosystems and document the temporal and spatial complexities of changing environments. In the context of Australia, acoustic ecology has evolved as a highly interdisciplinary, dynamic field with increased engagement across environmental sciences, urban planning, health, digital technology, creative arts, and humanities. The last decade has seen a strong emergence of socially engaged practice and sonic activism along with a rapid increase in education and academic research.

Biosphere soundscapes: Listening to UNESCO biosphere reserves

Biosphere Soundscapes is a large-scale international research project that works directly with the communities of UNESCO biosphere reserves to facilitate local acoustic ecology research projects that connect with a global network. The project was conceived in 2011 and launched on World Listening Day 2012 in Queensland, Australia in the Noosa Biosphere Reserve. The project encourages communities to actively listen to the environment through local workshops, masterclasses, sound walks, and international residencies that bring together artists and scientists. The project has developed various participatory methods for acoustic ecology and has facilitated capacity building frameworks that empower local communities to take ownership of the project and continue recording the changing soundscapes of UNESCO biosphere reserves over time. Biosphere Soundscapes has a strong focus on interdisciplinary research, education, and community engagement. The project has worked at the intersection of art and science to explore, advance, and develop accessible audio recording technologies that connect and empower local and global communities of UNESCO biosphere reserves to understand the value of listening to the environment (Barclay & Gifford, 2017). Biosphere reserves are ideal sites for acoustic ecology research as they are considered learning laboratories for sustainable development and interdisciplinary research. They are designated by UNESCO for the conservation of biological and cultural diversity and the World Network of Biosphere Reserves now includes 727 sites across 131 countries.

Participatory acoustic ecology research in biosphere reserves, as demonstrated through Biosphere Soundscapes (Barclay, 2014; Barclay & Gifford, 2017) allows communities to connect with their local environment through listening and offers various pathways to engage from both artistic and scientific perspectives. Biosphere Soundscapes has consistently maintained a balance between art and science, with the audio recordings providing scientific data for biodiversity analysis to advance the field of ecoacoustics in addition to providing source material creative works and ecological sound art that are shared with communities to bring awareness to these environments. While the project has shifted focus to various online delivery modes throughout the COIVD-19 pandemic, one of the most impactful outcomes from Biosphere Soundscapes has been immersive interdisciplinary residences with artists and scientists hosted with local communities in biosphere reserves. These have taken place in biosphere reserves across the Asia Pacific region and Latin America and have been the catalyst to facilitate collaborations, partnerships and ongoing community-based research.

In 2015, Biosphere Soundscapes hosted a residency in the Sian Ka'an Biosphere Reserve along the east coast of the Yucatan Peninsula in Mexico. The biosphere reserve comprises of tropical rainforests, underground river systems, mangroves and a diversity of coastal and marine ecosystems. The sonic environment is diverse with over 300 species of birds, 42 species of amphibians and reptiles, and wild cats including jaguarundi, jaguar, and pumas. The surrounding aquatic ecosystems include crocodiles, manatee, dolphins, 52 species of fish, and a diversity of aquatic life that is audible across the local reefs. Biosphere Soundscapes residencies are developed with permission from traditional custodians and in collaboration with the local community. The project partnered with Fonoteca Nacional de Mexico (the Mexican Sound Archive), CONANP (National Commission for Natural Protected Areas), and Amigos de Sian Ka'an (the local organisation who manage the biosphere reserve with the community) to host the residency in Mexico. Biosphere Soundscapes residencies are open for international participation but prioritise local engagement. The program was designed as an interdisciplinary laboratory focusing on the creative and scientific possibilities of listening and acoustic ecology in Mexico. The residencies focus on a small interdisciplinary group and often attract hundreds of applications meaning the group is selected and curated based on their interests and intentions to practise, use and share the knowledge they generate. As Sian Ka'an is an incredible location, the call attracted applications from high profile Hollywood film composers, popular musicians, and scientists with specific interests in the local ecosystem. The selected applicants focused on those who most aligned with the research intentions and would benefit from the opportunity, this resulted in a final group comprised mostly of early career researchers from both artistic and scientific perspectives.

The Sian Ka'an Biosphere Soundscapes Residency involved an intensive 10-day expedition with daily field recording sessions accompanied by interdisciplinary workshops and presentations with the local community. The Biosphere Soundscapes team ensure that equal time is spent on artistic and scientific perspectives and is guided by the local community in developing the thematic focus of the residency. The design of the project begins first and foremost with establishing a dialogue with local Indigenous communities to identify appropriate ways for participants to engage and learn about the biosphere reserve from an Indigenous perspective and seek permission to record at various field locations. The Sian Ka'an residency started with a field trip to archaeological sites in Muyil and visiting Mayan community leaders who introduced participants to Mayan culture through ceremonies, traditional medicine, local foods, and sharing Balché, an Indigenous Mayan drink extracted from the bark of the leguminous tree mixed with honey and water. The Mayan elders generously shared local knowledge about the environment and allowed participants to record a Mayan ceremony that revolved around purification. Starting the residency by grounding participants in culture, listening, and local knowledge is critical in establishing the ethical framework of Biosphere Soundscapes—that seeks to move away from the colonial approach of western field recording practices that often exoticise foreign soundscapes and fail to engage with the local community.

Biosphere Soundscapes builds relationships with the local community through listening and these experiences provide a starting point to explore notions of knowing a place through sound and the importance of acoustic ecology in the social, cultural and ecological health of a community and environment. This draws on Steven Feld's concept of 'acoustomology', or sound as a distinctive method for knowing the world (Feld, 1996). Following the engagement with Mayan culture, the Biosphere Soundscapes residency in Sian Ka'an explored the biodiversity of the local environment through presentations and workshops with local scientists and conservationists from Amigos de Sian Ka'an, the local organisation that manages the biosphere reserve. The residency structure includes daily field recording sessions as a group and allows flexibility for participants to explore the environment from their personal perspectives. The group field recording trips involve travelling to a particular location in the biosphere reserve that has been predefined through the development phase with various field kits and equipment. Once on location, participants explore the location through collective listening and participatory exercise that are facilitated by different group members. The group then divides throughout the landscape, usually with a distance of at least 100 meters between each recordist. Terrestrial recording sessions usually occur at different times throughout the day with a focus on dawn and dusk to capture the sonic shifts and diversity in the soundscape, while aquatic sessions using hydrophones (underwater microphones) in freshwater and marine ecosystems occur at various times throughout the day and night. The group field recording trips range from two to three hours where the recordists will remain in situ with the equipment, actively listening to the environment while recording. The recording session is followed by reflections on the locations and listening experiences and the recordings are then shared with the group once back at the field lab.

The collective listening sessions allow participants to recognise subtle differences in recording approaches, including the placement and height of microphones or different approaches to using equipment. These group listening sessions provide an opportunity for dialogue on the scientific and aesthetic approaches to recording and production and often result in a diversity of perspectives and interdisciplinary learnings. The residency in Sian Ka'an included a diversity of participants with extremely varied recording experience, but all agreed that acute listening in the field and a deep engagement with the soundscape produced the most effective recordings. These workshops allowed participants to exchange equipment and approaches and revealed that the intentions, ethics, and aesthetics of the recordings had a greater impact on the outcomes than the quality of the recording equipment. Participants agreed that it was possible to create a rich and engaging soundscape with very basic equipment if the field recordist made conscious decisions in the field that were guided by listening. The listening workshops also allowed participants to share their soundscape compositions with the resulting recordings. These were designed as immersive listening experiences where the aesthetics of sound were interrogated and discussed. Some participants were deeply critical of extensive manipulation and editing of the recordings while others processed and composed their sound materials into unrecognisable abstract soundscapes. These aesthetic listening experiences are open to the community and generate extensive discussion around how different species hear, how we interpret and analyse sound in the environment, and how sound can evoke such powerful emotional responses during these collective listening experiences.

The in-situ recording sessions were accompanied by remote recordings with field recorders installed on location for several days. In these instances, different equipment is used that has the capacity to run for extended periods. These devices are also distributed to the local community and often left with the community after the residencies so they can continue participating in the project and ongoing research. These devices include small AudioMoth recorders and low-cost Zoom H2 recorders which have the flexibility for multi-channel recording. For biosphere reserves interested specifically in the bioacoustics and ecoacoustics approaches, we recommend the installation of devices such as the Frontier Labs Bioacoustic Audio Recorder (BAR) which is accessible for community use and will run for 80 hours of recording without replacing the battery. The device is lightweight, robust, and durable with built in GPS and the ability to schedule recordings for specific durations and times of day that could focus around the dawn chorus or particular species of frogs vocalising at dusk.

At the conclusion of the residency, all the resulting recordings were included in the Fonoteca national sound archive in Mexico and catalogued for the Biosphere Soundscape community database. In addition to compositions, installations and albums created by the participants, the residency inspired various scientific outcomes and online projects. As one example, Stéphane Pigeon created a generative online project titled 'A Bird's Paradise: Interactive Tropical Birds Soundscape' using the Sian Ka'an recordings captured by French sound artist Félix Blume during the residency that attracts hundreds of daily listeners. Through the Sian Ka'an Residency, Biosphere Soundscapes facilitated a range of partnerships and collaborations that have had ongoing outcomes and positive impact in the community. This example of a Biosphere Soundscapes residency in Mexico demonstrates how the project engages directly with local communities to design and develop long-term mutually beneficial outcomes.

Biosphere Soundscapes has focused on advancing the field of acoustic ecology as an accessible, inclusive, interdisciplinary field that can assist local communities in biodiversity monitoring and understanding the environment in deeper and more attuned ways (Barclay & Gifford, 2017). The project has been flexible and responsive to the needs of collaborating communities, and has developed and encouraged appropriate protocols for field recording which always involve seeking permission to record from traditional custodians and respecting Indigenous knowledge systems and protocols in local environments. The project has acted as a catalyst for bringing diverse voices together and highlighting the critical importance of Indigenous approaches in environmental management. The project brings together interdisciplinary knowledge, emerging science, new technology, and responsive community engagement to explore the social, cultural, and ecological soundscapes of biosphere reserves. Biosphere Soundscapes has successfully delivered workshops, education programs, acoustic monitoring and large-scale research projects in over 20 biosphere reserves and offers communities various pathways and tools to engage with acoustic ecology. The project is fundamentally focused on promoting the value of listening to the environment from artistic and scientific perspectives and encouraging engagement through participatory research, education, partnerships, and collaboration. While the project has focused on recording the changing soundscapes of biosphere reserves, Biosphere Soundscapes has expanded to live streaming audio from remote locations through a network of open microphones.

BIOM: Towards a Biosphere Open Microphone network

Biosphere Soundscapes launched the Biosphere Open Microphones (BIOM) in 2017 at the Sound + Environment Conference at the University of Hull in the UK. BIOM is a multi-partner acoustic ecology project establishing a network of live audio streams from sites of ecological significance in UNESCO biosphere reserves. The live audio streams are designed for public engagement and interdisciplinary research and are used for biodiversity monitoring and a diversity of global creative projects with the live streams managed by the local community. The project brings together a group of pioneering arts and science initiatives to establish new live streams from places of ecological and acoustic interest and builds on the existing work of Biosphere Soundscapes in collaboration with SoundCamp (London), Locus Sonus (Marseille, France), Cyberforest (Tokyo, Japan), and the UNESCO World Network of UNESCO Biosphere Reserves (Paris). BIOM provides a network for research, monitoring, artistic work and community engagement through a real-time acoustic observatory reflecting the diversity of terrestrial and aquatic biomes in UNESCO biosphere reserves across the world. The project focuses on community engagement and partners with local organisations to develop and maintain the audio streams at each location. As an extension of Biosphere Soundscapes, the community engagement in the development, design and maintenance of the audio streams makes them a hub for learning, collaboration, citizen science, and interdisciplinary creativity with sound.

BIOM extends community engagement to virtual environments and is concerned with spatial and temporal scales of our sonic environments. The project seeks to connect listeners in local and global communities through the live streams and explore different perspectives on listening and methods of engaging with real-time audio online. BIOM has developed a series of pilot sites across Australia,

Europe, North America, and Mexico and builds on a series of existing streams with the project partners including Cyberforest at the University of Tokyo with multiple streams across Japan. The BIOM streams are hosted on an open access global sound map managed by Locus Sonus in France. The map displays satellite imagery of the world with microphone icons located each stream site on an intuitive interface that allows listeners to navigate between audio streams.

Public sound maps have proliferated the acoustic ecology community in recent years and have been an increasingly popular way of publishing crowd sourced audio. While sound maps raise various challenges and tensions in perspective and perception (Carlyle, 2014), this methods of interacting with the streams online remains a familiar and accessible approach for community engagement where the listener has a sense of agency moving between streams and can decide on the listening durations with each stream. Locus Sonus has presented the streams with various viewing options and includes a basic list of server mountpoints available as an option from the map. Each stream has an information panel with information about the site and images. The map interface has a predictable levelling effect, with all the streams appearing together on the same virtual plane, as if on the surface of the earth allowing listeners the navigate in and out of listening regions. The map shows clusters of streams in parts of the planet and reveals some of the imbalance in the project with continents such as Africa and Antarctica currently silent. The coming years will see priority placed on establishing BIOM streams in these locations with further efforts around managing streams in remote locations and extreme environments.

During the initial phase of pilot streams in 2017, a solar powered BIOM site was established in Mexico's Monarch Butterfly Biosphere Reserve, a site with rugged forested mountains and diverse ecosystems roughly 100 km northwest of Mexico City. The stream was designed to listen to the soundscapes of migrating Monarch Butterflies, with around 70 percent of the migratory eastern population of monarchs passing through the reserve. The stream attracted wide interest among the network of Monarch Butterfly researchers and advocates, as a tool for community engagement, collective listening and generating awareness of the habitat's importance, fragility and ecological value. The site is a pilot for an extended project to link the butterflies' migration routes from Mexico to Canada led by Dr Rob Mackay to draw attention to the need for collaborations across disciplines and borders for biodiversity conservation. The stream acted as a valuable learning experience and was significant in bringing together a diverse team who embraced the technical challenges of the installation in a remote environment and found innovative ways to communicate and trouble shoot challenges and repairs. The process of prototyping and field testing in a mountain location was challenging; equipment and large batteries were brought up the mountain by horseback and the mobile connection was often unpredictable.

The resulting live audio stream presents a rich and engaging sonic environment, though the fragility of the environment is reflected in the audio, with regular disconnections and unpredictability. The migrating butterflies do not dominate the soundscape—the fluttering of wings can be heard subtly at times, but this requires a sustained engagement with the stream, similar with field listening experiences. Engagement with the local community has been integral to establishing and maintaining the live stream, including arborists Pato Moreno, who became a core part of the project team. The stream has informed the development of BIOM streams in remote locations in Australia and approaches for ongoing community engagement strategies. While still in the early stages of the project, BIOM has the capacity to be a valuable acoustic observatory for a wide spectrum of artistic, scientific and activism purposes through community engagement with acoustic ecology. Biosphere Soundscapes and SoundCamp have developed various resources for communities to participate that include building and installing stream kits and creating live audio streams from mobile devices. As an interdisciplinary acoustic observatory, BIOM provides global open access to the live streams and encourages durational deep listening. The aesthetic engagement with the streams have resulted in performances and installations at various galleries and museums including the Smithsonian. There is an element of compelling intangibility while listening to the live audio. The unpredictable soundscapes draw listeners into a sonic portal that could change at any second.

River listening: Connecting communities to freshwater ecosystems

Biosphere Soundscapes has been the catalyst for a range of community-based research projects that focus of the soundscapes of specific ecosystems including coral reefs, wetlands, and rivers. These projects use hydrophones (underwater microphones) as the core method to engage communities in listening to the changing soundscapes of aquatic ecosystems. Hydrophones have been used in the context of sound art since the 1970s, with composers including Maggi Payne, Jana Winderen, Douglas Quin, Chris Watson, David Monacchi, and Ros Bandt renowned for their creative projects that use underwater recordings as creative source material to bring attention to aquatic environments. Freshwater conservation urgently needs more effective techniques and remains challenging (Linke et al., 2018) when we often cannot see beneath the surface. The conservation of freshwater ecosystems is critical as we continue to see the catastrophic impacts of climate change and biodiversity loss.

Listening to rivers with hydrophones can help local communities understand freshwater biodiversity in accessible and engaging ways (Barclay et al., 2018; Gilmurray, 2017). This was the inspiration for River Listening, an ongoing research project initiated in 2014 through a Synapse Residency awarded to the author and the Australian Rivers Institute. River Listening works at the intersection of art and science by investigating the cultural and biological diversity of freshwater ecosystems through real-time listening and underwater recording used for biodiversity monitoring and community engagement. The project has developed various ways for communities to engage with the recording and production of underwater audio and has been the catalyst for establishing the scientific discipline of freshwater ecoacoustics (Barclay et al., 2018).

River Listening has advocated for community led research models that encourage communities to be actively involved in the creative and scientific outcomes. The community process for each River Listening location has involved permission from traditional custodians, community consultations, and interdisciplinary workshops prior to hydrophone recording, live streaming, analysis, and mapping. The River Listening workshops are accessible for conservation groups, specialists, and local school groups with each workshop responsive to the group's interests. The youth workshops involve students making their own hydrophones, which they can keep to continue recording and uploading to the sound maps after the workshop. In similar design to Biosphere Soundscapes, River Listening has focused on creating pathways and opportunities for young people to engage with acoustic ecology through participatory experiences and active engagement in the research process. This interdisciplinary framework draws on participatory action research that works directly with local communities and is scalable and responsive to each community and river system. The adaptable nature of the project means it has been effective in a range of international contexts. In addition to major river systems in Australia, the project has worked with freshwater ecosystems throughout Europe, North America, and the Asia Pacific facilitating major creative projects and community engagement.

River Listening has also advocated for virtual community engagement, particularly as travel restrictions have limited field work during the COVID-19 pandemic. These efforts include a customised digital platform, virtual sound maps with live streaming hydrophones, and a mobile application that allows communities to record, upload, and compare sounds in a global database (Barclay, 2016; Barclay et al., 2020). These digital outcomes underpin the major creative projects and facilitate community engagement in the collection of recordings for scientific monitoring. The central tools for public engagement around River Listening have been creative outcomes, including performances and installations interactive sound walks that require audiences to visit their local river to experience the work.

The River Listening sound walks involve a custom-built mobile application that transforms the listener's phone into a sonic compass to guide them along the riverbank and explore the cultural and biological diversity of the river through sound (Barclay & Gifford, 2017). The installation triggers geo-located soundscapes that include hydrophone recordings, local voices, and composed soundscapes as listeners walk along the riverbank. The sounds included in the sound walks are created through participatory workshops with the local community and evolve into collaborative soundscapes and compositions that can adapt and evolve along the river system over time. The communities feel a sense of ownership of the work when they can hear the sounds they recorded in the final experience (Barclay et al., 2020). The accessibility and multi-platform nature of the creative outcomes has been essential for community engagement in River Listening and the participatory nature of these experiences has been successful in engaging communities in their local river systems. Communities who engage with these experiences are often enthralled and shocked by the soundscapes beneath the surface of the water. The audio production and composition of the sound walks is essential in engaging communities through an aesthetic listening experience. The sound walks are designed to provide communities with a new way to connect with their rivers and inspire freshwater conservation through accessible mobile technologies.

River Listening has navigated various challenges throughout the community engagement processes and has always ensured the project was conducted ethically and respectfully, with permissions, and collaboration where possible, with traditional custodians. The project has facilitated opportunities for Indigenous knowledge holders to take a leadership role in the projects and has brokered partnerships with cultural leaders and scientists to look at new approaches to freshwater conservation. The River Listening project has had the great privilege of learning from Indigenous leaders such as Lyndon Davis, Brendan Kennedy, and Anne Poelina in Australia. Anne Poelina has been leading the conservation and protection of the Mardoowarra (Fitzroy River) in Western Australia and has advocated for the river to be recognised as a living ancestral being, following the recognition of the Whanganui River in Aotearoa (New Zealand) in 2017. Poelina believes there are Indigenous ways of knowing, being and doing based on non-linear time, a located temporality, which connects people, rivers, and place (Martuwarra RiverOfLife et al., 2021) that can help us listen to rivers. Poelina shares a similar view to many First Nations leaders in Australia, who are advocating for water management to engage with Indigenous-led restorative research and practices to ensure Indigenous values are recognised (Martuwarra RiverOfLife et al., 2021). Poelina believes that healing river ecosystems is a cultural endeavour and that listening, hearing and voicing rivers should be guided by Indigenous knowledge of place. Listening to rivers from a diversity of perspectives provides unique opportunities for bridging traditional knowledge systems, emerging science, and creative practice when working directly with communities in freshwater management.

Conclusion

As the world continues to face the ramifications of the COVID-19 pandemic and the catastrophic impacts of climate change, communities urgently need more accessible and affective methods to participate in global conservation efforts. Encouraging communities to listen to the environment through interdisciplinary acoustic ecology research can revalue Indigenous leadership and traditional knowledge and create pathways for genuine collaborations, climate action, and participatory conservation. The projects featured in this chapter are drawn from a large body of interconnected research that is embedded in participatory acoustic ecology that values diverse knowledge on listening from artistic, cultural, aesthetic, and scientific perspectives. The projects seek to move beyond the uncomfortable tones of exoticism, imperialism, and exploitation that have often accompanied a colonial approach to field recording and frame acoustic ecology as a socially engaged, inclusive, accessible, interdisciplinary field that can inspire communities to listen.

Biosphere Soundscapes, BIOM, and River Listening are designed with a sense of hope and optimism, and focus on facilitating connections and collaborations between art, science, and technology to mobilise local and global communities to listen, share, and collaborate. The accessibility of new audio technologies and emerging scientific interest in ecoacoustics and acoustic ecology has provided an opportunity for communities to actively engage in shifting the climate narrative by listening to the changing soundscapes of our global ecosystems. The resulting creative works realised through compositions, installations and performances can engage and immerse listeners in changing ecosystems and inspire a sense of environmental empathy and action. Interdisciplinarity is critical to advancing participatory acoustic ecology and this research calls for further collaborations between artists, scientists, and technologists to record, document, and share the changing soundscapes of the planet to help us understand and respond to climate change and engage community in the process.

References

Barclay, L. (2014). Biosphere soundscapes. *Leonardo, 47*(5), 496–497. 10.1162/LEON_a_00820

Barclay, L. (2016). River listening. In F. Bianchi & V. Manzo (Eds.), *Environmental sound artists* (pp. 68–75). Oxford University Press.

Barclay, L., & Gifford, T. (2017). Acoustic ecology in UNESCO biosphere reserves. *The International Journal of UNESCO Biosphere Reserves, 1*(1), 23–65. 10.25316/IR-386

Barclay, L., Gifford, T., & Linke, S. (2018). River listening: Acoustic ecology and aquatic bioacoustics in global river systems. *Leonardo, 51*(03), 298–299. 10.1162/leon_a_01516

Barclay, L., Gifford, T., & Linke, S. (2020). Interdisciplinary approaches to freshwater ecoacoustics. *Freshwater Science, 39*(2), 356–361. 10.1086/709130

Burtner, M. (2011). EcoSono: Adventures in interactive ecoacoustics in the world. *Organised Sound, 16*(3), 234–244. 10.1017/s1355771811000240

Carlyle, A. (2014). The God's eye and the buffalo's breath: Seeing and hearing web-based sound maps. In R. Castro & M. Carvalhais (Eds.), *Proceedings from invisible places, sounding cities: Sound, urbanism and sense of place.* Viseu, Portugal: Jardins Efémeros. http://invisibleplaces.org/2014/pdf/ip2014-carlyle.pdf

Deichmann, J., Acevedo-Charry, O., Barclay, L., Burivalova, Z., Campos-Cerqueira, M., d'Horta, F., Game, E., Gottesman, B., Hart, P., Kalan, A., Linke, S., Nascimento, L., Staaterman, E., & Aide, T. (2018). It's time to listen: There is much to be learned from the sounds of tropical ecosystems. *Biotropica, 50*, 713–718.

Duan, S., Towsey, M., Zhang, J., Truskinger, A., Wimmer, J., & Roe, P. (2011). Acoustic component detection for automatic species recognition in environmental monitoring. In *2011 Seventh International Conference on Intelligent Sensors, Sensor Networks and Information Processing (ISSNIP)* (pp. 514–519).

Farina, A., & Gage, S. (2017). *Ecoacoustics: The ecological role of sound.* Wiley.

Feld, S. (1996). Waterfalls of song: An acoustemology of place resounding in Bosavi, Papua New Guinea. In S. Feld & K. H. Basso (Eds.), *Senses of place.* School of American Research Press.

Fuller, S., Axel, A. C., Tucker, D., & Gage, S. H. (2015). Connecting soundscape to landscape: Which acoustic index best describes landscape configuration? *Ecological Indicators*, 58, 207–215. 10.1016/j.ecolind.2015.05.057

Gagliano, M. (2018). Planetary health: Are we part of the problem or part of the solution? *Challenges*, *9*(2), 38. 10.3390/challe9020038

Gilmurray, J. (2017). Ecological sound art: Steps towards a new field. *Organised Sound*, *22*(1), 32–41. 10.1017/S1355771816000315

Hill, A., Prince, P., Snaddon, J., Doncaster, C. P., & Rogers, A. (2019). AudioMoth: A low-cost acoustic device for monitoring biodiversity and the environment. *HardwareX*, *6*, e00073. 10.1016/j.ohx.2019.e00073

Krause, B. (1987). Bioacoustics, habitat ambience in ecological balance. *Whole Earth Review*, *57*, 14–18.

Linke, S., Gifford, T., Desjonquères, C., Tonolla, D., Aubin, T., Barclay, L., Karaconstantis, C., Kennard, M., Rybak, F., & Sueur, J. (2018). Continuous monitoring of freshwater environments using underwater passive acoustics. *Frontiers in Ecology and the Environment*, *16*(4), 231–238. 10.1002/fee.1779

Martuwarra River Of Life, Unamen Shipu Romaine River, Poelina, A., Wooltorton, S., Guimond, L., & Sioui Durand, G. (2021). Hearing, voicing and healing: Rivers as culturally located and connected. *River Research and Applications*, *38*(3), 422–434. 10.1002/rra.3843

Monacchi, D. (2013). Fragments of extinction: An eco-acoustic music project on primary rainforest biodiversity. *Leonardo Music Journal*, *23*, 23–25. 10.1162/LMJ_a_00148

Monacchi, D. (2016). A Philosophy of eco-acoustics in the interdisciplinary project fragments of extinction. In F. Bianchi & V. J. Manzo (Eds.), *Environmental sound artists* (pp. 159–168). Oxford University Press.

Ribeiro, J. W., Sugai, L. S. M. & Campos-Cerqueira. M. (2017). Passive acoustic monitoring as a complementary strategy to assess biodiversity in the Brazilian Amazonia. *Biodiversity and Conservation*, *26*, 2999–3002.

Schafer, R. M. (1977). *The tuning of the world*. Random House.

Staaterman, E., Ogburn, M. B., Altieri, A. H., Brandl, S. J., Whippo, R., Seemann, J., Goodison, M., & Duffy. J. E. (2017). Bioacoustic measurements complement visual biodiversity surveys: Preliminary evidence from four shallow marine habitats. *Marine Ecology Progress Series*, *575*, 207–215. 10.3354/meps12188

Sueur, J., & Farina, A. (2015). Ecoacoustics: The ecological investigation and interpretation of environmental sound. *Biosemiotics*, *8*(3), 493–502. 10.1007/s12304-015-9248-x

Sueur, J., Pavoine, S., Hamerlynck, O., & Duvail. S. (2008). Rapid acoustic survey for biodiversity appraisal. *PLoS ONE*, *3*(12), e4065. 10.1371/journal.pone.0004065

Tucker, D., Gage, S. H., Williamson, I., & Fuller, S. (2014). Linking ecological condition and the soundscape in fragmented Australian forests. *Landscape Ecology*, *29*(4), 745–758. https://eprints.qut.edu.au/70754/

Voegelin, S. (2014). *Sonic possible worlds: Hearing the continuum of sound*. Bloomsbury Academic.

15

AURAL COUNTERPUBLIC RESISTANCE: NOISE, SILENCE, AND ACOUSTICAL AGENCY IN PROTEST TACTICS

Nimalan Yoganathan

Introduction

Within dominant discourse related to social movements, there is a tendency to focus on the visual tactics and aesthetics employed by protestors. Mainstream media coverage of the Black Lives Matter (BLM) protests that spread across the Unites States and globally during the summer of 2020 circulated viral news images of mass demonstrations, protestors intensely staring into the eyes of police as a mode of refusal (Mirzoeff, 2018) as well as racist monuments being toppled. These visually striking protest images are indeed crucial for shifting public discourse. Such spectacular protest tactics were one of many catalysts that brought the policy demands of BLM organisers into the public sphere. Discourse surrounding the defunding and abolition of police was able to infiltrate mainstream debates in part because the engaging visual protest aesthetics caught the attention of the dominant gaze (see Osterweil, 2020).

However, news outlets and politicians were quick to focus their gazes on incidents of 'rioting' and 'looting' during the BLM protests (Nguyen, 2020; Tracy, 2020). Trump tweeted racist narratives framing protestors as 'thugs' and 'outside agitators' (Ali, 2020). Such criminalisation of protestors of course has historical precedents. For instance, amidst the 2005 'riots' in the Paris banlieue (low-income estate housing) of Clichy-sous-Bois following the police chase and death of two teens, Zyed Benna and Bouna Traoré, then French Minister of the Interior Nicolas Sarkozy referred to the Arab and Black youth who were rioting as *racaille* ('scum'). French news outlets repeatedly disseminated images of burning cars.

Numerous media scholars have examined how visual depictions of protests in mainstream news influence public discourse. McLeod and Hertog (1992) have studied how media framing determines whether the public will perceive a protest as legitimate or not. They argue that the riot and confrontation frames in news

DOI: 10.4324/9781003164227-26

often serve to delegitimise social movements and negatively affect public opinion, thus helping to maintain the status quo. In contrast, the debate frame used by journalists moves beyond the spectacle of protests and attempts to investigate the socio-political contexts in which protests take place. This news frame emphasises the political agendas and demands of protestors (McLeod & Hertog, 1992). However, Kilgo and Harlow (2019) address how the debate frame appeared infrequently for American news coverage in 2016 and 2017 of the Indigenous-led Standing Rock protests, BLM protests, as well as the NFL national anthem protests led by Colin Kapaernick and other Black players. The Standing Rock and BLM protests in particular were more likely to be framed in the news as disruptive and violent. Such negative framing of protests is partly due to reporters defaulting to official sources like police and politicians for statements (Kilgo, 2020). The implicit racial biases of reporters and the overwhelming whiteness of American newsrooms (Goodwin, 2021) additionally contribute to the negative news framing of social movements.

Aside from the visual spectacle of protests, it is also critical to understand how the act of listening influences how the dominant public engages with social movements. The dichotomy between 'legitimate' forms of protest and criminal 'rioting' is arguably reinforced by a failure on the part of dominant society to listen critically and ethically to social unrest.

In her analysis of the Clichy-sous-Bois uprising, Bassel (2017) insists on the need for a politics of listening that can challenge norms of intelligibility. She suggests the need for ethical forms of listening that break down binaries between 'Us', the audible, and 'Them', the silent Others (p. 6). The French media and state were preoccupied with morally condemning rioting but refused to listen to what rioters were struggling to tell them for many years concerning police brutality and systemic employment discrimination. Bassel thus theorises rioting as a political event in itself challenging the assumption that rioting youth are merely violent, ignorant, and depoliticised. Furthermore, she insists that it is essential for marginalised groups to develop practices of resistance so that unheard people have the chance to oppose dichotomous discourses and tell their own stories on their own terms. The concept of the counterpublic responds to Bassel's call for spaces where marginal voices can speak on their own terms.

Fraser (1990) conceives of *subaltern counterpublics* as 'discursive arenas where members of subordinated social groups invent and circulate counter-discourses' (p. 67) that challenge dominant narratives. In this chapter, I propose the concept of *aural counterpublics* to articulate how communities at the margins of society amplify their grievances and right to exist through the use of performative sound.

It is critical we move beyond confining the theorisation of protest tactics to the visual domain and instead listen to how protestors employ sound practices as creative methods for challenging the news media's delegitimisation of their grievances. In this chapter, I study how social movements on a large-scale and even individual political actors engage aural counterpublics to exploit the often underheard sonic registers of political resistance. These sonic registers subvert the status quo in ways that ocular-centric protest tactics may not. I pose the question, what roles do sound and listening play in the mobilisation of social justice movements?

This chapter is organised as follows. The second section briefly surveys literature on the politics of listening so as to lay out the theoretical foundation for the anti-racist, ethical, and self-reflexive modes of listening I explore here. In the third section, I define my concept of aural counterpublics and clarify how such alternative discursive spaces work to unsettle white normative listening modes and what Dreher (2009) describes as hierarchies of attention. In the fourth section, I return to the Clichy-sous-Bois uprising of 2005 but listen to it alongside the Indigenous-led Idle No More movement's use of traditional round dances and drumming in shopping malls as a mode of sonic disobedience. I highlight how these subversive deployments of noise via aural counterpublics help elucidate philosopher Michel de Certeau's (1988) concept of the *tactic*. For de Certeau, subjugated groups may use tactics as everyday acts of resistance to construct spaces of agency in opposition to strategic power. I am particularly interested in the potential of *acoustical agency* which sound studies scholar Tom Rice (2016) examines to understand how communities actively resist and reshape everyday oppressive soundscapes by producing their own noise. While Rice concentrates on noise as a tactic, I apply his concept of *acoustical agency* to resistive forms of silence as well.

In the final section of this chapter, I explore the expression of acoustical agency through NFL players Marshawn Lynch and Malcolm Jenkins' performative silence and strategic refusal to speak to news reporters. I read the creative sound practices of these football players as performance art committed to a politics of listening in which silence can work to agitate the status quo. I posit Lynch and Jenkins use their large public platforms to construct aural counterpublics that quietly yet subversively circulate counter-discourses regarding racism perpetrated both by the news media and society at large.

Through this chapter, I aim to disrupt the false dichotomy of 'respectable' versus 'uncivil' modes of protest but through a sonic perspective. Rather than arguing whether confrontational noise or silent resistance is more effective at attracting attention to social justice issues, I instead suggest that a diverse array of creative tactical interventions, both loud and quiet ones, are needed to rupture racist dominant narratives.

Towards a politics of listening

In Western societies that prioritise the liberal notions of individualism and 'free speech', dominant discourse and political debate are seldom focused on self-reflexive and ethical forms of listening that decentre whiteness and its links to power and privilege. Our dominant culture is one that speaks rather than listens (Back, 2007). Such an inequitable economy of listening is driven by hierarchies of attention that shape not only what is and is not heard and valued but also who is and is not heard (Dreher, 2009). Marginalised groups thus often struggle to find discursive spaces where they can speak on their own terms while being heard by those with power and privilege.

Spivak argues that it is not a matter of the subaltern being incapable of speech (that is, lacking agency) but rather that there is a systematic inability or refusal of the discursively privileged to hear that speech (Fernandez, 2018). As I will suggest throughout this chapter, tactical uses of noise and silence carry the potential to unsettle established economies of listening.

I propose that such established economies of listening are deeply rooted in whiteness. Whiteness, as I apply it in this chapter, refers to the way that white people, their customs, culture, and beliefs operate as the standard and racial default by which all other groups are compared (National Museum of African American History and Culture, n.d.). This phenomenon of racial normalisation is commonly termed white normativity—that is, the cultural norms and practices that make whiteness appear natural, normal, and right (Ward, 2008).

Stoever (2016) proposes the concept of the *white listening ear* to articulate the epistemological function of normative listening as a modality of racial discernment. This listening ear is an aural complement to an interlocutor of the white gaze. The sonic practices of resistance I study in this chapter attempt to force the dominant public to hear beyond the white listening ear and instead listen through an intersectional framework.

Several scholars have outlined the need for a politics of listening that unhouses us from our normative listening impulses. Robinson (2020) insists on the need for critical listening positionalities that adopt a self-reflexive questioning of how race, class, gender, sexuality, ability, and cultural background intersect and influence the way we are able to hear sound, music, and the world around us.

Bickford's (1996) notion of 'political listening' speaks to Robinson's call for self-reflexive listening. For her, oppression works, in part, through 'not hearing certain kinds of expression from certain kinds of people' (p. 129). Political listening assigns an ethical responsibility to those who are discursively privileged to be self-conscious about how they listen and what they hear. Building on Bickford's work, Dreher (2009) proposes the concept of 'listening across difference' to suggest how critical and ethical listening modes must entail 'not merely polite conversation or consensus but also risk and conflict, the possibilities of discomfort and difficulty' (p. 451).

The counter-hegemonic modes of listening discussed so far aim to interrupt the monopoly over discursive spaces and attention economies the dominant white public holds. The following section demonstrates how aural counterpublics link these alternative forms of listening to the performance and aesthetics of protest.

Sonic agitation through aural counterpublics

Within hegemonic listening contexts, marginalised communities are often expected to limit themselves to the arena of 'civil' discourse so as to persuade the dominant white public to sympathise with their concerns. One of white supremacy's more covert and insidious effects is how it is able to conceal unequal distributions of power and privilege using the cloak of meritocracy. In particular, dominant discourse within Western liberal 'democracies' suggests that we all, despite our intersectional identities, have been afforded equal social and economic opportunities. In reality, white supremacy is a political, economic and cultural system in which whites overwhelmingly control power and material resources (Gillborn, 2006). As a result, racialised groups demanding social justice are often forced to appeal to political bodies and news media institutions that are primarily controlled by whites (Heider, 2014). This means that social justice demands are expected to conform to and appease white cultural norms.

However, some social movement scholars have highlighted the limits of solely relying on incremental reform and persuasion as tactics. Daum (2017) argues that counterpublics like BLM and the Standing Rock protestors have creatively devised methods of cultivating their own power by rejecting the dominant discourse, respectability politics and norms of appropriate decorum that do little in challenging the system of white supremacy.

Kilgo (2020) posits that journalists pay little attention to protests that are not dramatic. Knowing this, protesters find creative ways to capture media and public attention. This often includes visually striking tactics. Protests are often performative in nature in that protestors perform their existence through resistance: they demand recognition, embody visibility, and articulate a political voice. Furthermore, the aesthetics of protest can rupture the conventions of doing politics (McGarry et al., 2020).

McGarry et al. (2020) examine the role of aesthetics in political life and social change. They conceive of protest aesthetics as forms of communication used to disseminate protestors' ideas and demands, while ensuring their voices are heard. Protestors produce and document the choreography of protest through aesthetics such as slogans, art, graffiti, and gestures. Aesthetics can be understood as political practices with the power to transform existing structures, ideas, and orthodoxies.

However, even once protestors attract media attention by performing their protest aesthetics, news coverage often remains superficial and delegitimising. News media tend to focus solely on the tactics and disruptive nature of protests while excluding discussions about protestors' demands (Emmerich, 2020; Kilgo, 2020). Despite the news media's selective attention, counterpublics still maintain that their protest aesthetics can be politically transformative as long as they circumvent normative codes of behaviour and discourse.

Counterpublics may find agitation to be a more effective tactic for disrupting the status quo. This is primarily because engaging the public in the vernacular of dominant discourse is unlikely to dismantle existing institutions and power arrangements (Daum, 2017). I employ the concept of aural counterpublics to articulate how marginalised groups claim their own space within everyday soundscapes and force their counter-narratives into the public sphere in an attempt to unsettle the existing political order. I do not limit my inquiry to traditional forms of speech (i.e. voice). After all, politics is not produced solely by the vocalised claims or demands of protestors but by their action, and sometimes their inaction (McGarry et al., 2020), which in the case of my study in this chapter can be heard as silence.

I focus on political uses of sound in itself—articulations of noise and silence by subjugated groups that can be read as radical 'speech acts' (Butler, 2015) and subversive practices of refusal and agitation. More specifically, I suggest that aural counterpublics encrypt counter-narratives, subjugated knowledges, and expressions of agency within their noise and silence so that they not only force their visibility into mainstream discourse, but their audibility as well.

Listening politically to noise as tactical resistance

The politics of listening I engage in this chapter is one that listens to how entrenched power structures shape speaking and listening relationships. In this section, I study the use of noise as a protest tactic that challenges established hierarchies of attention while simultaneously permitting marginalised groups to express their agency through sound. I start by listening critically to riots.

Hearing aural counterpublics in riots

In his essay, *Banlieue Sounds, or, The Right to Exist*, Tchumkam (2019) examines how rioting within France's low-income and predominantly racialised housing estates known as the *banlieues* is employed as a tactic to denounce social injustice. The primarily Maghrebi and Sub-Saharan African immigrants and their children who live in banlieues have, since the 1980s, been attempting to make dominant white French society listen to their grievances concerning high unemployment and police brutality by way of peaceful marches and political hip-hop performance. These peaceful means of protest have not led to significant systemic changes

Tchumkam (2019) therefore argues that amidst the highly publicised riots that started in Clichy-sous-Bois and spread throughout numerous other French banlieues in November 2005, a new form of community emerged, a youth resistance culture, in which the noise of rioting can be understood as a sonic tactic that directly responds to stigmatisation and silencing by the state and mainstream media. This resistance culture understands that the spectacular performance of lighting cars on fire and throwing Molotov cocktails at buildings is both a mode of existence and mode of speech. As Butler (2015) warns:

> We have to ask whether it is right that verbalisation remains the norm for thinking about expressive political action. Indeed, we have to rethink the speech act in order to understand what is made and what is done by certain kinds of bodily enactments: the bodies assembled "say" we are not disposable. (p. 18)

When banlieue youth assemble to riot, they are not apolitical. The visually confrontational and loud protest aesthetics through which their own bodies are employed as mechanisms to destruct property are where the expressive political action and 'speech' lie. Rioting does not use clear verbalisations of demands, but rather embodied performativity as an articulation of resistance.

Furthermore, I suggest we understand the noise of rioting as a means of constructing an aural counterpublic for the banlieue youth. Noise tactics help create a new counter-hegemonic acoustic territory for these youth—that is, a space where they can circulate their own counter-discourses related to daily life and use the speech acts of rioting as a method of affirming their own subjugated knowledges.

Fraser (1990) argues that counterpublics permit marginalised groups to formulate oppositional interpretations of their identities and needs. The counterpublics formed via the noise-based sound practice of banlieue rioters is one means of translating their anger, frustration, and grievances into a language that the French state and dominant public has no choice but to listen to, whether or not they listen with outrage and condemnation.

The aesthetics of protest is acutely important for minority and marginalised voices that are often unheard (McGarry et al., 2020). Surely, the visual aesthetics of rioting, including images of burning cars repeatedly circulated in 2005 French news coverage, play a key role in drawing mainstream attention to the banlieues. However, when only looking at rather than also listening to riot aesthetics, do we lose some of the ideas these aesthetics are attempting to communicate publicly? I suggest we think through Butler's (2015) argument that the aesthetics of protest reveal how democracy is constituted through a complex interplay of performance, images, and acoustics. Seeing a 'mute' photo of a car in flames on the front page of a newspaper or having a news reporter talk over the soundscapes of youth screaming and smashing glass does not provide significant insight into the political role of sound itself. Whether or not French youth are intentional in their use of sound in the deafeningly loud agitation of rioting, the combined performance of both visual and aural aesthetics helps to further amplify their grievances.

To fully understand the French riots as political events we must examine the unique ways in which white supremacy operates in France. Tchumkam (2019) listens to the riots within the wider context of what he calls France's 'colonial obsession' (p. 188). The descendants of former colonial subjects, even those born in France, continue to be regarded as bodies at the disposal of the state. However, France's colonial obsession is not only manifested via visible modes of discrimination (such as banning of Islamic headscarves). We may also listen to its sonic manifestations. Black and Arab youth are Othered and reduced to silence in the same way their grandparents were silenced in former colonies. This systematic silencing in conjunction with *laïcité* (French secularism) narratives and Sarcozy's thinly veiled racist discourse at the time about banlieue 'scum' transforms banlieue youth into what de Souza (2018) terms 'expert listeners' (p. 464)—that is, racialised groups and individuals who, through lived experience, become acutely attuned to the sonic traces and histories of racial injustice.

Butler's (2015) theory of performativity understands how when marginalised bodies assemble on the street, they are exercising a 'plural and performative right to appear, one that asserts and instates the body in the midst of the political field' (p. 11). Perhaps the performance of banlieue rioting is one way for many racialised 'expert listeners' to come together in order to collectively resist injustice. This urban assembly produces visually striking aesthetics in conjunction with collective noise to function as what Butler describes as a 'bodily demand for a more liveable set of economic, social, and political conditions no longer afflicted by induced forms of precarity' (p. 11).

These banlieue expert listeners 'speak back' to state power by devising their own politics of refusal. As an aural counterpublic, the rioters capitalise on their outsider status and reject dominant modes of dialogue and discourse and their affiliated norms of appropriate decorum. Daum (2017) argues that counterpublics must position themselves as agitators and locations of resistance as opposed to persuasive agents working within sites of assimilation. Doing so draws attention to the counterpublic, its discourse and identities, and works to decentre the centre.

The rioters' sonic agitation and decentring of dominant discourse may be understood as a tactic to force dominant French society into engaging with political listening. Such a critical listening positionality helps invert the roles of speakers and listeners (Bassel, 2017) so that banlieue communities reclaim a discursive space where they wield their own disruptive power.

The rioters force their visibility and audibility onto French society, no matter how much France pretends not to see them or hear them (Tchumkam, 2019). Their sonic tactics help foster a transformative politics of listening that, as Dreher (2009) suggests, has the potential to shift some of the onus and responsibility for change from marginalised voices on to the discursively privileged and powerful. The noises of cars burning, glass smashing and rioters shouting at police allow the banlieue youth to serve as audible political agents who have shifted the hierarchies of attention previously controlled by the media and political elite.

Additionally, rioters subvert and transform their existing environmental acoustic spaces in order to assert a sonic form of agency. Rice (2016) proposes the concept of *acoustical agency* to articulate how prisoners are not merely passive listeners to their everyday oppressive soundscapes (such as the frequent sound of slamming metal gates). Instead, they actively resist these acoustic environments through their own noise to shape the sonic space they inhabit. Prisoners assert their own acoustical agency by playing music loudly on portable stereos or through cell-to-cell shouting.

Rice's understanding of acoustical agency is helpful in framing how rioters 'make do' (de Certeau, 1988) with whatever tactical materials they have access to, finding creative ways of manipulating the constraining order of a space to make it work to their benefit. The tactic of noise empowers rioters with an acoustical agency facilitating the quite literal amplification of their grievances, subjugated knowledges and right to exist. Additionally, the youth resist and reshape their daily oppressive environmental soundscapes of cramped living conditions and police surveillance by inserting their own loud sound into urban spaces. Therefore, the community-centred aural counterpublic fostered by riot noise creates both alternative discursive and environmental acoustic spaces where banlieue youth can 'speak back' to dominant narratives and the soundscapes of being Othered within France's colonial continuum.

Acoustical agency through sonic disobedience: The audible tactics of Idle No More

We may also hear acoustical agency in tactics of sonic resistance that are perhaps not as spectacular and extreme (in an aesthetic sense) as rioting. Take for instance, the Idle No More Indigenous-led protests held in numerous shopping malls throughout Canada and the United States during the Christmas holiday season of 2012. The protests aimed to draw attention to the Canadian federal government's then proposed Bill C-45 that favoured resource-extraction industries and was understood as a threat to Indigenous treaty rights (Nardone, 2016). Protestors occupied the malls through the performance of massive traditional ceremonial round dances. The noise and rhythms were at times loud enough to actually disrupt the acoustic architecture of the building. At a Regina, Saskatchewan mall, the sounds of singers and drummers masked the mall's ambient Muzak, cancelling it out. Nardone eloquently describes the sonic experience: 'The architecture shakes. Impossible to ignore or avoid, the music's vibrations affectively claim the space, sonically consume it"' (p. 92).

Such sonic disobedience allowed Indigenous bodies to create both a physical and acoustic barrier halting consumers' movements. Idle No More formed an acoustic equivalent of the physical blockades used by Indigenous activists to interrupt oil pipeline construction and corporate development on sacred lands. The round dance protests articulate a refusal to passively accept the environmental soundscapes linked to 'quotidian circulations of capital' (Nardone, 2016, p. 95). As an aural counterpublic that rejects dominant procedures and norms of respectability, Idle No More forced their own anti-colonial counter-narratives into otherwise mundane and homogenous commercial acoustic spaces.

In her examination of the phenomenology of whiteness, Ahmed (2007) argues 'spaces are orientated "around" whiteness, insofar as whiteness is not seen which makes non-white bodies feel uncomfortable, exposed, visible, different, when they take up this space' (p. 157). Because whiteness is unmarked and assumed to be given, it may function as a form of public comfort by allowing white bodies to extend into spaces that have already taken their shape. To build on Ahmed's analysis, I would ask what does it sound like when spaces take the shape of normative whiteness? I have argued that the loud protest aesthetics of banlieue rioters and Idle No More activists function as assertions of acoustical agency by which marginalised groups reshape the oppressive urban sonic spaces they inhabit. This literal reshaping of soundscapes can be heard as a disruption of the 'public comfort' (Ahmed, 2007, p. 158) that everyday spaces grant to white bodies and a forcible attempt on the part of racialised and Indigenous bodies to both visibly and sonically 'extend into' normative spaces where, otherwise, they are accustomed to surveillance and exclusion.

The rioters and protestors discussed thus far in this chapter confront systems of power and create spaces of agency using the performance of tactical noise. Sometime this includes literal noise (such as rioting). In the cases of Idle No More, aural counterpublics use a symbolic form of noise—sound perceived by the dominant public as 'noise' and a nuisance because of its potential to interfere with daily operations and commerce.

Sounding out respectability politics: The loud/quiet protest dichotomy

So far in this chapter, I have underscored how groups at the margins of society capitalise on the potential of urban noise to agitate the status quo. Noise in the case studies I have examined does not assimilate into the norms of respectability and decorum that political elites and the dominant public often impose on social justice movements. There are numerous examples of social movements that have refused respectability politics including the Black Panther Party and the Mohawk land defenders of Kanehsatà:ke during the 1990 'Oka Crisis' in Quebec, Canada. These counterpublics were essentially labelled as enemies of the state if not terrorists in part because of their confrontational and 'indecorous' protest aesthetics.

To be clear, I am not suggesting that respectability cannot in some instances be used as a carefully devised tactic by protestors. In fact, Anderson (2017) highlights how respectability as an aesthetic and practice greatly informed the non-violent Civil Rights Movement of the 1950s and 1960. She states:

> Blacks in the movement worked hard to curate an image as God-fearing, hardworking, law-abiding, and family loving Americans. The point was to skilfully use the new medium of television so that the brutality that rained down on Black people—especially respectable ones—would shock the conscience of the white public and lawmakers. (para. 10)[1]

It is however critical to push back against the binary logic that associates 'civil' protests with 'quietude' and indecorous agitation with 'noise'. Even Civil Rights organisers understood that solely relying on quiet tactics would not be sufficient to make significant racial justice gains. Black protesters during the Birmingham campaign of 1963 were not strictly non-violent. They sometimes fought back against the police by throwing rocks at them and breaking windows. Robert Kennedy, afraid that the increasingly riotous atmosphere in Birmingham would spread across Alabama and the South, convinced John F. Kennedy to deliver his speech calling for a civil rights act (Osterweil, 2014).

It is clear that a diverse array of tactics, both loud and quiet, are needed within social movements. With this in mind, I now shift my attention to the other end of the audible spectrum—that is, to the barely audible hums and quiet frequencies produced by aural counterpublics. How can silence, like noise, be deployed within quotidian practices of resistance and refusal that disobey respectability politics?

I will problematise the assumption that the performance of protest and expression of agency are solely dependent upon clearly articulated demands and vocalised claims by activists.

Listening for resistance in silence

Campt (2017) listens to quiet and quotidian articulations of agency in Black resistance. She sounds out the 'lower frequencies' in the seemingly quiet mug shot photographs of young 'Freedom Riders' during the American Civil Rights movement. Around 328 young Black men and women were arrested in Mississippi on buses, at lunch-counters and parks for defying Jim Crow segregation laws. Campt sees and, most interestingly, hears confidence and intention in the mug shots. The Freedom Riders are wearing dignified suits and dresses. They 'highjack the booking process' (p. 94) through the simple yet radical gesture of subtly grinning for their photos. These are not merely mute subjects without agency. They are resistive 'quiet subjects who refuse the silence imposed upon them' (p. 91) by the state who attempts to reduce them to criminals. They achieve this without having to utter a word to their oppressors. We may find parallels between this silent tactic and recent examples of young Palestinian protestors smiling while being arrested or violently attacked by Israeli security forces (Kuttab, 2021). These smiles which are recorded on cellphones and disseminated via social media are one mode of asserting Palestinian dignity and resistance.

The quiet tactics of the Freedom Riders exemplify how the performance of silence as a protest aesthetic can serve as a technique for 'micro-political resistance' (Wagner, 2012, p. 100) that asserts one's humanity. Listening to the disruptive hum emanating from the mug shot photos helps articulate Fernandez's (2018) proposal that silence need not only be interpreted as an absence, a lack, or a negation. This Black radical tradition of silent embodied resistance that Campt references is still practiced today.

Sounding out agency and counter-narratives in Marshawn Lynch's silence

In 2013 and 2014, Black NFL football player Marshawn Lynch engaged in a long-term practice of refusing to speak to the media. At one press event, Lynch responded to the NFL's contractual obligations that require players to engage with the press by stating, 'I'm here so I don't get fined. You can ask me all the questions that you want to. I'm going to answer with the same answer. I've got nothing for y'all'. He then ceased speaking for the remainder of the event.

In *Lynch: A History* (2019), filmmaker David Shields provides a critical reading of Lynch's embodied performance of silence. Through a collage method, he suggests links between Lynch's politics and the wider context of American race politics. For instance, Lynch's touchdown runs and interviews are intercut with horrific clips of American violence such as a white police officer shooting Walter Scott, an unarmed Black motorist (Hsu, 2019).

Shields attempts to position Lynch's tactical silence as a response to how the American white gaze consumes Black masculinity and Black labour. 'Marshawn is an incredible performance artist', Shields comments. 'I wanted to show what he was doing wasn't what people thought. It wasn't to be funny or playful or shy or petulant. I wanted to correct a false media narrative that made him some cartoonish character' (Williamson, 2020).

Lynch's press conference performances are noteworthy for their discursive tactics. When reporters ask mundane questions about how he felt a game went or persistently probe him about the motivations behind his silent protest, Lynch remains calm and silent, occasionally providing them with vague and intentionally nonsensical responses: 'You know why I'm here', 'maybe', or 'I'm thankful'. Such sonic agitation subverts the mainstream media and NFL's imposed norms of Blackness (Vats, 2016) that expect Black athletes to always be at the disposal of the predominantly white press. Lynch's protest aesthetics are sophisticated in how he satirically performs what the corporate media imagines Black respectability looks and sounds like. However, he does so through an ostensibly polite and quiet demeanour.

Here, I am particularly interested in the meanings hidden within the long silent gaps between Lynch's sparse comments to the press. As Campt (2017) suggests, 'Quiet is a modality that surrounds and infuses sound with impact and affect, which creates the possibility for it to register as meaningful' (p. 4). I also listen out for what Fernandez (2018) describes as 'hidden resistance', (p. 199) where the potential for contestation is sought in the interstices of the established order, rather than through direct confrontation. Without explicitly attacking the NFL and the corporate media, Lynch deploys silence as a refusal to speak and perform labour in the ways forced on him by the NFL. It is a means for him to assert control over how his body is propertised (Vats, 2016). Such reclamation of control is significant especially considering the NFL once fined Lynch $25,000 for having performed his signature 'crotch-grabbing' celebration dance during a game (Shields, 2019).

We may also hear acoustical agency in Lynch's quiet intensity. The media insist that he deliver clichéd sound bites during press conferences and perform respectability on their terms. Rather than passively abiding by the NFL's speech protocols, Lynch weaponises his silence to speak back to the overwhelming soundscape of reporters expecting him to entertain them not only on-field, but off-field as well. By simultaneously working within and against the corporate media and the NFL, Lynch speaks to de Certeau's assertion that tactical manoeuvres need not always confront systems of power from the outside. These manoeuvres can originate from within the systems (Sawchuk, 2013).

Fuentes (2015) points out the intertwined relationship between performance, aesthetics and politics. Anchored in the society of the spectacle, demonstrators put in practice a variety of communicative styles and mobilising techniques that include strategic uses of non-linguistic, embodied actions as statements. Lynch creatively re-appropriates and 'hijacks' the press conference room and transforms it into a stage where he can perform his silent embodied resistance. His performance aims to critique the artificiality and white normativity of the spectacle that is the corporate press conference format.

Furthermore, Lynch's performed refusal to speak can be understood as a form of Black fugitivity. Campt (2017) hears fugitivity in the quiet mug shots of Freedom Riders grinning. However, such fugitivity is not fleeing or escape but rather a practice of refusing the terms of negation. It is a practice of refusal honed by the dispossessed in the struggle to create possibility within the constraints of everyday life.

A clear articulation of Lynch's performed fugitivity can be heard in one scene of *Lynch: A History* in which he tells a crowd of reporters, 'Ya'll shove microphones and cameras down my throat. So now for these next three minutes, I'll just be looking at ya'll the way that y'all looking at me'. He then remains silent and once his cell phone timer rings after the three minutes, he abruptly stands up and leaves the stage. This film clip is juxtaposed with a commentary from former Black NFL player Alan Grant discussing the challenge, as a 'big dark-skinned athletic man', of feeling like you are an 'exhibit', subject to someone else's gaze. He notes how fans and the media look at you like zoo animals. 'What they fail to realise is that you can see them, you can hear them, you can formulate your own thoughts and reach your own conclusions about them' (Shields, 2019). Lynch thus rejects the media's terms of negation that attempt to characterise him as uncivil, apolitical and devoid of agency. As Hsu (2019) suggests about Lynch, a Black athlete's insubordination, particularly when it is public, is rarely apolitical. Many wondered if there were deeper motivations behind Lynch's silence that he was declining to elaborate.[2]

Lynch, during this press conference, was indeed abiding by his NFL contractual obligations and agreeing to sit in front of cameras for three minutes rather than opting to not show up at all and pay the fines, if not quit the NFL. Some may interpret this choice as him still being 'on exhibit' and placating the corporate media. I insist that he is asserting his agency and refusing to passively submit to having his body be commodified by the league and media. He finds ways to work within and against a constraining order, in the de Certeauean sense, as opposed to a wholesale attack on the NFL and the media.

Lynch's defiant 'looking back' at reporters works to expose the ideology of whiteness and its desire to consume and exploit Blackness for entertainment. He deploys what Yancy (2012) terms a Black counter-gaze that centres on Black lived experiences and sees beyond the supposed invisibility of whiteness. This counter-gaze challenges the cultural norms and practices that define white normativity. By choosing a silent 'looking back' rather than speech, Lynch refuses norms of decorum and speech codes that the press attempt to impose on him.

I also propose we read Lynch's performed fugitivity through Jiwani's (2011) theorisation of *reflectionism* as a possible tactical intervention that activists and artists can employ. This involves 'reflecting back' (p. 345) and mirroring or mimicking those who are in a dominant position, thus working to invert the gaze.

Lynch's reflectionism also operates through a sonic register. He challenges the white listening ear (Stoever, 2016) by echoing back to reporters the banality and absurdity of their questions. During a post-game locker room press conference, reporters bombarded a visibly uncomfortable Lynch with questions such as, 'What was the song of the day on the way to the game, Marshawn?' Reporters were perplexed when Lynch repeatedly responded with, 'I'm thankful'. Lynch's performed aural mimicry is perhaps informed by his ability to hear as an expert listener (de Souza, 2018) attuned to the sonic traces and histories of racism. The previously discussed banlieue rioters have become accustomed to decoding the racist dog whistle remarks of Sarcozy. Whereas, Lynch is attuned to the sonic registers of anti-Blackness within mainstream media discourse that attempts to police the manners in which Black male athletes are allowed and not allowed to speak to the public (see Brown, 2017).

Lynch capitalises on his status and the media exposure he receives to construct an aural counterpublic. His tactical use of silent body gestures and calculated responses to questions works to circulate alternative epistemologies that disrupt hegemonic conceptions of Black labour and Black respectability. Furthermore, Lynch's silence can be read as a wielding of his own power—'a way of controlling his own narrative by refusing to offer one' (Hsu, 2019, Para. 2).

Lynch uses his intentional (un)voicing to take up space and inspire a community of other Black athletes to reject the expectation that they need to perform palatable and predicable sound bites that satisfy white capitalistic norms. In one press conference, Lynch, in an unusual turn, spoke much more than usual. However, his statement was completely unrelated to the question posed by a white reporter. Lynch chose to speak but only to the Black listening ears of current or future young Black football players. He gave them advice on how not to be financially and emotionally exploited: 'Take care of y'all money African, cus that shit don't last forever. While ya'll in it take care of y'all bread, and take care of y'all mentals, so when y'all done, you go ahead and take care of yourself' (NFL Season, 2020).

You aren't listening

Shields (2019) suggests, 'Lynch is leaving a legacy of the eloquence of silence, echo, and mimicry as key tools of defiance'. Black football player Malcolm Jenkins built on the legacy of Lynch's silent tactics during a 2018 locker-room press conference. Jenkins silently held up a series of large white poster boards with hand-written statistics related to police brutality and mass incarceration in the US. When repeatedly asked by reporters whether he was protesting Trump's disinvitation of his team the Philadelphia Eagles from a White House event due to the NFL national anthem protests, Jenkins remained silent and responded by holding up a sign that read 'YOU AREN'T LISTENING'. Reporters were not attuned to silence and its ability to carry counter-meanings that critique the political status quo and systemic racism. They're only response: 'What aren't we listening to?'

The media were expecting dramatic political slogans and sensational attacks on Trump, but Jenkins' silence itself is where the political work lay.

Berlant (2011) argues that the aim of performative silence in social movements is to 'interrupt normative strategies of affective orchestration—that is, to viscerally move spectators into emotional spaces that diverge sharply from the "normative noise" of mainstream politics' (p. 238). Jenkins demonstrated on live television the fraught ways in which sports is central to how we understand and misunderstand racial issues today (Sheppard, 2018). He provokes the media to engage in ethical modes of 'listening across difference' that expose networks of power and privilege and one's own location within them (Dreher, 2009, p. 451). As Dreher suggests, 'For those who enjoy the prerogative of not listening, it means giving up that privilege' (p. 451). Billias and Vemuri (2017) posit, 'The act of attention is the beginning of ethical behaviour because it is through paying attention to the other that I move beyond myself into the space between us. These in-between spaces are filled with silence' (p. 4). Therefore, the silent act of attention is essential to the meaning-making process.

Because the press is fixated on the normative noise of Trump and his tactics of distraction, Jenkins chooses to decentre his own celebrity voice and instead amplify the data and lived realities linked to systemic anti-Blackness in America that are often rendered inaudible in dominant discourse. He remains silent rather than taking on the labour of educating the press about social justice issues by answering their questions. Doing so is perhaps a prompt for the reporters to remain silent as well so that they may take on the role of ethical agents willing to at least temporarily surrender their privileged position within the hierarchy of attention and read the documented evidence of systemic racism Jenkins cites for them.

Conclusion

Daum (2017) argues that when counterpublics retain their outsider status and reject the centre's attempts to control their discourses and identities, they not only maintain their internal power, but also wield it as a disruptive force by directly challenging the public in its own spaces and venues. In this chapter, I have suggested how the banlieue rioters as well as the Idle No More movement deploy noise as a tactic of resistance to reclaim a counter-hegemonic discursive space where they can wield their own power and unsettle the established roles of dominant speakers and marginalised listeners. These rioters and protestors construct aural counterpublics as a response to the systematic silencing imposed on them by the state and corporate media.

Lynch and Jenkins are certainly not silenced in the same manner. They have major televised platforms that allow them to speak and stage their resistance. However, their intentional and calculated withholding of speech works to disrupt the expectation that Black athletes always be at the disposal of the predominantly white media to perform for them.

In this manner, the quiet embodied resistance of Lynch and Jenkins carries the potential to agitate the status quo as much as confrontational noise tactics do.

However, while the banlieue rioters and Idle No More attempt to force their audibility and counter-narratives into urban public spaces, Lynch and Jenkins disrupt dominant narratives by way of tactical inaudibility—that is, by the mere act of not even engaging with these narratives. They deploy an agential form of silence akin to what Keating (2013) describes as 'resistive silence' (p. 25) where there is a rejection to insert oneself in speech engagement that would reinforce oppressive power relations or practices.

I suggest we understand the loud and quiet sonic protest tactics studied in this chapter as working in parallel with visual-centred tactics, as well as grassroots organising and strategies for demanding legislative changes. The sonic tactical interventions I have outlined are just a few of the many creative ways subjugated groups 'make do' with the resources at their disposal in order to assert agency and launch counter narratives.

Notes

1 However, Anderson warns that respectability offers little to no protection against anti-black violence. She points to the ease at which 12-year-old Tamir Rice and 17-year-old Trayvon Martin could be 'dirtied up' and their humanity stripped from them (Anderson, 2017) by the media and attorneys defending police.

2 Lynch once stated that the media is rarely present when he is speaking in his hometown of Oakland, California supporting local Black community initiatives (Brown, 2017). However, beyond this, he has yet to disclose his specific reasons for his silent protest.

References

Ahmed, S. (2007). A phenomenology of whiteness. *Feminist Theory, 8*(2), 149–168. 10.1177/1464700107078139

Ali, S. S. (2020, September 27). Not by accident: False 'thug' narratives have long been used to discredit civil rights movements. *NBC News*. https://www.nbcnews.com/news/us-news/not-accident-false-thug-narratives-have-long-been-used-discredit-n1240509

Anderson, C. (2017). *Respectability will not save us*. [Literary Hub] https://lithub.com/respectability-will-not-save-us/

Back, L. (2007). *The art of listening*. Berg.

Bassel, L. (2017). *The politics of listening*. Palgrave Pivot.

Berlant, L. (2011). *Cruel optimism*. Duke University Press.

Bickford, S. (1996). *The dissonance of democracy: Listening, conflict, and citizenship*. ICornell University Press.

Billias, N., & Vemuri, S. (2017). *The ethics of silence*. Palgrave Macmillan.

Brown, D. D. (2017). Ballers without Blackness: The suppression of Black cultural agency in the NFL. In D. J. Leonard, K. B. George, & W. Davis (Eds.), *Football, culture and power* (pp. 44–58). Routledge.

Butler, J. (2015). *Notes toward a performative theory of assembly*. Harvard University Press.

Campt, T. (2017). *Listening to images*. Duke University Press.

Daum, C. (2017). Counterpublics and intersectional radical resistance: Agitation as transformation of the dominant discourse. *New Political Science, 39*(4), 523–537. 10.1080/07393148.2017.1378492

de Certeau, M. (1988). *The practice of everyday life*. University of California Press.

de Souza, P. (2018). What does racial (in)justice sound like? On listening, acoustic violence and the booing of Adam Goodes. *Continuum, 32*(4), 459–473. 10.1080/10304312.2018.1488524

Dreher, T. (2009). Listening across difference: Media and multiculturalism beyond the politics of voice. *Continuum, 23*(4), 445–458. 10.1080/10304310903015712

Emmerich, L. (2020). *The media and "riot frames": Shaping the 2020 protest narrative.* Carnegie Council (Report No. 2). https://www.carnegiecouncil.org/education/2020-internship-program/_res/id=Attachments/index=0/CarnegieCouncilSummerInternship2020ResearchReports.pdf#page=4

Fernandez, B. (2018). Silence as a form of agency? Exploring the limits of an idea. In T. Dreher & A. A. Mondal (Eds.). *Ethical responsiveness and the politics of difference* (pp. 187–204). Palgrave Macmillan.

Fraser, N. (1990). Rethinking the public sphere: A contribution to the critique of actually existing democracy. *Social Text,* (25/26), 56–80. 10.2307/466240

Fuentes, M. A. (2015). Performance, politics, and protest. *What is performance studies?* https://scalar.usc.edu/nehvectors/wips/performance-politics-and-protest

Gillborn, D. (2006). Rethinking white supremacy. *Ethnicities, 6*(3), 318–340. DOI:10.1177/1468796806068323

Goodwin, S. (2021). Reckoning with whiteness: The limits of desegregation in America's newsrooms from the 1960s to the present. *Comparative American Studies: An International Journal,* 17(3-4), 271–295. 10.1080/14775700.2021.1892445

Heider, D. (2014). *White news: Why local news programs don't cover people of color.* Routledge.

Hsu, H. (2019, June 14). The profound silence of Marshawn Lynch. *The New Yorker.* https://www.newyorker.com/culture/cultural-comment/the-profound-silence-of-marshawn-lynch

Jiwani, Y. (2011). Pedagogies of hope: Counter narratives and anti-disciplinary tactics. *Review of Education, Pedagogy, and Cultural Studies, 33*(4), 333–353. 10.1080/10714413.2011.597646

Keating, C. (2013). Resistant silences. In S. Malhotra & A. C. Rowe (Eds.). *Silence, feminism, power: Reflections at the edges of sound* (pp. 25–33). Palgrave Macmillan.

Kilgo, D. (2020, May 30). *Riot or resistance? The way the media frames the unrest in Minneapolis will shape the public's view of protest.* [Nieman Lab] https://www.niemanlab.org/2020/05/riot-or-resistance-the-way-the-media-frames-the-unrest-in-minneapolis-will-shape-the-publics-view-of-protest/

Kilgo, D. K., & Harlow, S. (2019). Protests, media coverage, and a hierarchy of social struggle. *The International Journal of Press/Politics, 24*(4), 508–530. 10.1177/1940161219853517

Kuttab, D. (2021, May 11). New generation of Jerusalem Palestinians emerges. *Al-Monitor.* https://www.al-monitor.com/originals/2021/05/new-generation-jerusalem-palestinians-emerges

McGarry, A., Erhart, I., Eslen-Ziya, H., Jenzen, O., & Korkut, U. (2020). *The aesthetics of global protest: Visual culture and communication.* Amsterdam University Press.

McLeod, D. M., & Hertog, J. K. (1992). The manufacture of public opinion by reporters: informal cues for public perceptions of protest groups. *Discourse & Society, 3*(3), 259–275. DOI: 10.1177/0957926592003003001

Mirzoeff, N. (2018). Tactics of appearance for abolition democracy #BlackLivesMatter. *Critical Inquiry.* https://criticalinquiry.uchicago.edu/tactics_of_appearance/

Nardone, M. (2016) Skirmish at the oasis: On sonic disobedience. *Leonardo Music Journal, 30,* 92–96. DOI:10.1162/LMJ_a_00983

National Museum of African American History and Culture. (n.d.). *Talking about race: Whiteness*. https://nmaahc.si.edu/learn/talking-about-race/topics/whiteness

NFL Season. (2020, January 14). *Marshawn Lynch drops jewels on young NFL players: "Take care of your chicken"* [Video]. Youtube. https://www.youtube.com/watch?v=eGrTAF-KIc0

Nguyen, T. (2020, June 2). There isn't a simple story about looting. *Vox*. https://www.vox.com/the-goods/2020/6/2/21278113/looting-george-floyd-protests-social-unrest

Osterweil, V. (2014, August 21). In defense oflooting. *The New Inquiry*. https://thenewinquiry.com/in-defense-of-looting/

Osterweil, V. (2020, June 12). Burning down the 3rd police precinct changed everything. *The Nation*. https://www.thenation.com/article/activism/police-precinct-minneapolis/

Rice, T. (2016). Sounds inside: Prison, prisoners and acoustical agency. *Sound Studies*, *2*(1), 6–20. 10.1080/20551940.2016.1214455

Robinson, D. (2020). *Hungry listening: Resonant theory for Indigenous sound studies*. University of Minnesota Press.

Sawchuk, K. (2013). Tactical mediatization and activist ageing. *MedieKultur: Journal of Media and Communication Research*, *54*(1), 47–64. https://tidsskrift.dk/mediekultur/article/view/7313/6765

Sheppard, S. N. (2018). Close-up: Sports, race, and the power of narrative. *Black Camera: An International Film Journal*, *10*(1), 156–161. DOI:10.2979/blackcamera.10.1.09

Shields, D. (Director) (2019). *Marshawn Lynch: A history* [Film]. Cargo Film & Releasing.

Stoever, J. L. (2016). *The sonic color line: Race and the cultural politics of listening*. New York: NYU Press.

Tchumkam, H. (2019). Banlieue sounds, or, the right to exist. In G. Steingo & J. Sykes (Eds.), *Remapping sound studies* (pp. 185–202). Durham: Duke University Press.

Tracy, M. (2020, June 6). Top editor of Philadelphia Inquirer resigns after 'Buildings matter' headline. *The New York Times*. https://www.nytimes.com/2020/06/06/business/media/editor-philadephia-inquirer-resigns.html#:~:text=Stan%20Wischnowski%2C%20the%20top%20editor,by%20dozens%20of%20staff%20members.

Vats, A. (2016). Marking disidentification: Race, corporeality, and resistance in trademark law. *Southern Communication Journal*, *81*(4), 237–251. 10.1080/1041794X.2016.1200128

Wagner, R. (2012). Silence as resistance before the subject, or could the subaltern remain silent? *Theory, Culture & Society*, *29*(6), 99–124. DOI: 10.1177/0263276412438593

Ward, J. (2008). White normativity: The cultural dimensions of whiteness in a racially diverse LGBT organization. *Sociological Perspectives*, *51*(3), 563–586. DOI: 10.1525/sop.2008.51.3.563.

Williamson, B. (2020, July 16). *Marshawn Lynch's silence was former Raiders' RB 'just doing it in his own way,' documentary director says*. [Silver and Black Pride] https://www.silverandblackpride.com/2020/7/16/21326738/marshawn-lynch-documentary-history-director-david-shields

Yancy, G. (2012). *Look, a white!: Philosophical essays on whiteness*. Temple University Press.

16

STREET HASSLE: NOISE, ART, AND ACTIVISM

Mitchell Akiyama, in conversation with Dont Rhine and Syrus Marcus Ware

When I was invited to have a solo exhibition at Trinity Square Video in the summer of 2019, I decided I didn't want to do it alone. I had proposed to create a 30-channel sound system for the gallery, which is housed in Toronto in Canada's oldest media arts centre, but I conceived the apparatus as a platform for collaboration, for inviting other artists and the public to help me continually recreate the exhibition. I wanted to know what an exhibition could be if I intended it to be a question rather than an assertion. *Everyonce* included three (re)installations of the sound system: one version resulting from a public workshop I gave on field recording; a second in collaboration with Dont Rhine (a founding member of the activist art collective, Ultra-red) and Syrus Marcus Ware (a Toronto-based artist and organiser); a third with Anishinaabe artist, Maria Hupfield; and a concert featuring Diné composer, Raven Chacon. It is the second of the installations, an instalment we called *Street Hassle*, a work addressing the experiences of front-line workers in the current opioid crisis, that will eventually be the focus of this text.

As an idea, *Everyonce* had been in gestation for years. I had been thinking about coordinating a group of people to simultaneously record a site, creating a synchronous sound map, which I would then play back in a gallery, scaled down, allowing listeners to hear a vast stretch of space as though their ears spanned kilometres. But I hadn't followed through with this because it had felt like an empty, formalist experiment; there wasn't a whole lot at stake in the gesture. I was also working on my dissertation at the time—a critical history of field recording that addressed how media are always understood in terms of their relations to site (Akiyama, 2014). For example, when we hear an old, crackly recording of Balinese gamelan with roosters crowing in the background, we instinctively know that there was a microphone somewhere, someplace where something *real* happened. Conversely, when we hear the disembodied voice of a radio news anchor in our speakers, her words broadcast from a soundproof, treated vocal booth, we buy into

DOI: 10.4324/9781003164227-27

the fiction that this sonic moment is placeless and exists only for the listener. In my research, I kept encountering practitioners of field recording—ethnographers, biologists, and composers—who treated the microphone like a window, a transparent portal that allowed a listener to traverse time and space, to be *there* in another moment and place (Akiyama, 2010). This struck me as profoundly problematic in that this seemed to essentialise their subjects while neither acknowledging the artifice of the practices nor owning up to the inequities of power, privilege, and access created by encounter between recordists and their Others.

In the first work in the trilogy, my eponymous piece *Everyonce*, I tried to work through the limitations and possibilities for recording in the field that, until that point, had been largely theoretical. I had been thinking about point of view and how the microphone typically represents the single-point perspective of a solitary recordist. Played back on stereo speakers or headphones, a recording of wind rustling through the leaves of an old growth forest seems to put *you* there, where some*one* else was. But what, I wondered, would happen if this point of view were kaleidoscopic, distributed, and the result of community and collaboration rather than stoic solitude? I organised a workshop at a public space situated under a highway underpass in downtown Toronto called The Bentway, where I was artist in residence at the time (see Figure 16.1).

Using their phones, participants started recording at a point in the middle of the site and then made their way to 30 designated places, roughly equally distributed across The Bentway. I asked the recordists, once they had reached their sites, to be still and to listen deeply to everything going on around them. And then, after 10 min,

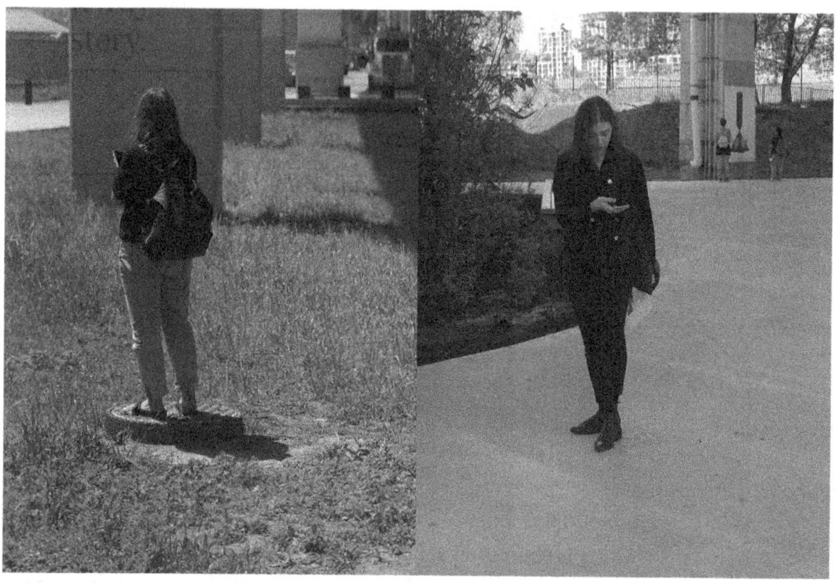

FIGURE 16.1 Field recording workshop participants, The Bentway, Toronto. Photograph by the author

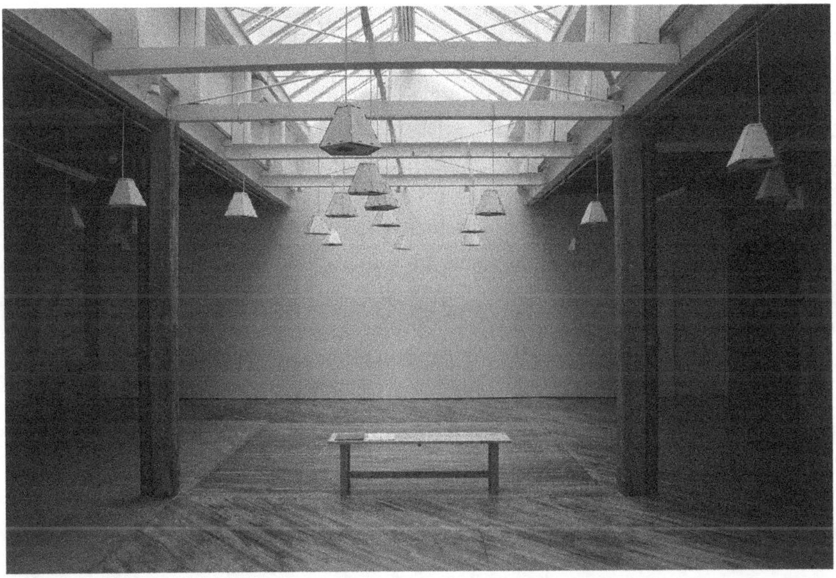

FIGURE 16.2 Installation view of *Everyonce*, Trinity Square Video, Toronto. Photograph by the author

each was to say, 'returning', before making their way back to the starting point. For the gallery installation, I attached speakers to the ceiling, each positioned to approximate where one of the recordists had stood in relation to each other. At The Bentway, the participants radiated up to 300 m from the centre point; at Trinity Square Video, this distance was compressed down to around 4 m from the middle of the gallery (see Figure 16.2).

As the recordings played back in this sonic scale model, strange and interesting things happened. A car entering into earshot of a recordist at the east end of The Bentway would travel into the recording, then into range of the person just west of her, and so on. If it took that car 10 s to travel a kilometre, it took 10 s for the sound to travel less than 10 m through the gallery. It was as though the sluggish ghost of an automobile was drifting through the room, detailed and dimensional enough to almost feel material. More importantly, though, this contraction of space and this expansion of time could only have been achieved by a group of individuals working in concert. We had built a model of a multi-nodal, sensing body with ears 600 m apart and everywhere in between; a mediated body that offered both hopeful and unsettling possibilities for shared subjectivity or surveillance.

Mega Sonic Jingle, Maria Hupfield's intervention with the installation, was similarly invested in the possibilities that sonic mediation might offer for intimacy and community, albeit at a very different scale and with a shift in political valence. Describing herself as 'an off-rez Anishinaabe [belonging] to Wasauksing First Nation', (Hupfield, 2021) Maria's work is multifaceted, spanning performance, installation, and video. She transmutes grey industrial felt, two-by-fours, plywood, and other common construction materials into assemblages and supports for performances that vibrate and pulse with her commitment to decolonisation and Indigenous resurgence.[1]

For *Everyonce*, Maria worked to create a disembodied soundscape to further her investigations into sound through the relationship between traditional Anishinaabe jingles and the lived experiences of Anishinaabe women. A jingle is a small piece of metal shaped into a cone that, when sewn to a dress or boots and then activated through dance, transforms the garment and dancer into a sounding, rhythmic body whose every movement announces its presence. Traditionally, jingles are fashioned from rolled disks of metal repurposed from discarded tins—chewing tobacco, flour, snuff. A jingle is potential, it vibrates subtly with an imminent vitality (Bennett, 2010), but as Maria writes, 'The jingle cannot make noise on its own, it needs others to make a sound, to connect and resonate with. A jingle is formed through intention' (Hupfield, 2019). The intention to sound is also the intention to commune; jingle dancing is an important ceremonial healing practice for Anishinaabe, Ojibwe, Dakota, and other Indigenous peoples.

For her work, Maria invited her sister, Johna Hupfield, and her sister-in-law, Deanne Hupfield to dance and reflect on their relationship to jingles and jingle dress dancing. For this iteration, Maria and I reinstalled the speakers on the ground, the long cables gathered into loops that prompted visitors to navigate stationary eddies of wire. We placed all of the sound apparatus on the floor, the electronic guts exposed and artifice withheld, and in front of the amplifiers and audio interfaces, a pair of Maria's felt jingle boots worn during the audio recording sat, as though waiting to be activated in sympathy with the recorded sound (see Figure 16.3).

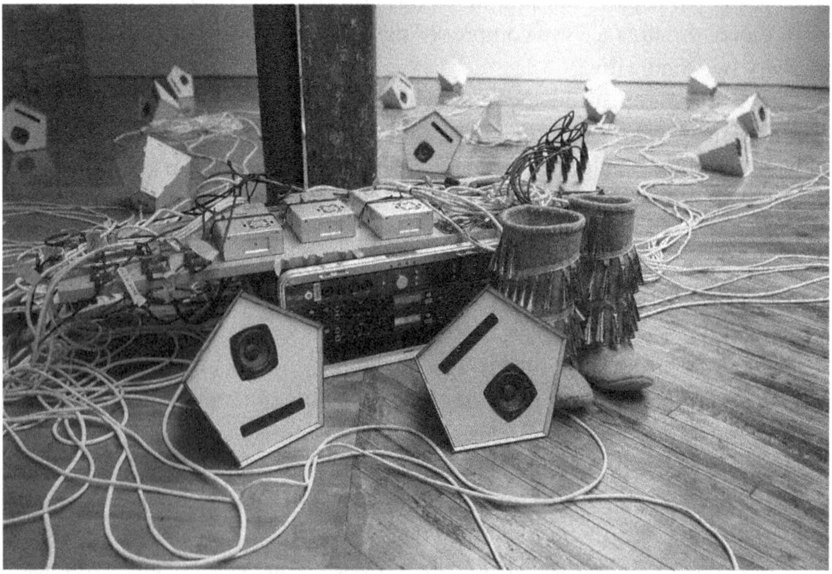

FIGURE 16.3 Detail of *Mega Sonic Jingle*, Maria Hupfield in collaboration with Mitchell Akiyama, Trinity Square Video, Toronto. Photograph by the author

The recording session happened behind closed doors, and the three women's familiarity and intimacy are palpable in the composition. In the sound piece Maria and I produced, the three women speak of healing and of their community, and even though we are eavesdropping on a private moment, their confidence and joy feel freely given to the audience. At the end of the work, a plaintive melody played by Deanne on an Ojibwe flute envelopes the sisters and the room, fading away until the jingle starts again.

Street Hassle, the work created in collaboration with Dont Rhine and Syrus Marcus Ware, also dealt with tensions between privacy and exposure, intimacy and mediation. I had been following Dont's work with Ultra-red for years; in fact, it was their unflinching political commitments and activism that had played a big part in my conception of *Everyonce* in the first place. Ultra-red is a group of activists and artists, scattered across the globe. Founded in 1994 by AIDS activists Dont Rhine and Marco Larsen, Ultra-red produced a series of albums of experimental electronic music on the German label Mille Plateaux in the late 1990s and early 2000s. I had first heard their work when I was making my start as producer and DJ in my early 20s. Their CDs were full of glitchy recordings of protests, of men cruising and having sex in public parks, of myriad expressions of political struggle. At the time, I didn't really get what they were doing—I was more interested in losing myself in music than in using it to immerse myself in unfamiliar conflicts. But years later, I would circle back to these recordings after discovering Ultra-red's writings, which called for a completely different politics of representation than the ones I had encountered in the literature on sound recording. Most of the writing on field recording I had come across posited that it was the job of the recordist to intervene as little as possible, to 'accurately' or 'authentically' capture their source. Ultra-red (2008), on the other hand, advocated for what they called a practice of 'militant sound investigation', in which 'Listening is organized on behalf of the microphone'. Ultra-red rejected neutrality and embraced the idea that recordings are pure artifice; recordings don't contain the world, they create it. Or, as Ultra-red (2000) writes, 'Separating sound from context produces the most artificial results: a utopia so to speak' (p. 4).

I invited Dont to collaborate with me on the *Everyonce* exhibition, and he accepted, but on the condition that the work not be produced solely with an artworld audience in mind. I told him that this was precisely what was motivating my decision to approach the exhibition as a social platform, and after some planning, we decided that our collaboration would focus on the opioid crisis and the response from the harm reduction community in Toronto. Dont had previously worked on an Ultra-red project in Toronto, mounting an exhibition in 2006 at the Art Gallery of Ontario (AGO). The exhibition, *SILENT|LISTEN*, was an effort to build a public record documenting impact of the HIV/AIDS crisis through sonic practice.[2] While working on the exhibition, Dont met Syrus Marcus Ware, with whom he reconnected and invited to collaborate on our project. Syrus is a prolific artist with a wildly eclectic output; he creates life-sized portraits of activists, works for theatre, and installations, all of which emerge from his profound commitment to confronting systemic racism and ableism. Syrus's work with Black Lives Matter, his advocacy for trans rights, and harm reduction—in addition to his written output as a scholar and essayist—have made a deep impact on Toronto, and we were thrilled and grateful to have him work with us, for him to connect us to people working on the frontlines of the opioid crisis in the city.

Our first action was to host a series of listening sessions with community organisers and healthcare workers. We structured these meetings according to the protocols for 'sound investigation' developed by Ultra-red (2011, p. 3). We held three sound investigations, and in each we played recordings collected by Ultra-red—'sound objects' documenting the group's work in harm reduction—for participants and then asked them, 'what did you hear?' Stories of police persecution, of deaths and of lives saved from overdose, of racist drug policies, of solidarity and community poured out. We recorded these powerful, raw testimonies, which we used to create a sound work for the exhibition.

For the reinstallation of the sound system (see Figure 16.4), we placed a table at the centre of the room, upon which we set four speakers that played back the sound objects that we had used as prompts.

We positioned the rest of the speakers, which played participants' responses, on six tables, each draped with 'African print'/Dutch wax fabric, arranged in a ring. At the very centre, we installed a live microphone that was automated to occasionally send signal to speakers at random, creating intermittent feedback while sonically implicating the audience in the work. Voices would announce themselves throughout

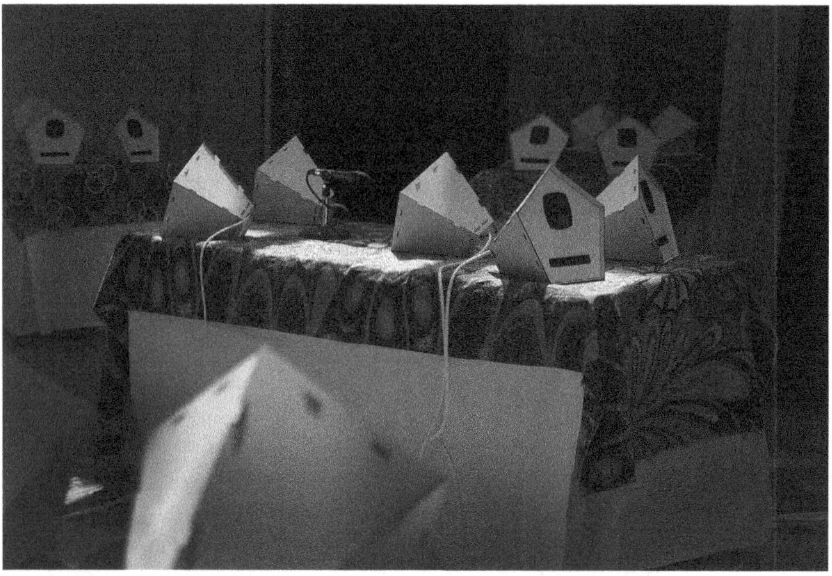

FIGURE 16.4 Detail of *Street Hassle*, Dont Rhine and Syrus Marcus Ware in collaboration with Mitchell Akiyama, Trinity Square Video, Toronto. Photograph by the author

the room, sometimes talking alongside each other, creating polyphonic textures that prompted listeners to move through the room, to gravitate towards the voices that spoke to them. It was powerful to hear those reflections in that space, in a gallery situated in a long-gentrified area of the city that is now a shopping and entertainment district.

Almost a year after the exhibition and only two months into the COVID-19 pandemic, Dont, Syrus, and I reconnected on a video call to discuss, to consider, and to reconsider *Street Hassle*. It was a freewheeling conversation, and we were free with our associations. The following dialogue has been edited for clarity.

Dont: [*Street Hassle*] was an accumulation of conversations; there were the conversations that preceded the actual installation amongst the three of us, and between Syrus and me. But something that I always find interesting is that there is a kind of formality to getting people in the same room at the same time, setting the recording device on the table asking pre-considered questions, but for the people who are in the room, who are involved in those conversation oftentimes, that exchange happens in an ongoing way. The particular configuration makes it a bit unique, but for the people who are having the conversation, it's oftentimes *not* very unique because it's just part of an uninterrupted flow of discussion, debate, and so forth. I am mindful that we end up working with the recording in a way that the recording becomes precious for all sorts of reasons, but for the people who are recorded, it's another exchange.

Mitchell: Dont, I wanted to ask you about how you saw this collaboration working into the trajectory of the work that Ultra-red's been doing. Was this part of that trajectory, or was it a swerve in any way? Have the protocols changed over time? Did this experience add something to the protocols?

Dont: Definitely. In my short time there in Toronto, the discussion we had about safe injection and Moss Park[3] have since impacted the contribution that I am making to the discussions we're having out here. To be specific, like a lot of harm reduction interventions, L.A. Community Health Project[4] has been very institutionalised so that it's protected from the law, protected from NIMBYism, and from developers. There's something that's also lost with that institutionalisation. One of things that's lost is the sense that you can do things even if you haven't figured out all the possibilities in advance. Instead, like with a safe injection site—we have to figure everything out before we enter into that adventure. From my time in Toronto, I was reminded of how this whole thing started; it didn't start because everything was figured out in advance! And so, I could use the example of Toronto in relationship to older histories of mutual aid. Communities do this for themselves; they don't wait for the state to tell them it's OK.

Syrus: When I was living in Vancouver in the nineties, there was a safe injection space that was illegal, and the people operating it saved a lot of lives. And then it was only 20 years later or so that they ended up opening up Insite[5] and started all the important work they've done there. And then here, in Toronto, with Toronto Overdose Prevention Society (TOPS), they're organising in Moss Park, and they just did it.

Dont: To bring it back, Mitchell, to your question about the protocols, this collaboration definitely deviated from previous sound investigation work in Los Angeles. That work has a very clear purpose, and it's not visible to an art audience whatsoever. It takes place entirely within the context of the needle exchange program. So, stepping out of that was kind of awkward, but I was grateful to have the chance to work with you, Mitchell, and to work with you, Syrus. For me it was a detour into a more experimental and speculative process. That has always been useful, particularly when the work is completely absorbed within use-value and there's nothing left over. Having this speculative space where the use value is rather minimal is helpful. It helps me think about things. And in this case, with the conversations that we were having, I didn't go into those conversations with an agenda, so it allowed me to listen in a different way to the exchanges that were happening around the table.

Syrus: Having participated in the protocols in 2006 with the intervention at the Art Gallery of Ontario, and then getting to do this experimental project now, it's just so amazing. One of the things that I think is the continuation is that I build a sense of trust in your practice and trust in you through that work, and that carried through to the point where you could call me up and be like 'Hey, do you want to do this thing?' Of course, because I really trust Dont. I think that is a testament to the work that you and Ultra-red have done, that your politics have been so solid and so resonant in the work. For someone like me who's sort of a sceptical activist who wants to be hopeful, it makes me trust just in this process. To be able to experiment and go in wild directions, and experiment with fabric and textiles and ladders and different ways of putting things together that sort of reference a kind of Black or BIPOC Toronto aesthetic, an activist aesthetic felt really nice.

Dont: At the same time I think it's useful to acknowledge what steps could have been taken next that weren't. From popular education (Freire, 1970), we know that after you produce the codification in relationship to your fieldwork, then the encounter that community members have *with* the codification isn't just reflection for its own sake; it's reflection that results in some sort of collective analysis that then gets tested in action. But the whole apparatus of contemporary art pretty much forecloses that next step—that is from analysis to action. And that foreclosure is repeated in every institutional situation, from the academy to the gallery, and from the funder to the curatorial. It permeates everything, which is the definition of what bourgeois art is. Community members often like an experience that defamiliarises that relationship to their own experience, and the gallery affords them that. It's a different kind of environment, a different space. The protocols are strange, and that strangeness can help people experience things differently and critically. But then the next step: to turn those reflections into a shared analysis that could be tested, and then the whole thing starts over again. All of us doing cultural work in a contemporary art context, that's the challenge we face.

Mitchell: I think what was interesting about this exhibition was that its manifestation as an installation in a gallery was just a small fraction the whole process; the trust-building and collaborative work that had to happen to even do those interviews was super important. I ended up working with people I didn't know very well in advance, and for me at least, there were all these moments of learning and engagement in very micro kinds of ways. It was clear to me that I didn't want to just create some work and put it in the space and then deinstall the whole thing, and then for everything else that followed—documentation, etc.—to be parasitical on the installation. I don't feel like it ever really manifested as a stable thing. It was always changing, and it taught me a lot about what it means to make things and who they're for and what they're for and when they're for. But Dont, you were saying that the listening sessions in Toronto were a deviation given the artworld context, and it's interesting for me to hear you say that because I know your work pretty much exclusively through art circles, especially music. And having gotten to know you, I really respect and appreciate that you think of the work that you're doing as political work. But when I think of what you do, all the evidence I have of your work is these recordings and these texts that, whether or not you think of them as artworks, they're taken up as artworks. How does that sit for you?

Dont: I guess part of it has to do with the fact that the river never stops moving. I'm fifty-seven now, and most of the work that people know from an art context or a music context is anywhere between 10 to 20 years in the past. This experimentation phase of Ultra-red's work occurred largely, but not entirely, in contexts where there weren't strong consequences for trust. Like making an installation in Germany with material that comes out of Los Angeles—if that installation fails at an aesthetic or even at a political level, it's not going to have an impact on the trust-building process back home. Sometimes that separation can be helpful because you can experiment with forms, processes, and practices in a way that can be liberating. But there comes a point where you have to put up! For example, bringing the sound investigation protocols into the work of the needle exchange in 2012 represented a significant shift in that process of experimentation. Once we began making site recordings at the needle exchange, interviewing community members who use the exchange, and organise the listening sessions, at that point the consequences for trust were front and centre. If the people who I'm playing the recordings for feel exploited in that situation, or feel mocked, or feel like they're serving someone else's agenda or someone else's entrepreneurial goals, it could be pretty damaging. Yes, the experimentation continues. I'm not suggesting that there's no experimentation in conducting a sound investigation in the context of the harm reduction movement. But the stakes are higher now that the protocols are embedded in an explicitly political context. While there might be an art context that you and others have become familiar with, that context for me functioned more like a moment, a particular period within a larger project.

Mitchell: For me, this raises important questions about the utility of even looking to art as a category. If you're making recordings and using them to engage with people, if that's within a community where there's a use-value to what you're doing *with* those recordings, I don't know why it's important to even ask the question: is or isn't that art? But then I'm trying to figure out what motivates the question at all. Is it better to just suspend or call a moratorium on the idea of art? Or is it more useful to think of everything as art or as aesthetic experiments?

Syrus: I can remember when Dont and I first met. I was working at the Art Gallery of Ontario, and I had been an activist for decades, and I was starting to see how people were bringing activism into their practices. And I was fascinated by it because I was like, oh, is that art? Because if it is, I'm doing those things too but thinking of it as activism. Now, of course, my artistic practice directly engages with those considerations of activism and activist aesthetics, of what it is to engage in organising. And I am an organiser who uses my artistic skills to be able to do creative activism. When I describe my work, I say I'm an artist and an activist, and I've been organising for about 25 years. I started doing them both actively around the same time, so for me they've always been interwoven.

Dont: To the present moment, I feel that the work of sound investigation is more grounded now, both in terms of the work that Robert [Sember, another member of Ultra-red] has done in New York City, and the work that we've done out here in Los Angeles. The invitation was particularly sweet, Mitchell, because it gave a chance for you and me to have a different relationship, and to spend time together. You were the most gorgeous host ever! And it was lovely to reconnect with Syrus. I think all of this speaks to something that we should really emphasise, and that's the idea that things need time. Relationships need time. For projects and practices to emerge and to go through cycles, there needs to be a long-term investment. That kind of investment is difficult for institutions to wrap their heads around because the way that contemporary art works, it's a mill that's grinding people through all the time.

Mitchell: That's something that doesn't really come up a lot—even in art practices that aren't obviously connected to a gallery system or creating art commodities; they still manifest in a way that leads you to believe that the results are fully formed works. Even if a work is process-based, given the container of the exhibition, I think you can get the impression that everything was leading to that moment, and then it's over. And I really appreciate what you're saying because every step of the process of creating this work was as important and as generative as every other step, including but not limited to its manifestation as an installation. Conviviality and trust and relationship—all of those things are required to make work, but they don't get framed as parts of the work. But being a good host or being able to have hard conversations—these are also important features of making work. But to bring us back to sound, I'm wondering how we can talk about what it meant to make this sound installation in a gallery space, about what it means to talk about sound as a focus.

Syrus: For me, the opportunity to get to sit and listen to the recording of that woman, [Renée Edgington, the first director of the Los Angeles needle exchange—then called Clean Needles Now—who, in the recording is receiving a citation from the police for being in possession of drug paraphernalia, while the exchange is taking place around them on the street], so fierce and so powerful, getting to hear recordings like that connected us through time and space. In that way sound was so powerful because of what she was talking about—it may as well have been 2019.[6] The same kind of things are still happening, and when we asked people what they heard, a lot of them said yes, the same things are still happening.

Dont: Yeah, you can talk about the apparatus whether the apparatus is materialised in technology or internalised through practice. But the apparatus causes a kind of defamiliarisation of our context so that we become mindful of it, we become mindful of the situation. That mindfulness gives us access to the intention to transform the situation. That, to me, is a kind of language that partly comes from the avant-garde, partly from Russian futurism and formalism, and partly from situationism and different forms of abstraction. That abstracting process is how we critically engage the world around us. We abstract it, we abstract an experience. The challenge, I think, has always been that next step: what do we do with the analysis that arises out of that abstraction? How do we test it through practice in the world in ways that aren't just a constant deferral of the hard questions, but actually embracing the consequences, embracing the risks involved?

Mitchell: I think there's a synthesis of what you were both saying—the possibility of having juxtapositions and superimpositions of different temporalities, of different moments in time, different conversations with different groups of people who are speaking to similar things, but who aren't present with each other; but they can become present with each other through sound in that space. As silly as it sounds to talk about, but the political significance of how many channels you use in a sound installation, the fact that there were thirty channels, and that these conversations scrambled time, as you were saying Syrus, it really stopped becoming important what was 1996 and what was 2019 because what fused everything together was the concern and the political moments that have changed but still need attention.

Dont: Maybe it's a difference between a flattening or an eternal present versus the challenge to think about historical memory. I think the latter is super important for the politics of the work. For instance, if a recording from the early 1990s could just as easily come from today, then that says something about the fundamental conditions of this particular era. There are certain things that seem to be conditional, whether it's early 1990s or 2019. We also have to be mindful of what is specific about this moment. Because we need that distinction in order to figure out a strategy. I think what was coming up strongly at that point in the summer of 2019 was a fundamental sense of moral exhaustion. Even understanding the necessity of a safe injection site, there were still so many deaths as a result of overdoses. That is a very different lens and a different set of conjectural terms from when Renée [the woman in the previously mentioned sound object] was arrested in 1995. Being able to hear both what is consistent, but then also being able to hear those differences, is super important for political strategy.

Mitchell: To use the word that you were using earlier, Dont—defamiliarisation—it's doing both things at the same time; it's historicising that struggle, and it's also collapsing the time of it, because things change for the better, things change for the worse, things take one direction and slide backwards. And not having the luxury of a total linear historicisation makes you really have to attend to what of the past is occupying the present, and what kinds of echoes you hear that go back from the present into the past. And I feel like the piece ended up being really defamiliarising, even sensorially, but also as a conversation, as people speaking to each other across space and time.

Dont: I would maybe draw a distinction between work that flattens time and work that teases out what Gramsci (1971) would call organic time or organic conditions. It's a weird phrase, and it never quite sits right with me, but he's basically describing how under capitalism there are certain operations that will be in effect whether it's the early 1900s or the 2020s. Those operations will be in effect, but the conjuncture in which we live, in the present moment, possesses certain aspects that are fundamentally different from the early 1900s. Politically, how do we tease those things out so that we can understand, say, something that is fundamental to harm reduction, which is an adversarial relationship to the police? We could advocate for participation in police training in harm reduction, but that work doesn't transform the fundamental reality that the function of the police is to protect private property with all the implications for racism, gender violence, and class violence that comes with private property systems. That's important to acknowledge because it helps us understand the class position of injection drug users. But still, we have to ask; what's the conjunctural quality of our moment? If we look at how the state is responding to the safe injection site and the contradictions it manifests, then I wonder if the political demands of the harm reduction movement are truncated in contrast. Our work is to think about how the listening sessions can help us understand something about what is fundamental to capitalism around the crisis of opiate overdose and drug criminalisation. We have to make those distinctions, otherwise we're going to be stuck with strategies that are irrelevant to our current moment or stuck with strategies that are *only* in the present and miss the larger fundamental questions.

Mitchell: I wasn't suggesting that what we did was *flatten* time. I think the potential usefulness and power of this kind of creative work is creating new epistemologies and new ways of thinking about how these politics can operate in different modes of time. This nonlinear way of organising present and past material can show you both through-lines that exist, but also the ruptures that exist at the same time. I think a lot about the politics of genre, and when you reach for an available genre, you're always letting the politics of how that genre has become a thing dictate how you organise your material. So, for example, we could have made a documentary. It could have been a linear work that laid out the history of safe injection and harm reduction, which then moves from the 1990s into the present to show how things have progressed and/or devolved. For me, it's like transforming the analysis in the way that one creates a work of science fiction. I'm thinking about it in the way that Jameson (2005) does—it's making the present seems strange, even though there's no doubt that you're still in the present.

Syrus: Walidah Imarisha, in *Octavia's Brood* (Brown & Imarisha, 2015), says that all activism, all organising is a speculative fiction because it's daring to dream or imagine another world being possible. Which is what you do in speculative fiction, which is also what you do in activism. And I've been thinking a lot about how that conversation would have unfolded if we were having it right now. Because now, suddenly, there are other things on the table like safe supply. If all that we are engaging in is 'daring to dream that another world is possible', i.e. creating speculative fiction, then what is that other world? And how do artists play into this practice and world building? I've been making work that is set in the near future that allows me to dream into what might happen in the next few years if we continue on this path, and also to dream far into the future and imagine what might happen if we veered off of it and started something new, something abolitionist in its place. If we do the work of imagining a freer world, a more just world, a world away from carceral logics, we must also imagine in this world a safe supply and an end to the drug war. We must root this work in harm reduction for this is life, this is how we will make it to the future, together. In my creative work, I've imagined a future where drug supplies are rationed, as in my play *Antarctica,* given freely but of substandard quality to the main characters from the government/company. Speculative fiction as a creative practice has helped me to understand what we are building towards and where we need to go.

Dont: We also want to be careful because if we lose sight of what are the more baseline conditions, we could easily fall into non-profit consciousness, which is about addressing but only to reform and remediate what's right in front of us. A safe supply conversation without a discussion about capitalism, racism, and police repression sets itself up for all sorts of just completely bananas shit.

Syrus: What I would hope for is that we're in this revolutionary moment and that we're actually going to finally once and for all overthrow capitalism. That's what I'm hoping for during my pandemic moments. There's potential for rapid change in this current moment.

Dont: And these are the moments where, within Ultra-red, we'll often use this expression, 'when in crisis, slow down'. We get so caught up in responding to the leadership of someone else's politics that we miss the opportunities that are actually here, right now. You can see this with everyone fixed on federal politics, on Trump, on Washington, DC. The real opportunities that exist for us at the moment slip through our fingers because everyone is asking, what will Trump do next? If listening to a recording allows us to slow down and to begin to develop the muscle to see the opportunities that exist right now, then that's important work. I don't think that there's anything that says that can't happen in a gallery or some other kind of space. It's more a matter of what that kind of space can contribute to such a process—the 'revolutionary moment' that Syrus is referring to. For a lot of working-class and poor folk who've never been to a museum or who haven't been to that particular gallery, it is such an odd and unfamiliar environment that makes possible hearing some things. It's up to us to make sure that any kind of violence that might also accompany that experience is mediated.

Syrus: This is the trouble. As a long-time museum educator, it's like you're reaching your hand out and trying to get a squirrel to come to a nut, and you're trying to not make big sudden moves, and you spend time building relationships, and you do all of the work and get everything sorted. And then the person comes who you've always hoped would come, and the security screams at them to take the bag off, or screams at them for their umbrella, or screams at them because they're Black or whatever. That can make someone jaded and undoes our work. This is one of the things that was so troubling about working within the institution.

Mitchell: On the flip side of that, when you bring difficult, politically engaged work into those kinds of institutions, it can also open a door for white, bougie people. I was thinking of my mom being at the opening and meeting you, Dont, and experiencing the work. She's an upper-middle class white lady who hasn't taken drugs in her entire life, who doesn't know users. And it was just completely mind-blowing for her to hear those stories, and she was going in there with an openness and a lack of judgment that probably came from being my mom, first and foremost. But in a sense, it's the opposite situation, because there's no reason why my mom wouldn't feel safe or entitled to be there, which makes it an important moment to share those stories with someone like her.

Syrus: And that's the thing about safety: it can be taken away very quickly, as quickly as it's been given. But I do think that it's amazing that I was able to witness times where community really felt like they were taking up space and felt at home, even for a moment. And I think that one of those times was when we were all sitting around, and we were listening to the recording of Zoe Dodd [co-organiser of TOPS] talk about harm-reduction organising. And everyone was sitting on the floor and just making it a familiar space, our space. We own this space right now: this is our space.

Dont: I think there's something about this conversation that brings me back to an understanding of what noise is; not noise as a kind of metaphysical category, which there might be a tendency to assume. Noise in different contexts is going to have very different functions, and so that moment when you're in a contemporary art gallery, and the bodily experience of noise, beyond just the cochlear experience, is 'oh my god, am I going to lose control? Is this space going to lose control? Who's going to come in and stop this? Who will keep this from going too far?' People really experience that below the skin. Now, in a noise concert, of course it's not going to function that way, but in other spaces—you get in a conference room and suddenly there's feedback. 'Where's that sound person? Make this stop!' And I think there's something about that moment that is so fucking useful for us, and particularly in relation to the politics of harm reduction.

Mitchell: I like to think of noise as everything all at once. Michel Serres (1982), instead of thinking about noise as being the thing that obscures the signal, thinks of noise as being the *foundation* of signal, that everything is noise, and signal is what we sculpt *from* the noise. I think being attentive to how one wants to filter the noise—and then paying attention to what those filters are and how they operate—can tell us a lot about the kinds of political intuitions that we have, that aren't even necessarily ones that we are committed to. Experience is really noisy, and the filters we inherit limit that all-at-onceness and only let certain things pass. Take white noise in music or in sonic terms; if you run noise through a filter, you can pull a tempered scale out of it very easily. But that tells you about the desire to hear a kind of order in a chaotic system. So, this has me thinking, what's social noise? And what's the analog of a bandpass filter that gives you a tempered scale out of social conditions? I know that Dont, you wanted to talk about the feedback from the mic. I'd love to hear you riff on that as a sort of formal, metaphorical device in that space.

Dont: I haven't resolved it. But to me it's an echo of that 2006 performance (see Figure 16.5) we did at the Art Gallery of Ontario in the education department, where we couldn't control the microphones.

There really wasn't anyone who was in a position to control the microphones. There was a room with so many fucking hot microphones. It was bananas. And then when we actually processed the speech back using real-time digital sound processing, it was just complete bananas. Every facilitator at the table had their own composition that they played. We have a recording of the room and it is so fucking chaotic. It's fantastic. So that moment 14 or 15 years later, when we're in Trinity Square Video, for me there was a kind of pleasure in that echo. I'm thinking about Mitchell at the table trying to create chaos that was usable, and how much effort that took. If all technology is accumulated labour, the sheer amount of accumulated labour necessary to create an experience of chaos is fascinating.

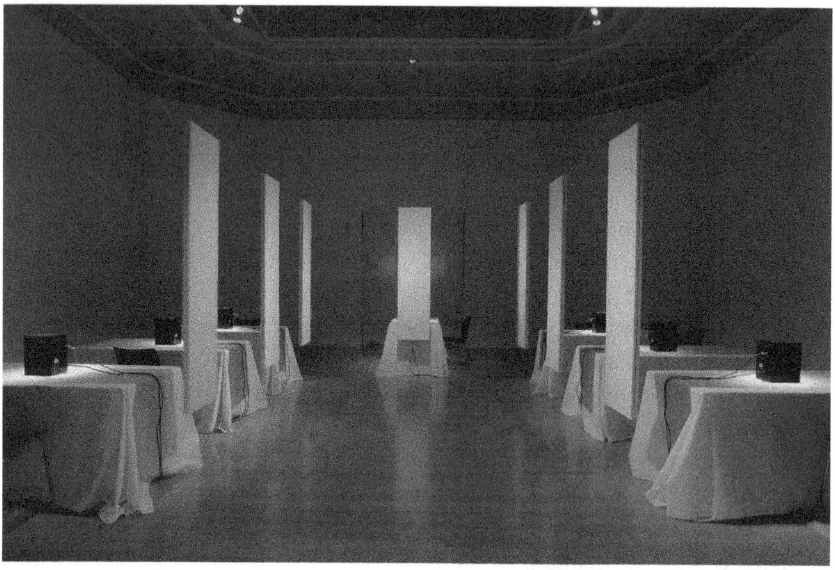

FIGURE 16.5 Installation view of *Silent|Listen: The Record* (2006), Ultra-red, Art Gallery of Ontario, Toronto. Photograph by the Art Gallery of Ontario

Mitchell: What I loved about the feedback was that it made it really difficult to just settle into a flow. It would spike and be uncomfortable, and then you had to start listening again. You couldn't get comfortable listening, you couldn't let yourself flatten those stories because as soon as you got comfortable with a certain mode of listening, with that content being available there for you, the feedback would just kind of kick you in the ear. And then you had to find a new way of orienting yourself to the piece.

Dont: Someone came up to me and said, 'Oh you know, you're having some feedback problems,' and I was like, 'Yes, I know. Isn't it gorgeous!'

Mitchell: It also shows how paradoxical noise is. If you're going to go with a very strict definition of what noise is as some sort of physical fact, well, feedback is exactly the opposite of noise because it is just one frequency being amplified over and over again to the exclusion of noise. So then it makes defining noise, well, really noisy because is noise unwanted sound, or is noise an excess of all frequencies all at once? It's kind of both at the same time.

Dont: So in this case, the traditional noise/signal binary is inverted. In this case we're experiencing noise *as* signal. At the same time if we draw on Serres's (1982) argument that noise is ground to a figure, there are some interesting things going on there, of defamiliarising your relationship to an undifferentiated surface. I'm reminded of a strange little book from the nineties: Susan Bickford's *The Dissonance of Democracy* (1996). She's basically asking, what is solidarity? She has this idea that all too often we want to think that politics emerges out of consensus, out of resonance. When in fact it actually emerges out of dissonance. How do we then position a relationship to dissonance? And she talks about how authentic engagement—I don't think authentic is her word—is where we make of ourselves the background to someone else's figure. Because if we enter into a listening relationship of empathy then we're just only listening for where there's commonality, we're not listening for dissonance. How do we attune our ears to hear contradiction and not always consensus?

Mitchell: I've always turned to Chantal Mouffe to think about dissensus. Mouffe (2007) has this short essay about art and consensus where she says that it's problematic to make a political art practice out of the idea that you're going to connect others; it's more that art has an important role to play in *suspending* the idea that consensus is what we should all be searching for. Because obviously consensus is going to boil down to whoever has the power to secure consensus, which is never going to be equitable. Everybody is coming with different needs and concerns and being comfortable in dissension and having art be that sort of noise, signal, oscillation—art has a really important role to play showing the limits of consensus and creating uncomfortable situations. What I'm always trying to do is in terms of a political experience is to figure out how you create a *dissensus* of the *senses*. How can you create a situation that is grating and inviting at the same time, so that even what you think you know about what should be aesthetically pleasing is upended and thrown into question?

Dont: The challenge with political philosophy is that it often doesn't have a politics. By which I mean talking about the abstract without an object, talking about solidarity without knowing who the subject of solidarity is. I think in our installation, we're clear on who we're in solidarity with. And so, if we're trying to create a dissensus, that dissensus is not disinterested. I think that project has a privileged political subject. So how does dissensus empower that political subject and empower our solidarity with that subject?

Mitchell: I think that frontloads the idea that you're going to have to do some work. There's a Bruno Latour (2004) essay in which he talks about matters of concern versus matters of fact. We talk about matters of fact as though there's something transcendental that we could all agree to. But that's unrealistic and stupid, and framing things as matters of concern can be a way of connecting to people who you don't necessarily agree with. The concern *here* is that people are dying. The concern *here* is that people don't have adequate housing, and I think that it seems just like an almost majoritarian and sensible position; people don't want people to die, and I would hope that people don't want others to be unhoused. But how do you get to the kinds of actions that can address these issues? Rather than asking, how did we get here? What is the state doing to fix this? Those aren't the right questions. First let's identify what's important, and what we care about, and go from there.

Syrus: It's such a thing—we always turn to the state as if it was somehow broken, as if it was something that needed to be fixed when it's working exactly as it's supposed to. In our organising now and our organising in the past and our organising going forward into the future, how do we not fall into the idea that the state we live under will exist forever. Like right now, this is the longest conversation that I've had with human beings since March 11th that hasn't been about COVID. All I talk about is about how to get people out of prison because of COVID. How to deal with targeted policing of people in the streets because of COVID. How to get people to do rent strikes because of COVID. That's all we're talking about. So, it's interesting to consider, in the absence of thinking in relation to COVID, what is it that we would love to see in the future that we heard coming out of the work?

Notes

1 On decolonization, see Tuck (2012). On Indigenous resurgence, see Simpson (2011).
2 Ultra-red's multi-platform initiative titled *SILENT|LISTEN* (2005-2008) investigated the affective landscape in the present conjuncture of the AIDS pandemic. Taking the form of performance-based inquiries in art museums, *SILENT|LISTEN* turned its focus on the present moment of the AIDS crisis where direct action and mass mobilisation has been replaced with the state and non-profit corporate administration of HIV and AIDS illness. In the summer of 2006, in the weeks leading up to the Sixteenth International AIDS Conference in Toronto, Dont had the opportunity to hold conversations with a range of groups and individuals organising HIV/AIDS advocacy and services in the city. It was during these interviews that Syrus and Dont met for the first time. Syrus was working in the education department at the Art Gallery of Ontario while his activism navigated the various HIV and harm reduction interventions in the city, particularly those organized by the Black queer community. On August 9, Ultra-red staged a *SILENT|LISTEN* performance at the Art Gallery of Ontario. Syrus was one of the participants in that performance where he presented a pre-recorded statement from an HIV+ inmate in the Canadian prison system. A week later, on August 14, Ultra-red held a large-scale performance of *SILENT|LISTEN* for delegates of the AIDS Conference with participants from each of the prior seven iterations of *SILENT|LISTEN* across the U.S. and Canada. Syrus participated once more as a representative of the Toronto event the week before. Documentation of the multi-site initiative was then mounted in an exhibition at the Art Gallery of Ontario. Ultra-red released two albums of electroacoustic music based on the *SILENT|LISTEN* performances; The Minutes and The Minutes XVI International AIDS Conference (Public Record, 2006). https://ago.ca/exhibitions/present-tense-ultra-red.
3 Dont is referring to the Moss Park Overdose Prevention Site. https://www.srchc.ca/programs/harm-reduction/consumption-and-treatment-service-moss-park/.
4 https://chpla.org
5 Insite was the first sanctioned supervised injection site in North America. https://www.phs.ca/program/insite/
6 The listening sessions took place June 25–26, 2019, at the offices of the Black Coalition for AIDS Prevention (Black CAP) and at the Daniels Faculty of Architecture at the University of Toronto.

References

Akiyama, M. (2010). Transparent listening: Soundscape composition's objects of study. *RACAR: Revue d'art canadienne/Canadian Art Review*, *35*(1), 54–62. 10.7202/1066802ar

Akiyama, M. (2014). *The phonographic memory: A history of sound recording in the field* (Publication No. 28254220) [Doctoral dissertation, McGill University]. ProQuest Dissertations Publishing.

Bennett, J. (2010). *Vibrant matter: A political ecology of things*. Duke University Press. 10.1515/9780822391623

Bickford, S. (1996). *The dissonance of democracy: Listening, conflict, and citizenship*. Cornell University Press. 10.7591/9781501722202

Brown, A. M., & Imarisha, W. (Eds.). (2015). *Octavia's brood: Science fiction stories from social justice movements*. AK Press.

Freire, P. (1970). *Pedagogy of the oppressed* (M. B. Ramos, Trans). Continuum.

Gramsci, A. (1971). *Selections from the prison notebooks*. (Q. Hoare & G. N. Smith, Trans.). International Publishers.

Hupfield, M. (2019, July 13–27). *Mega Sonic Jingle* [soundscape]. Trinity Square Video.

Hupfield, M. (2021, October 24). *Maria Hupfield Artist Talk at Centre3 [Video]*. Youtube. https://www.youtube.com/watch?v=58mu6KetNiM

Jameson, F. (2005). *Archaeologies of the future: The desire called utopia and other science fictions*. Verso.

Latour, B. (2004). Why has critique run out of steam? From matters of fact to matters of concern. *Critical Inquiry*, *30*(2), 225–248. 10.1086/421123

Mouffe, C. (2007). Artistic activism and agonistic spaces. *Art & Research*, *1*(2), 1–5. www.artandresearch.org.uk/v1n2/mouffe.html.

Serres, M. (1982). *The parasite*. (L. R. Schehr, Trans.). Johns Hopkins University Press.

Simpson, L. B. (2011). *Dancing on our turtle's back: Stories of Nishnaabeg re-creation, resurgence and a new emergence*. Arbeiter Ring.

Tuck, E. (2012). Decolonization is not a Metaphor. *Decolonization: Indigeneity, Education & Society*, *1*(1), 1–40.

Ultra-red. (2000). *Constitutive utopias: Sound, public space and urban ambience*. https://temporaryservices.org/served/wp-content/uploads/2014/07/constitutive_utopias.pdf

Ultra-red. (2008). *Some theses on militant sound investigation, or, listening for a change*. https://inthemiddleofthewhirlwind.wordpress.com/some-theses-on-militant-sound-investigationor-listening-for-a-change/

Ultra-red. (2011). *Five protocols for organized listening*. https://www.ultrared.org/uploads/2012-Five_Protocols.pdf

CONCLUDING ACKNOWLEDGEMENT

Sarah Woodland and Wolfgang Vachon

In the Introduction to this volume, we established a frame for understanding the aesthetics and ethics of community engaged sound and audio practices that embraced the diversity of approaches represented throughout. We began with a move towards decolonising sound studies, setting up 'Listening to Country' as a model of holistic, community engaged sound practice that privileges First Nations' Knowledges and embodies ethical, generative cross-cultural collaboration. We believe such approaches must become the norm, rather than the exception, as we face the crises of the current moment. The subsequent contributors generously drew us into their thinking and practices around sound and community that extended our understanding of the different ways this wave of social engagement might carry us forward. The contributions highlighted what we see as being some of the most pressing functions of sonic engagement: sonic knowing, sonic assembly, sonic disruptions, and sonic resistance. And they offered us multiple perspectives on how we might use and think about these approaches in our own work. Some of the authors here are well known in sonic circles, others are newcomers to this 'broad church', and we think the volume is better for this—offering a new direction for the conversation that is bound by the tensions and nuances of community engagement.

In drawing the work to a close, we are compelled to return to the questions we posed in the Introduction. What are our ethical obligations as listeners when we enter the relational spaces created by these sonic worlds? How might we be altered or transformed in the process? And what actions does the experience call us towards, as we move on from the exchange? We hope that this collection has gone some way towards answering these, or at least provoked readers to consider them more deeply in their own work. In particular, the last question draws us towards thinking about how sonic engagement might create reverberations of justice that reach between and beyond ideological boundaries and echo chambers to effect

DOI: 10.4324/9781003164227-28

positive change. This is a conscious return to the idea of art, culture, and aesthetic engagement as being vital to human (and non-human) flourishing.

This book has taken a while to get off the ground. Casting our eyes back to the early emails between ourselves and then with Routledge leading to our first proposal in May 2020, we are reminded of how this also marked the rapidly unfolding COVID-19 pandemic and the concurrent escalation of unrest in the USA due to the violent death of George Floyd—both of which seemed to dominate the news cycle and produce a collective sense of shock and dismay across the globe. In Australia, we had just come through some of the worst bushfires in history, and it seemed that 2020 was delivering still more hardship. The pandemic wreaked havoc on both the arts and higher education sectors there, resulting in two years of cancelled projects and events, and tens of thousands of job losses. Canada was shaken by and forced to confront the ongoing legacy of colonialism and genocide through the discovery of thousands of unmarked graves, mostly children who died in residential school. Significantly for this book, these young bodies are continuing to be found through the use of high-frequency electromagnetic waves:a technology that allows politicians and others to hear the dead bodies when the words of survivors were too often ignored.

The ongoing effects of the pandemic were experienced across the globe in countries where the contributors to this volume were situated, and the suffering continued throughout 2021 with extensive lockdowns, serious illness, and mass deaths. The COP26 in October 2021 delivered still more dire news, with the magnitude of the climate emergency finally registering in the mainstream after 40-plus years of warnings. And as we pull together the final pieces of the volume, Sarah's home on the eastern seaboard of Australia is experiencing widespread and catastrophic flooding, the IPCC has reported that 'the window to secure a liveable future for humanity is rapidly closing', and Russia has initiated a war in Europe that yet again threatens global stability and raises the spectre of nuclear holocaust.

This grim inventory is not designed to depress readers or make us dwell on the negative, but rather to acknowledge the efforts of our contributors, the publishing team, and each other, under the various pressures of the past two years. As editors, we have been incredibly lucky to keep hold of our health, our lives, and our livelihoods through this time, which has enabled us to press on. Some authors who signed on at the start were unable to see it through, and we acknowledge them here, hoping that they will soon find the opportunities and resources to share their amazing work. Some of the contributors to this volume experienced illness, instability, and losses of their own, nevertheless responding to our editorial requests and feedback with patience and grace, for which we are hugely grateful. Of course, we also wish to acknowledge all those essential workers who have kept the rest of us safe and well, and those people for whom the major preoccupation of the past two years has been simply to survive.

In today's context, a project such as this can feel like a privileged indulgence, but we remind ourselves that producing art and knowledge at such times is an urgent, often instinctive, undertaking. Rather than becoming paralysed by feelings of helplessness, artists carry on, helping us make sense of the world as it is, and creating visions for what it might be. And community engaged arts play a crucial role in processes of recovery. In putting this volume together, we were conscious that we did not want it to be a 'pandemic book' that would become obsolete as soon as the threat had passed. But we see from the contributions here that it is that, and yet it is also so much more. In a time of intersecting and rapidly escalating crises, unfolding concurrently with rapidly developing sound technologies, the authors here provide nuanced and inspiring accounts of how we all might continue to move together towards creativity, justice, and hope. In that spirit, we express our gratitude to all those who have contributed here, and all those who will continue to make work in the face of whatever is to come.

INDEX

Note: **Bold** page numbers refer to tables and *italic* page numbers refer to figures.